GROWING GREEN

GROWING GREEN

The Economic Benefits of Climate Action

Uwe Deichmann
Fan Zhang

THE WORLD BANK
Washington, D.C.

Cover photo: © Frank Chmura/Getty Images; sunlight collectors at a solar panel station, Bohemia, the Czech Republic

Cover design: Naylor Design

Library of Congress Cataloging-in-Publication Data
Deichmann, Uwe.
  Growing green : the economic benefits of climate action / Uwe Deichmann and Fan Zhang.
    pages cm. — (Europe and Central Asia Reports)
  Includes bibliographical references.
  ISBN 978-0-8213-9791-6 (alk. paper) — ISBN 978-0-8213-9792-3
1. Environmental policy—Economic aspects—Europe. 2. Environmental policy—Economic aspects—Asia, Central. 3. Climatic changes—Economic aspects. 4. Climatic changes—Government policy. I. World Bank. II. Title.
  HC240.9.E5D45 2013
  363.738'745—dc23   2012047950

Contents

Boxes

Figures

Maps

Tables

Foreword

The world is heading for a rise in average temperatures of 4° Celsius by the end of the century and possibly more in higher latitudes. We could face a climate that has not been experienced in the millennia in which human civilizations have developed. Given the speed of climate change, adaptation can be costly and is unlikely to eliminate all risks, especially for the poorest and most vulnerable. As World Bank Group President Jim Yong Kim warned recently: *"Lack of action on climate change threatens to make the world that our children will inherit a completely different world than we are living in today. Climate change is one of the biggest challenges facing development, and we need to take action on behalf of future generations."*

Most countries in the Europe and Central Asia (ECA) region have been slow to respond to this challenge. In light of the prospect of more frequent and severe droughts, floods, heat waves, and wildfires, many ECA countries have teamed up with the World Bank and other development partners to explore adaptation options for coping with a warmer and more variable climate. Our 2010 regional report, *Adapting to Climate Change in Eastern Europe and Central Asia*, was written to inform these efforts. Adaptation will remain important as current heat-trapping emissions commit the world to further warming. But to prevent climate change that exceeds our adaptation capacity, climate action to significantly reduce emissions must become a greater priority for all countries. This report, along with two companion reports that distill the lessons from successful countries in increasing energy efficiency and in mitigating the welfare consequences of reductions in energy subsidies, shows how this can be made to happen.

There are legitimate concerns about the economic costs and social impacts of climate policies. But this report shows that well-designed climate action can bring numerous benefits, while its costs can be contained. It will be important for policy makers to be mindful of the following key considerations:

First, reducing emissions from energy consumption will require large investments but, given the high energy intensity of the ECA region, these investments offer attractive rates of return. This is particularly true for industrial energy use, which could be cut by half without loss of output. This is also true for reducing losses in power and heat generation, and for raising the efficiency of energy use by households and in public service delivery. By 2017, World Bank financed investments in energy efficiency in ECA will annually avoid the equivalent of today's CO_2 emissions of countries such as Bulgaria or Switzerland. There are many more cases where climate benefits are a small part of overall benefits. Improving sustainable transportation reduces congestion, local air pollution, and accidents. These local and immediate benefits dwarf those from lower greenhouse gas emissions. Air pollution from power generation alone causes almost $20 billion in health damages in ECA each year. In the rural sector, better land and forest management brings urgently needed productivity gains while also increasing the amount of atmospheric carbon captured in soils and trees. Restoring land abandoned in Western Russia since 2001 could yield 11 million tons of grain at a time of rising global demand for food. For some mitigation options costs still exceed immediate benefits—as is the case with some renewable energy technologies. Such options may not become a priority in many ECA countries for some time. But the costs of action have been falling as new technologies and experiences become widely shared, while the costs of inaction leading to dangerous climate change will continue to rise if mitigation is delayed. The time to act is now.

Second, climate action will require some difficult adjustments across the economy, but it will also bring new economic opportunities. Many firms have benefited from fiscally unsustainable and environmentally harmful energy subsidies. Protection from the true cost of energy has contributed to a lack of competitiveness in ECA compared to neighbors in Western Europe and East Asia. A modern industry should be able to cope with real input costs by using those inputs far more efficiently. The large shifts in energy and economic systems over the next few decades will also create entire new industries and businesses. The global market for renewable energy, for example, was $250 billion in 2011 and is growing fast. Opportunities exist in all countries and across the technology spectrum. Governments can encourage green growth by pursuing ambitious climate policies and by improving their business climate. Countries that have learned the lessons of experience in the ECA Region's recent assessment of the European growth model (*"Golden Growth"*) will be in a good position to benefit from the transition to a low-carbon economy.

Third, by raising the cost of energy, climate action can affect employment and household welfare. Greater energy efficiency and effective social protection systems can soften those impacts. Labor market reforms and active labor market policies can facilitate job transitions, and social safety nets can help those unable to find adequate new work. Similarly, rising energy bills would hit the budgets of the poorest households the most. Assistance for better home insulation and more efficient appliances can reduce those bills. And many ECA countries have already improved social assistance programs to provide additional support with energy costs of poor households.

Climate action is one of the ECA Region's three strategic pillars. As this report shows, it is closely linked to the other two—competitiveness and social inclusion. Climate policies should prioritize actions that will strengthen competitiveness and promote economic growth. There are many opportunities to do so in the ECA region. And they can be complemented by affordable policies that moderate the costs of climate action for the poor and vulnerable. By becoming leaders on climate action, ECA countries can "grow green."

Philippe Le Houérou   Indermit Gill
Vice President   Chief Economist
Europe and Central Asia   Europe and Central Asia

Acknowledgments

This report was prepared by Uwe Deichmann and Fan Zhang based on contributions from a large number of World Bank staff and external experts. This work was carried out under the direction of Indermit Gill, Chief Economist for the Europe and Central Asia (ECA) Region, who generously provided his guidance, insights, and encouragement. The report was sponsored by the ECA Regional Leadership Team under Philippe Le Houérou, Regional Vice President.

The report received substantial support from the ECA management team. The team would particularly like to thank the Sector Director for Sustainable Development, Laszlo Lovei, as well as Kulsum Ahmed, John Kellenberg, Henry Kerali, Ranjit Lamech, Dina Umali-Deininger, and Wael Zakout; Yvonne Tsikata, Sector Director, and Benu Bidani in the Poverty Reduction and Economic Management Unit; and Jesko Hentschel in the Human Development Sector Unit. Martin Raiser, Country Director for Turkey, was an early champion of this study and gave helpful suggestions throughout its preparation. Kseniya Lvovsky and Markus Repnik, country managers in Albania and Bulgaria, respectively, also provided useful inputs.

The study builds on more than 20 background papers and policy notes authored or co-authored by Brian G. Bedard, Brian Blankspoor, Hannes Böttcher, Hei Sing (Ron) Chan, Jacqueline Cottrell, Mame Fatou Diagne, Ariel Dinar, Mark A. Dutz, Eleanor Charlotte

Ereira, Carolyn Fischer, Alexander Golub, Mykola Gusti, Marcel Ionescu Heroiu, Gary Howorth, Erika Jorgensen, Matthew E. Kahn, Leszek Kasek, Olga Kiuila, Natalia Kulichenko, Florian Kraxner, Donald F. Larson, Sylvain Leduc, Michael Levitsky, Shanjun Li, Anil Markandya, Craig Meisner, Andrew Mitchell, Carolina Monsalve, Michael Obersteiner, Jung Eun Oh, Isil Oral, Caroline Plante, Louis Preonas, Irina Ramniceanu, Nina Rinnerberger, Lourdes Rodriguez-Chamussy, Indhira Santos, Dmitry Schepaschenko, Siddharth Sharma, Maria Shkaratan, Anatoly Shvidenko, Jas Singh, Ahmed Slaibi, Govinda Timilsina, Sebastian Vollmer, Krzysztof Wojtowicz, and Tomasz Zylicz.

The report benefited greatly from coordinating closely with two related studies: *Energy Efficiency: Lessons Learned From Success Stories* by Gary Stuggins, Yadviga Semikolenova, and Alexander Sharabaroff; and *Balancing Act: Cutting Energy Subsidies While Protecting Affordability* by Caterina Ruggeri Laderchi, Anne Olivier, and Chris Trimble. The relevant sections of this report draw on their work. These studies have been supported by the ECA Region's Regional Studies Program, coordinated by Willem van Eeghen. Marianne Fay, Vijay Jagannathan and Michael Toman were peer reviewers for this report and also provided valuable inputs and suggestions in the study's early stages. Elena Kantarovich and Rhodora Mendoza Paynor provided administrative assistance and helped in the production of the report. Sofia Chiarucci and Naotaka Sugawara carried out data analysis for various parts of the report and Irina Bushueva, Xu Chen, Tarik Chfadi, Ryan Decker, Po Yin Wong, and Kuangyuan Zhang provided research assistance. Gazmend Daci, Dmytro Glazkov, Marat Iskakov, Elena Klochan, and Artur Kochnakyan kindly provided data for energy subsidy estimation.

In addition to the contributors of the background papers and parallel regional studies, many people at the World Bank and in the region provided helpful comments, suggestions, and other inputs along the way. The team thanks Gabriela Elizondo Azuela, Benoit Blarel, Anna Maria Bogdanova, Pascal Boijmans, Ken Chomitz, Jane Ebinger, Daryl Fields, Antonina Firsova, Franz Gerner, Alexander Gershunov, Kathrin Hofer, Ron Hoffer, Peter Johansen, Stephen Karam, Sunil Kumar Khosla, Agi Kiss, Andreas Kopp, Holger A. Kray, Ryszard Malarski, Yuriy Myroshnychenko, Shinya Nishimura, Kari Nyman, Victor Olkov, Harun Onder, Salvador Rivera, Alexander Rowland, Jitenedra P. Srivastava, Claudia Ines Vasquez Suarez, William Sutton, Jari Vayrynen, Kryzysztof Blusz, and Michael Yulkin. We wish to apologize to anyone inadvertently overlooked in these acknowledgments.

Matthias Beilstein, Viktor Novikov, and Otto Simonett of Zoinet prepared most of the maps in this report and also provided valuable comments on its content. Michael Jones was the principal editor and Romain Falloux designed the Overview of the report. From the Office of the Publisher, Denise Bergeron, Susan Graham, Paola Scalabrin, and Dina Towbin managed the production of the full report. Preparation of several background papers was supported by the World Bank's Green Growth Knowledge Platform, the Research Support Budget, and the DEC Knowledge for Change Program.

Abbreviations

°C	Degrees Celsius
€/tCO$_2$e	Euros per ton of carbon dioxide equivalent
AMC	Advanced market commitments
BAU	Business as usual
bcm	Billion cubic meters
BEE	Biomass Energy Europe
BF/BOF	Blast furnace or basic oxygen furnace
BNI	Biological nitrification inhibitors
BRT	Bus rapid transit
BTA	Border tax adjustment on imports
BTU	British thermal unit
CAFE	Corporate Average Fuel Economy
CCS	Carbon capture and storage
CDM	Clean Development Mechanism
CER	Certified emission reduction
CH$_4$	Methane
CHP	Combined heat and power
CIS	Commonwealth of Independent States
CNG	Compressed natural gas
CO	Carbon monoxide
CO$_2$	Carbon dioxide

CO$_2$e	CO$_2$ equivalent
CRU/NEA	Climatic Research Unit of the University of East Anglia
CTF	Clean Technology Fund
DALY	Disability adjusted life year
DETAŞ	Turkish Railway Transportation Corporation
EBRD	European Bank for Reconstruction and Development
ECA	Europe and Central Asia
EE	Energy efficiency
EE/DSM	Energy Efficiency and Demand Side Management (Fund)
EJ/yr	Exajoules/year
ENPI	European Neighborhood Policy East
ENV-CC	World Bank Environment-Climate Change
EOR	Enhanced oil recovery
ESA	Energy service agreement
ESCO	Energy service company
ESMAP	Energy Sector Management Assistance Program
ESTD	Early-state technology development
ETS	Emissions Trading System
EU	European Union
EWEA	European Wind Energy Association
FDI	Foreign direct investment
FiT	Feed-in tariff
FLEG	Forest Law Enforcement and Governance
FSU	Former Soviet Union
gCO$_2$/km	Grams of carbon dioxide per kilometer
gCO$_2$eq/pkm	Grams of carbon dioxide equivalent per passenger kilometer
gCO$_2$eq/tcm	Grams of carbon dioxide equivalent per ton kilometer
GDP	Gross domestic product
GEF	Global Environment Facility
GHG	Greenhouse gas
GJ	Gigajoule
GJ/yr	Gigajoules/year
GtC	Gigatons of carbon
GW	Gigawatts
GWh	Gigawatt-hours
ha	Hectare
HFC	Hydrofluorocarbon
HV	High voltage
ICT	Information and communications technology

IEA	International Energy Association
IFC	International Finance Corporation
IPCC	Intergovernmental Panel on Climate Change
IPR	Intellectual property rights
IT	Information technology
IUCN	International Union for Conservation of Nature
JI	Joint Implementation
JRC-EDGAR	Joint Research Centre of the European Commission's Database for Global Atmospheric Research
kg	Kilogram
kgoe	Kilograms of oil equivalent
kt	Kiloton
kWh	Kilowatt-hour
LCOE	Levelized cost of electricity
LDV	Light-duty vehicle
LED	Light-emitting diode
lge/100km	Liters of gasoline equivalent per 100 kilometers
LLC	Limited liability company
LNG	Liquefied natural gas
LV	Low voltage
MAC	Marginal abatement cost
Mt	Megaton
$MtCO_2e$	Million tons of carbon dioxide equivalent
mtoe	Million tons of oil equivalent
MV	Medium voltage
MW	Megawatt
MWh	Megawatt-hours
NASA	National Aeronautics and Space Administration
NGL	Natural gas liquids
NGO	Nongovernmental organization
NH_3	Ammonia
N_2O	Nitrous oxide
NOAA	National Oceanic and Atmospheric Administration
NO_x	Nitrogen oxide
NPPs	Nuclear power plants
O_3	Ozone
OECD	Organisation for Economic Co-operation and Development
PC	Price cap
PES	Payments for environmental services
PFC	Perfluorocarbon
PJ	Petajoule

$PM_{2.5}$, PM_{10}	Particulate matter
ppm	Parts per million
PPP	Purchasing power parity
PSIA	Poverty and social impact analysis
PV	Photovoltaic
R&D	Research and development
RE	Renewable energy
REC	Renewable energy credit
REDD	Reducing Emissions from Deforestation and Forest Degradation
ROR	Rate of return
RPS	Renewable portfolio standards
SCI	Sustainable Citied Initiative
SEE	Southeast European
SME	Small and medium enterprise
SO_2	Sulfur dioxide
SO_x	Sulfur oxide
SRES	Special Report on Emissions Scenarios
T&D	Transmission and distribution
TAMT	Transport Activity Measurement Toolkit
tcm	Trillion cubic meters
TJ	Terajoule
Toe	Tons of oil equivalent
TRACE	Tool for Rapid Assessment of City Energy
TWh	Terawatt hours
UN	United Nations
UNFCCC	United Nations Framework Convention on Climate Change
USPTO	United States Patent Office
VAT	Value added tax
VSL	Value of a statistical life
WHO	World Health Organization
WRI	World Resources Institute
WRI-CAIT	World Resources Institute's Climate Analysis Indicators Tool
WTO	World Trade Organization
WWF	World Wildlife Fund

Key Country Groups (45 European Countries)

The following country groups are used throughout this report. These categories are broad and commonly used across all the chapters.

Commonwealth of Independent States

Azerbaijan, Armenia, Belarus, Georgia, Kazakhstan, the Kyrgyz Republic, Moldova, the Russian Federation, Tajikistan, Turkmenistan, Uzbekistan, and Ukraine

Eastern partnership countries

Armenia, Azerbaijan, Belarus, Georgia, Moldova, and Ukraine

EFTA

Iceland, Liechtenstein, Norway, and Switzerland

EU candidate countries

Albania, Bosnia and Herzegovina, Croatia, Kosovo, the former Yugoslav Republic of Macedonia, Montenegro, Serbia, and Turkey

EU-10

Countries that joined the EU in 2004: Bulgaria, the Czech Republic, Estonia, Hungary, Latvia, Lithuania, Poland, Romania, the Slovak Republic, and Slovenia

EU-12

Countries that joined the EU in 2004 or 2007: Bulgaria, Cyprus, the Czech Republic, Estonia, Hungary, Latvia, Lithuania, Malta, Poland, Romania, the Slovak Republic, and Slovenia

EU-15

Austria, Belgium, Denmark, Finland, France, Germany, Greece, Ireland, Italy, Luxembourg, the Netherlands, Portugal, Spain, Sweden, and the United Kingdom

EU-27

EU-15 plus EU-12

CHAPTER 1

Introduction

Main Messages

- In the first post-transition decade after the fall of Communism, Europe and Central Asia (ECA) moved its economy from plan to market. In the second decade, the 2000s, it moved from social division to inclusion. The region has an opportunity to use the third decade, the 2010s, to move from brown to green growth—making production and consumption more sustainable, increasing quality of life, and reducing impacts on the climate.

- Lowering climate change risks in ECA will involve many different actions that fall broadly into three areas: Some, like energy efficiency improvements, are often economically beneficial regardless of climate concerns. Others, like creating a good business environment for green enterprises, are investments that create new growth opportunities. Finally, actions like expanding wind and solar energy will have net costs for some time but are essential to tackling climate change.

- A simple framework helps guide climate action. The priorities are to use energy more efficiently, use cleaner energy, and manage natural resources better. Although price instruments like carbon or energy taxes tend to be most effective, climate action will also require regulations and investments such as fuel efficiency standards or research and development spending. Complementary growth, social, and environmental policies promote the broader benefits of climate action while limiting its costs.

A Third Transition

The ECA region has experienced remarkable changes over the past 20 years. The region spent the first decade after the fall of Communism freeing itself from the legacy of central planning and mending its economies in an often painful adjustment process. The reward has been economic recovery and growth built on integration with Western Europe and the rest of the world, as figure 1.1 illustrates. ECA spent the next decade, the 2000s, making growth more inclusive by improving public services, education, and social safety nets. The rewards have been a fall in poverty, an emerging middle class, and higher living standards overall as the benefits from the economic shift have been shared more widely, as shown in figure 1.2.

These two transitions—from plan to market and from division to inclusion—took enormous effort up front but eventually paid ample dividends. They are also by no means complete. There is still much that could be done to increase competitiveness and ensure opportunities for all.

The financial crisis has understandably distracted from the long-term view. Among the World Bank regions, ECA was hit hardest. Countries had to focus on crisis management rather than on broadening past gains. Despite the recent turmoil, policy makers should not lose sight of another transition that is necessary to make development in ECA truly sustainable: the region's third transition will be to reduce the adverse environmental side effects of economic growth. For example, greenhouse gas (GHG) emissions fell after the first decade of transition but steadily rose in the 2000s until the recent economic crisis put the brakes on economic growth, as shown in figure 1.3.

The consequences of neglecting the environment will affect ECA's most vulnerable citizens more than anyone. To thrive in the long

FIGURE 1.1
Median GDP Growth in Europe and Central Asia, 1990–2010

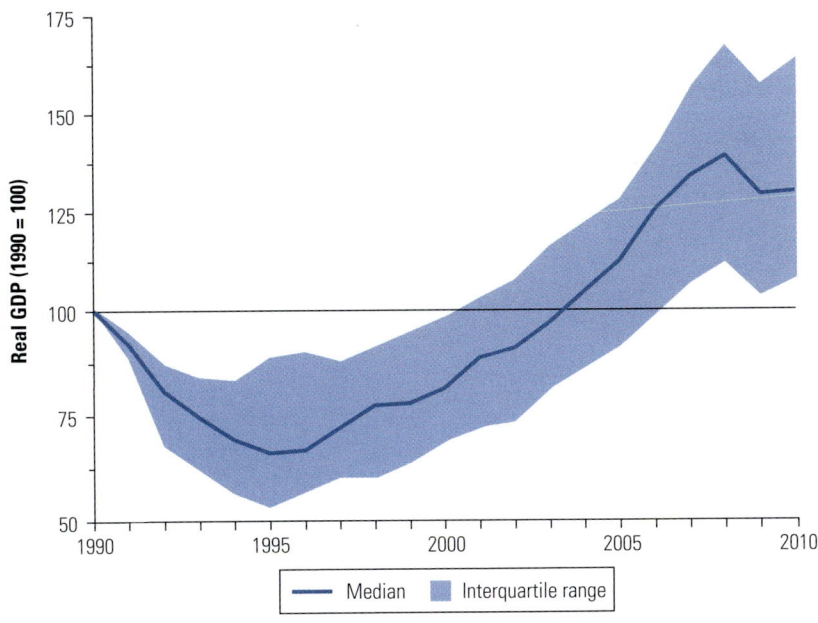

Source: World Bank, *World Development Indicators.*
Note: Bosnia and Herzegovina, Kosovo, and Montenegro are not included because data were not available. The figure also excludes Turkey because the focus is on the trend of transition countries.

FIGURE 1.2
Social Inclusion Trends in Europe and Central Asia, 1993–2008

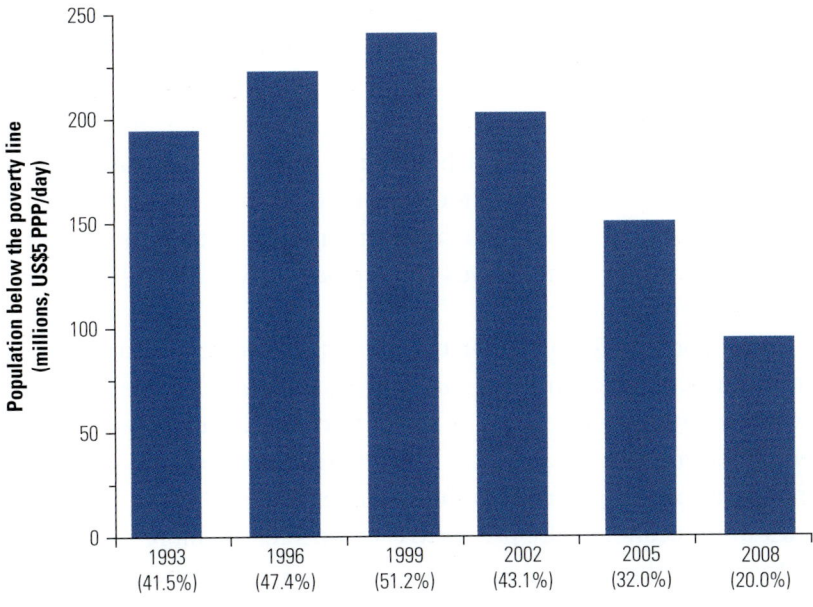

Source: World Bank, PovcalNet.
Note: Numbers in parentheses are poverty headcount ratios (as percentages of population). PPP = purchasing power parity.

FIGURE 1.3

Carbon Dioxide Emissions per Capita, Europe and Central Asia, 1995–2009

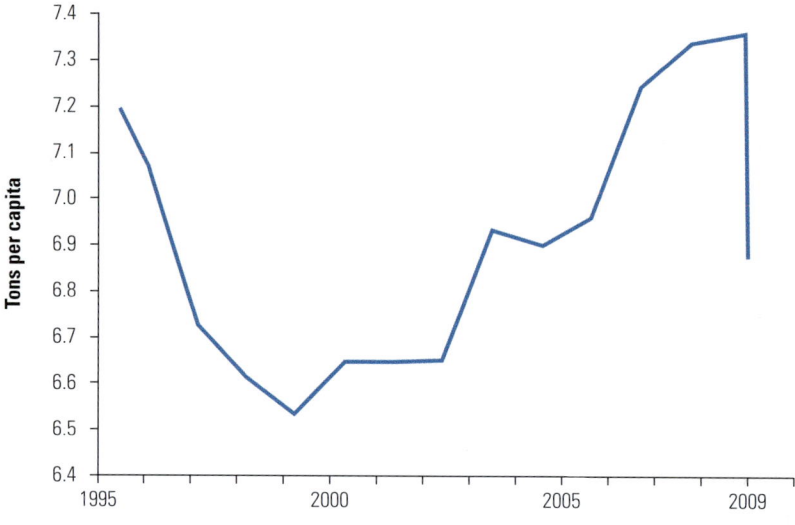

Source: IEA 2011.

term, the region's countries need to reduce acute, but often reversible, local pollution and ecological degradation by embracing greener approaches to growth (World Bank 2012). And they should join their Western European peers and tackle the danger of severe and irreversible climate change. The reward will be an economy that is far more efficient, clean, and green.

This third transition will be challenging. Globally, avoiding potentially dangerous climate change will require a reduction of per capita carbon dioxide (CO_2) emissions from 7 tons to 2 tons and an 80–90 percent reduction in harmful GHG emissions per unit of gross domestic product (GDP) (see Spotlight 1 at the end of this chapter). Some have called for an energy-industrial revolution (Stern 2009). ECA countries, which know about revolutions better than anyone else, will know that even ambitious climate action will not be anywhere near as disruptive or divisive.

This report will show that climate action in ECA would benefit from rapid progress in three main areas: (a) further, large improvements in energy efficiency; (b) a shift to cleaner energy systems that will also improve local health and energy security; and (c) better management of "green" natural resources that will make the countries more economically productive while keeping more carbon out of the atmosphere. The region has already achieved much progress in these areas, but much more could be done without jeopardizing other development objectives. Achieving these shifts will require smart and consistent policies because market prices for the use of energy and other resources do not reflect their impact on local populations and on the global climate.

As in any major transition, some firms and households face greater challenges than others. Well-balanced development strategies cushion the negative impacts of climate action to ensure social cohesion while promoting benefits to economic growth and competitiveness. The climate problem is difficult but not intractable. Without downplaying the urgency and magnitude of the challenge, what is needed more than anything else is a strong commitment and forward-looking policy making.

What are the main principles of a growth-promoting climate action strategy?

The first priority is to pursue measures that address existing inefficiencies in the use of resources, including energy and land. The benefits will generally exceed their costs, sometimes significantly. The best examples are energy efficiency investments such as improved air conditioning and heating systems, but such opportunities also exist through improved land use management, such as new farming techniques that increase soil productivity and also trap more carbon in the soil.

The second priority is to actively pursue green growth opportunities where climate action translates into new economic activities, jobs, and income growth. Manufacturing of green capital goods like wind turbines is one such growth opportunity, as are mostly nontradable energy services.

The third priority area is one in which low- and middle-income countries in particular should be most selective. It includes measures that are not financially viable without a subsidy but that are necessary to reduce climate change risk. Most renewable energy options still fall into this category, although their prices have been falling to an extent that they can soon become economically viable once climate, health, and energy security benefits are considered. Following these principles, climate action can be seen as a co-benefit, as an investment, and as an insurance policy.

Climate Action as a Co-Benefit

Advocates of climate action often seek to promote climate change mitigation policies by pointing out their co-benefits. Shifting electricity generation from coal-fired to lower-emission natural gas or renewables not only lowers GHG emissions but also reduces air pollution that causes respiratory health issues. Including these hidden costs of high-carbon energy options will make low-carbon options economically more competitive.

There are also climate-friendly measures that make economic sense even without considering climate benefits, which come as an added bonus. These "win-wins" are by no means abundant and by themselves will not be sufficient to achieve climate goals, but prioritizing policies that yield immediate and local benefits goes a long way toward a sustainable economy. In ECA in particular, large inefficiencies remain in the way energy is used. Energy use per unit of output in the region has been falling, but on average it is still more than twice that of Western Europe when using purchasing power parity (PPP) GDP (more than five times if measured using GDP at exchange rates). Waste of energy is particularly large in industrial production (chapter 8) and in buildings and urban services (chapter 9), which together account for 49 percent of GHG emissions in Europe and Central Asia. Energy efficiency gains could contribute as much as half of ECA's contribution to global emission reductions required to stabilize atmospheric greenhouse gas concentrations.

Energy efficiency is not the only area where economic gains exceed the costs of climate-friendly policies. The transport sector is the fastest-growing source of carbon emissions in ECA. A higher share of public transit and rail freight (relative to road transport) has benefits that are typically not reflected in market prices for transport services. Only about 5 percent of these side benefits from reduced road transport come from avoiding the damages of climate change. Ninety-five percent of the benefits come from reduced congestion, health issues, noise, and accidents (chapter 8). These benefits accrue immediately and justify sustainable transport policies that lower the price of transit or raise the costs of motorized transport where they are currently too low.

As a further example, the contribution of natural ecosystems to climate change mitigation is often overlooked. Forests in ECA store about 44 billion tons of carbon, and topsoil stores 55–120 billion tons. Combined, these carbon stores are equivalent to 8–12 times annual global GHG emissions. Deforestation is a far smaller problem in ECA than in tropical world regions, but poor forest management and increased frequency of fires jeopardize forest health. Further, about 685 million hectares of land are degraded in the region. With better management, ECA's natural ecosystems could store more carbon while also raising farm and forest productivity from levels that are generally lower than in other regions. Rates of return from land reclamation projects in Turkey and Uzbekistan, for example, have been around 20 percent even without including climate benefits (chapter 10).

Many of these seemingly attractive opportunities are ignored, and not just in ECA. One reason is that they compete with alternative investment opportunities that could have even higher payoffs. Another is that there are market, information, financial, and behavioral failures that cause firms, individuals, and governments to underinvest in projects that make economic sense and have climate benefits (chapter 2). Market signals are necessary but not sufficient, so governments would benefit from creating a policy environment that helps overcome these barriers. Even measures that pay for themselves within reasonable time frames may require a nudge. Doing so to take advantage of investments for which climate action is simply a co-benefit is a priority for all ECA countries.

Climate Action as Investment

Measures that reduce energy waste raise economic efficiency and thus contribute to growth. But policies that promote climate change mitigation can also contribute to economic growth more directly. The low-carbon transition is well under way and requires massive investments that generate economic opportunities and jobs. Currently this is happening mostly in the Organisation for Economic Co-operation and Development (OECD) countries and in China. These countries have ambitious domestic climate policies that promote the manufacture of energy-efficient capital and consumer goods, the development of clean energy infrastructure, and the emergence of green services such as energy services companies. Many start-up companies emerged to take advantage of new opportunities, but established companies are equally engaged. The engineering firm Siemens now generates a quarter of its turnover from green solutions and expects this share to grow. The solar and wind energy markets in Europe are now about €30 billion and growing at 15 percent per year. Globally as much as US$2.3 trillion could flow into renewable energy alone in this decade (Pew Charitable Trusts 2012).

As in traditional industries, green product leaders will concentrate on design and development. As these industries mature, they will seek locations with lower labor costs for other tasks and standardized manufacturing. ECA countries could capture a share of these economic opportunities. They have two main ways to do so: They can link up to the green part of "Factory Europe" by emulating other industries and becoming a part of green goods production networks. And they can develop domestic industries, initially mostly focused on lower-technology activities and green services. How can the region's countries attract domestic and foreign investments? A necessary condition is a good overall business environment. Many of these countries have made progress in this area and have climbed in the ranks of the World Bank Group's Doing Business indicators.[1] But others have lagged, and far more can be done to attract foreign and encourage domestic investment (Gill and Raiser 2012).

To attract *green* foreign direct investment (FDI), countries can go further (chapter 4). In traditional sectors, investments often followed historical strengths. Western European car manufacturers engaged in regions with long-established but uncompetitive automobile production. Volkswagen invested in the Czech Republic's Skoda, Fiat in Serbia's Zastava Automobiles and Turkey's TOFAS. Because most green-technology industries did not exist even 25 years ago, such historical links will not help. Instead, one way to attract such investments is to introduce strong and credible domestic climate policies. Such policies signal to investors that there is a commitment to climate action, a domestic market for green products, and strong support for an emerging domestic green economy sector. Some years ago, Costa Rica branded itself as a major hot spot for biodiversity by introducing comprehensive nature protection policies. This strategy yielded not just international publicity and goodwill but also a large increase in ecotourism and investments by bioresources companies, which are now mainstays of the economy. An ECA country that can brand itself as a leader in climate action will equally position itself as a preferred destination for green investment.

Besides reducing the environmental impacts of domestic production and consumption, growing green also means taking advantage of the economic opportunities presented by the large investments required for a low-carbon transformation. Climate policies—and related green policies not primarily motivated by climate concerns—will create some jobs and income opportunities in all countries. Going further and benefiting from global investment in green technology will be an opportunity that most countries in Europe and Central Asia can pursue.

Climate Action as Insurance

Some climate change actions will pay for themselves. However, solely relying on these interventions will not be sufficient to avoid potentially dangerous global warming. Achieving climate goals will also require measures that are initially more expensive than their conventional alternatives. Some are more expensive even when considering avoided future damages or immediate benefits such as lower public health impacts.

The shift from fossil fuels to an increasing share of clean energy in Europe, for example, requires support for renewable technologies through subsidies or incentives. Given the high price of clean energy technologies when policies such as feed-in tariffs were first introduced, these support policies were initially not justified by most estimates of the social cost of carbon emissions they avoided.[2] Instead they were motivated by the realization that climate change, if unchecked, could lead to catastrophic impacts as early as the second half of this century. The response to low-probability but high-impact events such as severe traffic accidents or a home burning down is to take out car or fire insurance. No insurance is available to protect against future catastrophic climate change (even though insurers and reinsurers are at the forefront of thinking about the implications of climate change for their businesses). Therefore, the subsidies required to make some climate change interventions (such as renewable energy technologies) competitive with traditional energy sources could be considered an insurance premium. As the use of these interventions increases, the industry will become more efficient, costs will fall, and the interventions will require less support.

The question is who should pay for what is essentially the global public good of low-cost clean energy or other mitigation technologies? Feed-in tariffs raised utility bills for ratepayers and taxpayers in Western Europe (chapter 6). The increases have been relatively minor, although they can add to hardship among poorer households even in wealthier countries. Low- or middle-income countries have understandably been reluctant to commit substantial resources to support technologies that are not competitive at current market prices. They do well in leaving the funding of initial cost reductions to high-income countries. But there are good reasons why they should not completely ignore developments among the climate action leaders.

One is that prices for green technology have been falling as demand—and therefore the scale of production—has increased. Solar panel prices have dropped by half in just the past two to three years. Wind energy in good locations is already cost-competitive with most conventional energy sources. It will be some time before such technologies will be able to consistently compete with fossil fuels at market prices, especially where conventional energy is subsidized. However, countries that make careful and early investments will develop the domestic expertise for quick and widespread deployment, while also promoting domestic jobs and energy security and lowering health costs from local air pollution. Smart power-sector strategies keep options open to invest in cleaner energy generation once prices reach parity. They will avoid lock-in to conventional generation, whose relative prices may well increase as renewables mature, fuel costs rise, or policy mandates emerge (chapter 3). European Union (EU) member states already face binding emission restrictions and clean energy targets, and EU-candidate countries will soon, too.

International support is available for countries willing to pursue clean energy investments. Several ECA countries can also take advantage of the Clean Development and Joint Implementation mechanisms under the United Nations Framework Convention on Climate Change to support clean energy investments. Also, the Clean Technology Fund, which channels funds through the multilateral investment banks, already supports investments in Kazakhstan, Turkey, and Ukraine. However, these funds are too limited to trigger large-scale changes and are best used strategically as additional incentives or for demonstration programs. Low- and middle-income economies should carefully evaluate domestic investments and support mechanisms for currently relatively expensive mitigation options (chapter 6). The benefits of well-designed programs will exceed the costs in many ECA countries.

Costs of Climate Action

Green growth is not free growth. Although win-wins exist, many mitigation measures will have net costs at least in the short to medium term, including transition or adjustment costs. These costs are the price for avoiding future damages that could be severe. What is important is to design climate policies that keep these costs as low as possible, which includes taking advantage of economic opportunities that a low-carbon transition brings.

Estimates of the annual cost of required climate change mitigation vary widely, from slightly negative to 3–4 percent of GDP. The European Commission expects its carbon cap to cost 0.3–0.7 percent of GDP to 2020. Most estimates, both globally and nationally, peg these costs at about 1 percent of GDP or less per year, with most predicting that net costs will turn into net benefits—for instance, when initial investments start to pay off in the form of reduced energy expenditures (World Bank 2011).

These cost estimates rely on economic models that include some notion of cost reductions due to technical change, but they are difficult to predict. Technical barriers may delay the low-carbon transition or make it more expensive than expected. Yet, in many instances, technical change happens faster and at lower cost than initially predicted when market signals or policy-induced demand growth create the right incentives. For instance, deployment of wind energy among the EU-15 countries[3] far exceeded the amounts forecast in the past, thanks to a combination of strong policies and rapid cost reductions, as shown in figure 1.4.

Besides the necessary investment costs of the low-carbon transition, climate action can put additional burdens on households—especially poor and middle-income families (chapter 5). Phasing out energy price subsidies makes economic sense and helps reduce wasteful energy use that harms the environment. Some countries in ECA have already moved to include some of the environmental costs of energy use in energy prices. This raises utility bills, which can create hardship among low-income households. Raising industrial energy prices or changing the price of high-emission fuels such as coal can also change the competitiveness of private firms and extractive industries, leading to employment losses that may be only partially offset by growth elsewhere. These kinds of impacts need to be anticipated.

FIGURE 1.4

Wind Energy Capacity in the EU-15, 2009 vs. Forecasts 1990–2002

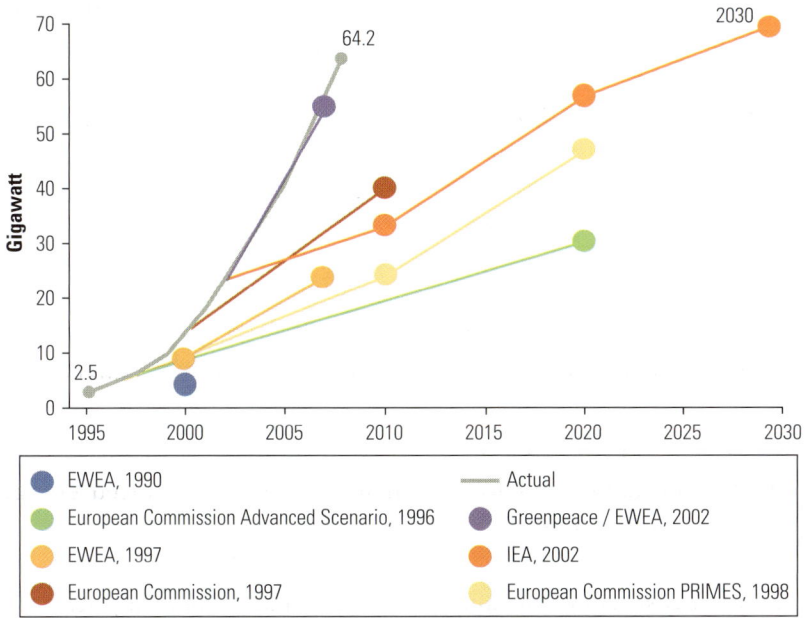

Source: Pieprzyk and Rojas Hilje 2009.

Note: EU = European Union; EWEA = European Wind Energy Association; IEA = International Energy Agency.

To win broad public support, climate and energy policies should be complemented by measures to reduce their social consequences (Ruggeri Laderchi, Olivier, and Trimble 2013). Policy-induced impacts of climate action are similar to those caused by other economic dynamics. Changes in trade regimes, commodity prices, or structural economic shifts due to changing prices for labor or inputs have triggered far greater disruptions in the past. The ECA region, therefore, has a lot of experience in dealing with such adjustments. It is important to ensure that energy services remain affordable for low-income households. It is also important to make it easier to invest in improved energy efficiency and to pursue labor market policies that help people to transition to new jobs in growing industries. Climate policies that are more inclusive will enjoy broader public approval.

The Elements of a Climate Action Strategy

From this brief overview emerge the elements of a climate action strategy, as shown in figure 1.5. Countries pursue economic, social, and environmental sustainability objectives. They pursue these objectives through various types of development policies: sectoral (or industrial) policies, social protection policies, and environmental safeguards. Climate action is often considered a subset of environmental

FIGURE 1.5
Elements of a Climate Action Strategy

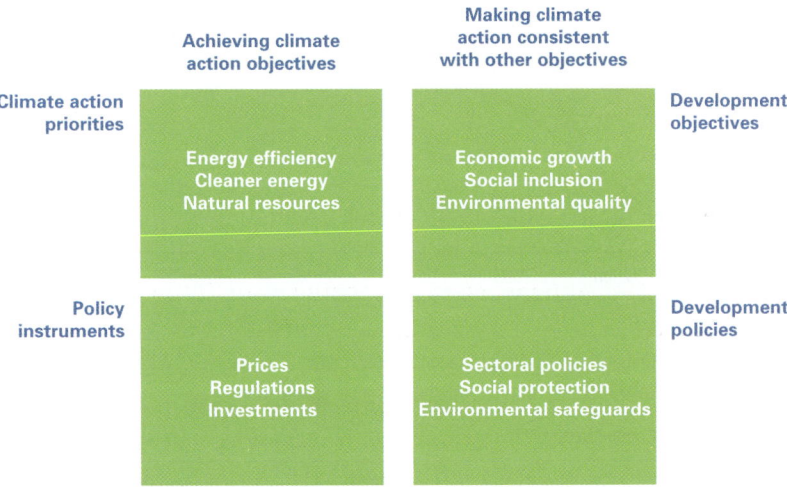

sustainability, but this report shows that it is closely linked to all three major development objectives. Climate policies therefore need to be coordinated with policy making in other domains.

There are three priorities for climate action: use less energy, use cleaner energy, and maintain carbon-storing natural resources. To achieve EU climate objectives, for instance, Europe's climate action leaders pursue all feasible emission-reduction options, but their relative importance and weighting will differ across countries. All countries need to continually improve energy efficiency where emission reductions come at lowest cost. There are large unrealized efficiency gains even in countries with low energy intensities such as Denmark or Ireland (Stuggins, Sharabaroff, and Semikolenova 2013). By switching to natural gas, further emission reductions are possible in countries that rely heavily on coal for electricity generation. Further, many countries in ECA have ample renewable energy endowments. These countries should be the regional leaders in renewable energy as the costs for generating clean energy continue to fall. Most of the region's countries can improve productivity in agriculture and forestry, which can also increase carbon sequestration. But it is in the very large ECA countries with enormous land and forest resources—such as Kazakhstan, Poland, the Russian Federation, Turkey, and Ukraine—that mitigation in the natural resource sector is of the highest importance globally.

Textbook environmental economics would suggest that, in an ideal world, a single policy instrument should be sufficient to achieve climate change mitigation: a general carbon price that reflects the damages that emissions will cause. That price triggers emission reductions in the economy, where they are most cost-effective. Yet, no country has introduced a general and sufficiently high carbon tax, and Europe's cap-and-trade system (a similar approach) extends to only a few sectors with an emissions cap that has so far been overly generous. Climate action therefore also requires other nonprice policy instruments to encourage energy efficiency, cleaner energy, and better natural resource management.

Price instruments should ensure that energy and natural resource use reflect market prices, whereas subsidies distort energy markets and encourage wasteful energy use. Ultimately, instruments should also reflect the indirect social cost of energy and natural resource use. The environmental economist Ernst Ulrich von Weizsäcker put it starkly: "Communism collapsed because it was not allowing prices to tell the economic truth, . . . capitalism may also collapse if it does not allow prices to tell the ecological truth"

(Von Weizsäcker et al. 2009). Price instruments include not only taxes and fees but also subsidies or preferential feed-in tariffs for renewable energy. Where price instruments are insufficient or not feasible, often for political reasons, regulations will be required. Almost all countries have fuel efficiency standards or restrict the release of harmful pollutants.

Governments can also support climate action more directly through *investments*. Most important, they should support knowledge creation and dissemination such as through information programs to help increase the adoption of energy efficiency measures. Where the private sector is unable to do so, public investments can also finance complementary infrastructure such as the grid expansion and management that support a higher share of renewables. Usually a combination of instruments is desirable or even necessary. For instance, when raising fuel taxes, governments must also make sure that people have alternative mobility options, which requires investment in public transit.

To ensure that climate action does not compromise other development objectives, climate policies need to be closely coordinated with growth, social, and environmental policies. To encourage growth benefits from climate policies, for example, governments can support national innovation systems and well-functioning labor markets. Removal of fossil-fuel subsidies raises heating costs and electricity prices for households. Poor households may need assistance to cope with such price shocks, either through the social safety net or assistance in energy efficiency investments. And environmental safeguards ensure that investments in climate change mitigation, such as hydropower or wind turbines, do not harm biological resources such as river ecosystems or birdlife. In sum, climate action needs to be closely integrated with policy making in other spheres of the economy and society.

Outline of the Report

The remainder of this report is divided into three parts: the first on priorities for a low-carbon transformation; the second on promoting growth and ensuring social inclusion; and the third on sectoral policies.

Part I: Priorities for Low-Carbon Transformation

Chapters 2 and 3 cover two of the main climate action priorities that cut across the issues discussed in later chapters: energy efficiency and cleaner energy sources.

- *Chapter 2*: Energy efficiency is still the highest climate change mitigation priority for all ECA countries and one that is relevant for all economic sectors. Energy efficiency enables a decoupling of economic growth from energy use and emissions; however, by itself, it will not yield sufficiently large reductions in emissions.

- *Chapter 3*: Large-scale emission reduction requires switching to low-carbon energy sources; the large uncertainties in power sector planning favor a greater focus on more decentralized and cleaner energy infrastructure.

Part II: Promoting Growth and Ensuring Social Inclusion

The second part of this report focuses on the link between climate action and the other main pillars of the ECA strategy: growth and social inclusion.

- *Chapter 4*: Green growth, in a narrow sense, concerns how countries can benefit economically from the low-carbon transition by encouraging green FDI, trade, innovation, and job creation in rising industries.

- *Chapter 5*: Several options address the potential negative impacts of climate action. The chapter examines how the increasing cost of traditional fossil fuels will affect competitiveness in mining and energy-intensive industries. In addition, it discusses labor market and household-level impacts and proposes strategies to buffer them, and it presents evidence of the direct social benefit of climate action in the form of reduced health impacts from fossil-fuel burning.

Part III: Sectoral Policies

The final part of the report discusses specific sectoral strategies and policy options for climate change mitigation.

- *Chapter 6*: Options for cleaner electricity generation include a switch from coal to natural gas in the short term and a much higher share of renewable energy in the medium to long term—whereas the potential contributions of carbon capture and sequestration and nuclear power remain uncertain.

- *Chapter 7*: This examination of production issues discusses options for realizing the large energy efficiency potential in energy-intensive industrial sectors.

- *Chapter 8*: Policies for more sustainable transport should take into account that the trend toward greater motorization will be hard to slow, let alone reverse. But ECA's compact cities and high rail share are a basis for sustainable mobility strategies.

- *Chapter 9*: Urban areas will be a focal point for climate action. The largest share of GHG emissions originates there, but cities also present a large and concentrated potential for emission reductions in buildings and public services as well as by managing urban form and its interaction with public transit.

- *Chapter 10*: Policies to reduce emissions and increase sequestration in rural areas include improved soil management, more sustainable livestock production, and better forestry practices.

Spotlights

Three short spotlights interspersed among this report's main parts provide context:

- *Spotlight 1*, which follows below, outlines the climate challenge. Given scientists' best assessment of risks if current emission trends continue, climate action is an urgent and formidable task.

- *Spotlight 2*, at the conclusion of chapter 3, summarizes current emission trends in ECA, over time and across sectors, and compares them with those in other world regions.

- *Spotlight 3*, at the conclusion of chapter 5, discusses why climate action is a particularly difficult task in the ECA region—not for technical or economic reasons, but because public perceptions do not favor immediate and ambitious mitigation.

Notes

1. For more about the Doing Business indicators, see http://www.doingbusiness.org/.
2. Environmental economics suggests that the subsidy for clean energy (or the penalty for dirty energy) should equal the marginal damage it avoids (or causes), as discussed, for example, in Aldy et al. (2010).
3. The EU-15 countries include Austria, Belgium, Denmark, Finland, France, Germany, Greece, Ireland, Italy, Luxembourg, the Netherlands, Portugal, Spain, Sweden, and the United Kingdom.

References

Aldy, Joseph E., Alan J. Krupnick, Richard G. Newell, Ian W. H. Parry, and William A. Pizer. 2010. "Designing Climate Mitigation Policies." *Journal of Economic Literature* 48 (4): 903–34.

Gill, Indermit, and Martin Raiser. 2012. *Golden Growth: Restoring the Lustre of the European Economic Model*. Washington, DC: World Bank.

IEA. 2011. IEA CO_2 Emissions from Fuel Combustion (database).

NRC (National Research Council). 2011. *Climate Stabilization Targets: Emissions, Concentrations, and Impacts over Decades to Millennia*. Washington, DC: National Academies Press.

Pew Charitable Trusts. 2012. "Who's Winning the Clean Energy Race? Growth, Competition and Opportunity in the World's Largest Economies." G-20 Clean Energy Factbook, The Pew Charitable Trusts, Washington, DC.

Pieprzyk, Björn, and Paula Rojas Hilje. 2009. *Erneuerbare Energien: Vorhersage und Wirklichkeit* [Renewable Energy: Prediction and Reality]. Berlin: German Renewable Energies Agency.

Ruggeri Laderchi, Caterina, Anne Olivier, and Chris Trimble. 2012. "Balancing Act: Cutting Energy Subsidies While Protecting Affordability." Energy study, Europe and Central Asia Region, World Bank, Washington, DC.

Stern, Nicholas H. 2009. *The Global Deal: Climate Change and the Creation of a New Era of Progress and Prosperity*. New York: PublicAffairs.

Stuggins, Gary, Alexander Sharabaroff, and Yadviga Semikolenova. 2012. "Energy Efficiency: Lessons Learned from Success Cases." Europe and Central Asia Region, World Bank, Washington, DC.

Von Weizsäcker, Ernst Ulrich, Karlson Hargroves, Michael H. Smith, Cheryl Desha, and Peter Stasinopoulos. 2009. *Factor Five: Transforming the Global Economy through 80% Improvements in Resource Productivity*. London: Earthscan.

World Bank. 2011. "Transition to a Low-Emissions Economy in Poland." Low-carbon growth study, Europe and Central Asia Region, World Bank, Washington, DC.

———— 2012. "Inclusive Green Growth: The Pathways to Sustainable Development." Sustainable Development Network report, World Bank, Washington, DC.

————. n.d. PovcalNet online database, World Bank, Washington, DC. http://iresearch.worldbank.org/PovcalNet.

————. Various years. *World Development Indicators*. Washington, DC: World Bank.

The Climate Challenge

Globally, heat-trapping emissions need to peak within a decade and drop by 80–90 percent by 2050 to reduce the risk of more frequent and severe impacts from flooding, heat waves, droughts, and forest fires in ECA.

Mikhail Ivanovich Budyko, who spent most of his career at academic institutions in Leningrad (today's St. Petersburg), was one of the fathers of climatology as a quantitative science. Following his pathbreaking work on global heat balances in the 1950s, he began to study the "earth snowball effect": a significant drop in atmospheric CO_2 could trigger a self-reinforcing process that might cause the earth to be completely covered in ice. Budyko did not believe this had ever happened in geological history, but by studying global cooling, he also clarified the role of CO_2 in warming the planet. He was one of the first to scientifically estimate the global warming effects of rising CO_2 levels, and he became convinced of humans' role in this process: "The conclusion is made that present-day climate appears to have changed as a result of man's inadvertent impact and this change may be considerably increased in the nearest decade" (Budyko 1977). It took almost half a century after his pioneering research for this insight to become the scientific consensus with the Third Assessment Report of the Intergovernmental Panel on Climate Change (IPCC 2001).

For the past 200 years, since the start of the Industrial Revolution, energy derived from coal, oil, and gas has fueled unprecedented economic growth and welfare gains. These carbon-rich resources had been safely stored in the ground for more than 300 million years in the case of coal. Burning them within the space of a few generations—each year we burn the products of 1 million years of photosynthesis[1]—has added far more carbon to the atmosphere than land and oceans can absorb. For the past 400,000 years, atmospheric CO_2 concentrations varied between 180 and 300 parts per million (ppm); in the past 10,000 years—the time in which modern civilizations developed—concentrations ranged between 260 and 280 ppm. Since the Industrial Revolution, they have increased to 394 ppm as of early 2012, a level the earth had last seen about 3 million years ago in the early Pliocene Epoch. Adding the effect of other greenhouse gases such as methane brings current concentrations to about 450 ppm of CO_2-equivalent or CO_2e. Atmospheric CO_2 and other gases trap incoming radiation from the sun, causing a gradual warming—the so-called greenhouse effect. Over the past 100 years, this phenomenon caused temperatures to increase by about 0.8°C, as shown in figure S1.1.

FIGURE S1.1
Rise in Global Land Air Temperatures, Deviations from the 1951–80 Mean

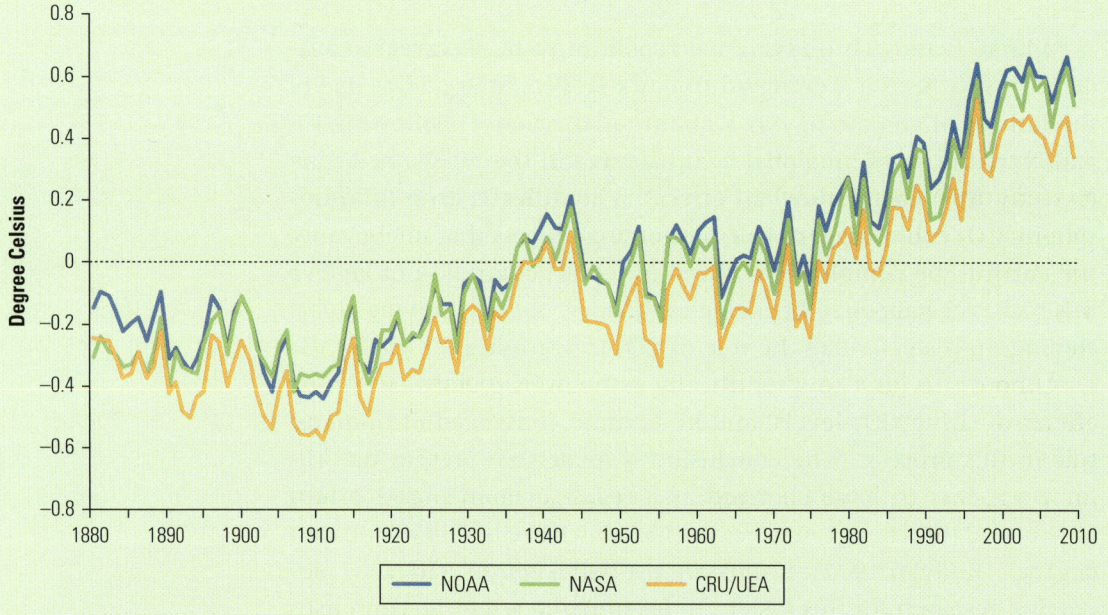

Sources: NASA, NOAA, CRU/UEA.

Note: NASA = National Aeronautics and Space Administration (http://data.giss.nasa.gov); NOAA = National Oceanic and Atmospheric Administration (www.ncdc.noaa.gov); CRU/NEA = Climatic Research Unit of the University of East Anglia (www.cru.uea.ac.uk).

The fundamental physics of this process have been known since Joseph Fourier's work in the 1820s, and in 1896, Svante Arrhenius established the link between human-induced emissions of GHGs and global warming. Today, basic science, climate models, and an increasing body of empirical evidence all confirm the existence of global warming. In response, the global community has adopted the scientific consensus that the increase in global temperature should remain below 2°C to prevent dangerous changes in the climate system. This is the goal that governments agreed to at the global climate summit in Copenhagen in 2009.[2] Without effective climate action, under current policies, temperatures will go up further—possibly by 5°C or 6°C above preindustrial levels by the end of the century. These are levels last seen about 30 million years ago, when snow and ice had largely disappeared and sea levels were 30–70 meters higher.

CO_2e concentrations are currently growing at about 2–2.5 ppm per year. But what drives warming is not the rate but the cumulative emissions that determine GHG concentrations and thus temperatures for at least 1,000 years after emissions stop (Solomon et al. 2009).[3] To have a good chance of keeping warming to 2°C, the world must limit cumulative emissions to 1.2 trillion tons of carbon (or about 4.4 trillion tons of CO_2e). About half of this has already been emitted since the Industrial Revolution. To achieve the Copenhagen goal, today's annual emissions of about 48 billion tons of CO_2e need to peak within the next 10 years and then drop quickly to about 20 billion tons by 2050 (Roeglj et al. 2011; NRC 2011). In per capita terms, emissions need to drop from today's 7 tons of CO_2e to about 4 tons by 2030 and 2 tons by 2050, when world population will be between 9 billion and 10 billion.

If economic output increases threefold as more countries achieve middle- and high-income status, emissions per unit of output will have to fall by 80–90 percent over the next four decades. This is a daunting task that must involve all countries in the world and one that will become harder the longer climate action is delayed. Countries with historically high emissions will need to cut more, and some developing countries may need to temporarily overshoot these targets.

Although the drivers of climate change are well understood, there is far more uncertainty about its future impacts. The earth's climate is already experiencing higher variability against a warming trend, as shown in map S1.1. Global temperatures are generally higher than in the middle of the 20th century. Temperatures vary from year to year and from place to place, but there is now a greater probability that any given place will experience higher than historically expected temperatures—sometimes much warmer, as in the heat waves in Western Europe in 2003 and Western Russia in 2010 that led to many deaths.[4]

MAP S1.1

Trends in Global Surface Temperature, 1955–2011

June-July-August surface temperature deviations relative to 1951–80 mean, °C

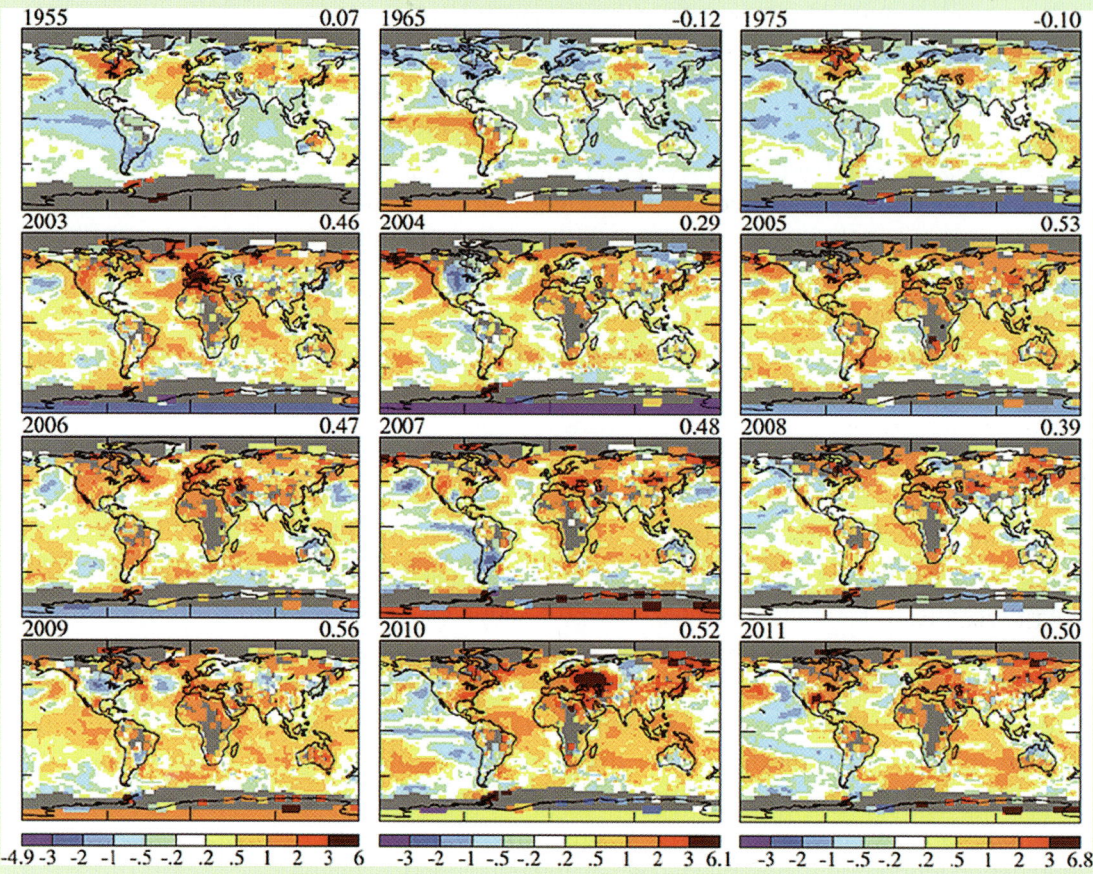

Source: Hansen, Sato, and Ruedy 2011.
Note: The number on the top right of each map shows the average annual temperature deviation from the 1951–80 mean.

Although it is impossible to definitively attribute specific events to climate change, such heat waves are consistent with climate model forecasts. In fact, changes such as shrinking glaciers and diminishing sea ice have occurred faster than models predicted. For the ECA region, these models predict changes in agriculture, hydrology, and temperature extremes that could bring more frequent flooding, thawing of northern permafrost, unreliable harvests, and severe fires in forests and drained peatlands. A previous World Bank ECA Regional Study summarized climate change impacts and adaptation in the region, and five more detailed adaptation pilots analyzed probable impacts and response options in Southeastern Europe and Central Asia (Fay, Block, and Ebinger 2010; Hoffer and Horvathova 2012; World Bank 2012b).

Notes

1. According to James Barber at Imperial College London. See http://www.gatsbyplants.leeds.ac.uk/TR/resource/uploads/Barber_J/player.html.
2. Some scientists doubt whether this is an achievable goal. New research also suggests that damages from such warming could be larger than previously estimated. See Anderson and Bows (2011).
3. Raymond T. Pierrehumbert makes this point forcefully: "It is the CO_2 emissions, and the CO_2 emissions alone, that determine the climate that humanity will need to live with for a time that stretches into the future at least as long as the time since the founding of the first Sumerian cities stretches into the past. The usual wimpy statement that CO_2 stays in the air for 'centuries' doesn't begin to convey the far-reaching consequences of the amount of CO_2 we decide to pump out in the coming several decades" (Pierrehumbert 2010). The scenario following this note in the text is adapted from Stern (2009).
4. The 2010 Russian heat wave coincided with 50,000 more deaths in the affected region than would have been expected under normal summer conditions, according to MunichRe Reinsurance's NatCatSERVICE database, "Natural catastrophes in 2010" (www.munichre.com/en/reinsurance/business/non-life/georisks/natcatservice/default.aspx). The 2003 Western European heat wave had a similar impact. There is a lively academic debate about whether the Russian heat wave can be attributed to climate change (Rahmstorf and Coumou 2011) or natural long-term variability (Dole et al. 2011). By the time there is statistical certainty about attribution of specific events to human-induced climate change, the world will likely have committed to warming beyond presumably safe limits.

References

Anderson, Kevin, and Alice Bows. 2011. "Beyond 'Dangerous' Climate Change: Emission Scenarios for a New World." *Philosophical Transactions for the Royal Society A* 369 (1934): 20–44.

Budyko, Mikhail I. 1977. "On Present-Day Climatic Changes." *Tellus* 29 (3): 193–204.

Dole, R., M. Hoerling, J. Perlwitz, J. Eischeid, P. Pegion, T. Zhang, X. Quan, T. Xu, and D. Murray. 2011. "Was There a Basis for Anticipating the 2010 Russian Heat Wave?" *Geophysical Research Letters* 38.

Fay, Marianne, Rachel I. Block, and Jane O. Ebinger. 2010. *Adapting to Climate Change in Eastern Europe and Central Asia*. Eastern Europe and Central Asia Reports. Washington, DC: World Bank.

Hansen, James, Makiko Sato, and Reto Ruedy. 2011. "Climate Variability and Climate Change: The New Climate Dice." NASA Goddard Institute for Space Studies, New York.

Hoffer, Ron, and Ivana Horvathova. 2012. *Adapting to Climate Change in Europe and Central Asia: Lessons from Recent Experiences and Future Directions*. Washington, DC: World Bank.

IPCC (Intergovernmental Panel on Climate Change). 2001. *Third Assessment Report: Climate Change 2001*. New York: IPCC of the United Nations Environment Programme and the World Meteorological Organization. http://www.ipcc.ch/publications_and_data/publications_and_data_reports.shtml#.ULz-kYawWtg.

NASA, NOAA, CRU/UEA

Pierrehumbert, Raymond T. 2010. "Losing Time, Not Buying Time." Real Climate blog, http://www.realclimate.org/index.php/archives/2010/12/losing-time-not-buying-time.

Rahmstorf, Stefan, and Dim Coumou. 2011. "Increase of Extreme Events in a Warming World." *Proceedings of the National Academy of Sciences* 108: 17905–09.

Roeglj, Joeri, William Hare, Jason Lowe, Detlef P. van Vuuren, Keywan Riahi, Ben Matthews, Tatsuya Hanaoka, Kejun Jiang, and Malte Mein-shausen. 2011. "Emission Pathways Consistent with a 2°C Global Temperature Limit." *Nature Climate Change* 1 (8): 413–17.

Solomon, Susan, Gian-Kasper Plattner, Reto Knutti, and Pierre Friedling-stein. 2010. "Irreversible Climate Change due to Carbon Dioxide Emissions." *Proceedings of the National Academy of Sciences* 106 (6): 1704–09.

Stern, Nicholas H. 2009. *The Global Deal: Climate Change and the Creation of a New Era of Progress and Prosperity*. New York: PublicAffairs.

World Bank. 2012. "Looking Beyond the Horizon: Adapting Agriculture to Climate Change in Four Europe and Central Asia Countries." Report AAA71-7E, Europe and Central Asia Region, World Bank, Washington, D.C.

PART I

Priorities for a Low-Carbon Transformation

Continuing current trends and patterns of energy use would lead to global warming of as much as 4 degrees Celsius (°C) by the end of this century even if the modest current mitigation commitments are kept. Some observers think we will run out of fossil fuels before cumulative carbon emissions have led to unsustainable climate change. That seems unlikely given recent large discoveries of unconventional fossil fuels and improved exploration and production techniques. The main climate action priority therefore must be to gradually transform the way we produce and use energy. The next two chapters discuss issues related to energy efficiency and cleaner energy that cut across the sector-specific policy priorities covered in chapters 6–10. Natural resource management, the third main priority, relates more directly to policies for agriculture and forestry, discussed in chapter 10.

Energy efficiency and cleaner energy would address severe market failures contributing to the climate change problem. One of these failures is that households and firms often use more energy to perform a given task than is necessary and economically beneficial. Lower or zero-emission energy more directly addresses the contribution of traditional fossil-fuel emissions to local air pollution and global warming.

As chapters 2 and 3 show, energy efficiency and cleaner energy objectives are interwoven in a number of areas. Both contribute to reducing environmental impacts relative to economic output. Energy efficiency can prevent growth in energy consumption and emissions that would otherwise result from economic growth. But achieving climate objectives will require absolute reductions in emissions, which will require both greater energy efficiency and a greater share of clean energy.

The cost-effectiveness of energy efficiency investments also depends on the price of energy. Where energy production—whether traditional or clean—is subsidized, energy efficiency measures become less profitable. Fossil-fuel subsidies are hard to justify even on social grounds because there are far more efficient ways to deal with afford-ability issues. By the same token, clean-energy subsidies should be scaled back as these technologies reach market competitiveness, so that incentives for greater energy efficiency remain strong.

Energy efficiency and clean-energy investments also share simi-larities when it comes to financing. Both require high up-front investments and relatively low operations and maintenance costs. Promoting them therefore requires similar financing instruments.

Finally, energy efficiency is also an important part of energy sector planning. Europe and Central Asia (ECA) will need to replace a large share of the energy capital stock and build some new generation capac-ity. Given the need for a low-carbon transition, the speed at which technology and prices have changed recently, and possible impacts of climate change on the power sector itself, improving energy efficiency can delay or even avoid costly and possibly suboptimal investments.

Chapter 2 begins by showing that because ECA countries signifi-cantly reduced energy intensity—the amount of energy that is used to produce a dollar of gross domestic product (GDP)—they avoided increases in total energy use that would have been expected with economic growth. Efficiency will continue to be an important com-ponent of climate action but will not by itself achieve sufficient reductions in energy use and emissions. The chapter discusses policy options for overcoming barriers to energy efficiency investments and concludes with a survey of financing options that are also relevant for clean-energy investments.

Chapter 3 discusses several sources of uncertainty that influence power sector investment planning. It argues that these uncertainties—in terms of regulations, technology development, and climate impacts—favor an increasing share of smaller-scale, decentralized, and clean-energy generation as ECA's power sector replaces and expands its energy infrastructure over the coming decades.

CHAPTER 2

Energy Efficiency[1]

Main Messages

- Energy use in relation to output in much of Europe and Central Asia (ECA) is still high. If the region produced output as efficiently as the Organisation for Economic Co-operation and Development (OECD) average, it could save energy equivalent to South America's entire consumption. But ECA's energy intensity has been falling—on track to gradually converge with Western European levels while also contributing to overall productivity gains.

- Introducing an economywide tax on emissions would promote energy efficiency but is politically difficult. Achieving higher energy efficiency will therefore involve numerous sector-specific policies. Price instruments, including revenue-neutral environmental taxation (ecotax) reform, can best trigger market-friendly solutions. Regulations and incentives will sometimes be necessary to overcome behavioral or financial barriers.

- Although a wide range of energy efficiency policies has been used, there is scope for further policy innovation, sharing experience across countries, and learning from economic and impact analysis.

- Economic growth—facilitated by higher energy efficiency—leads to more energy use, so total energy consumption and associated greenhouse gas emissions have not been falling as they must to achieve climate goals. Energy efficiency's contribution to emission reduction will thus be largely indirect by helping to create the wealth required for a transition to clean energy sources.

On September 4, 1882, the world's first electric power plant started its operation in New York City.[2] The Edison Electric Illuminating Co. had spent two years building the Pearl Street Station, where a steam engine drove six 27-ton "Jumbo" generators with a capacity of 100 kilowatts. The first underground electric mains had been installed the year before and enabled connection to individual customers in lower Manhattan. This system established the model of electricity generation, transmission, and distribution that persists to this day. But Edison's enterprise differed from today's electric utilities in one crucial respect: rather than selling kilowatt-hours, his first 59 customers purchased electric light. Rather than electricity, they bought the services that electricity provided. It was therefore in the interest of Edison's company not only to generate electricity as efficiently as possible but also to produce lightbulbs that required the least amount of energy. This model did not survive. Soon, electric utility profits started to depend on how much electricity they sold, and lightbulb manufacturers worried little about how much electricity their bulbs consumed because customers paid the utility bill.

Edison's business model of selling "illumination" rather than electrons highlights one of the most important aspects of energy efficiency: people do not want to consume energy; they want to consume the services that energy provides. So reducing energy consumption is not welfare reducing as long as it delivers comfortable temperatures in buildings, convenient travel, or the use of time-saving household appliances. Countries achieve similar living standards at levels of energy consumption that sometimes vary by an order of magnitude, as shown in figure 2.1.

FIGURE 2.1

Countries' Per Capita Energy Consumption Relative to Per Capita Incomes, 2009

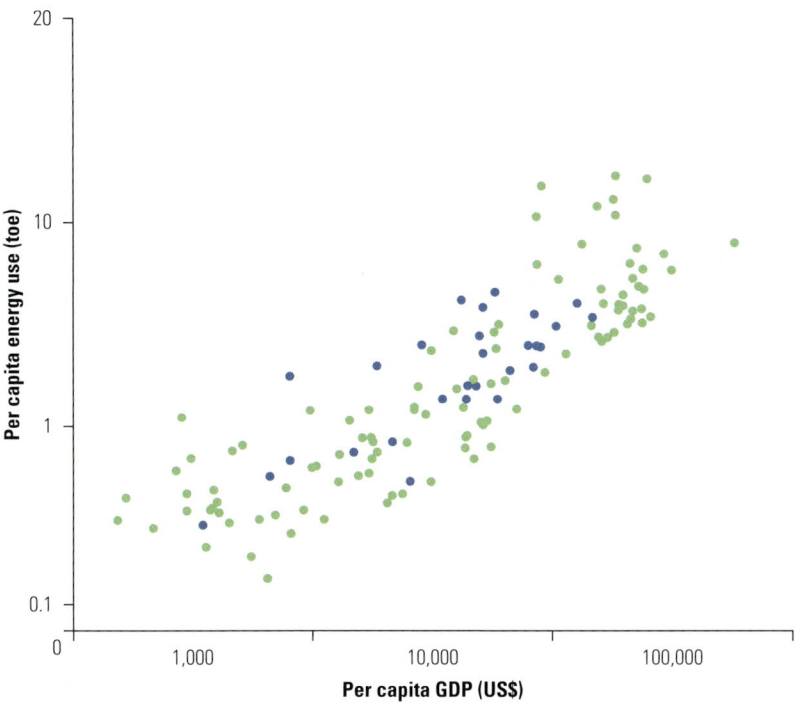

Source: IEA 2011.

Note: Blue dots indicate Europe and Central Asia countries; green dots indicate other countries; toe = tons of oil equivalent.

Some of these differences can be explained by climate (see, for example, Neumayer 2002). Very cold places need a great deal of energy for heating, and hot places need more energy for cooling. Some differences can be explained by economic structure: a large mining and basic manufacturing sector requires more energy input than a service economy. But the differences also have to do largely with how efficiently energy is used for household, commercial, and industrial uses. A main determinant of these remaining differences is the price of energy, which tends to be lower in countries with large deposits of oil, natural gas, and coal. Per capita energy use in countries with fossil-fuel reserves—including several in ECA—is about 80 percent higher on average than in countries with no reserves, and the larger the reserves relative to population, the larger the energy use, as figure 2.2 illustrates.[3]

ECA countries have substantial untapped potential to reduce energy consumption relative to economic output. If the region could match the OECD's average energy intensity, it would produce its current output using around 42 percent less energy, saving

FIGURE 2.2

Countries' Per Capita Energy Consumption Relative to Per Capita Fossil-Fuel Reserves, 2009

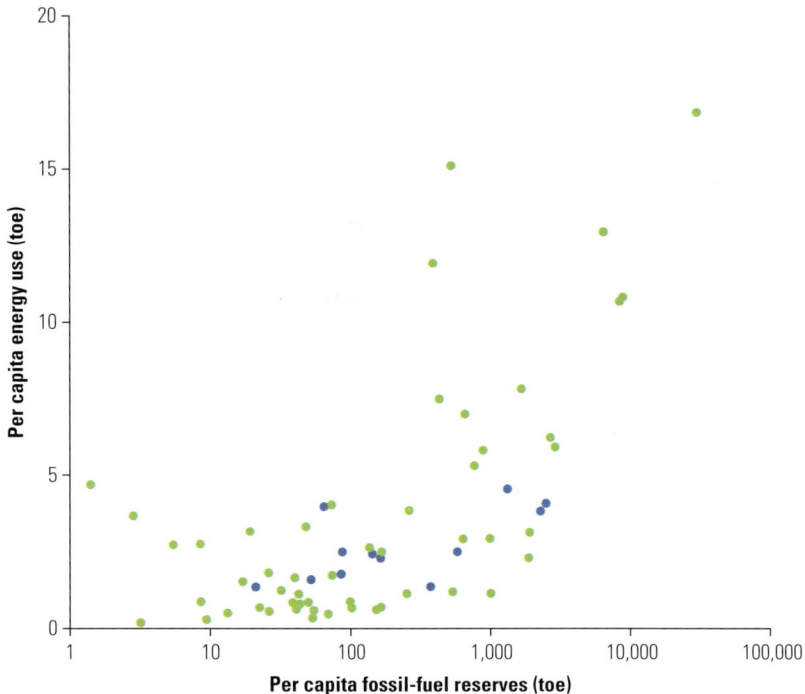

Sources: IEA 2011; BP 2011.
Note: Fossil-fuel reserves include oil, natural gas, and coal. Includes all countries with significant fossil-fuel reserves for which data were available. Blue dots indicate Europe and Central Asia countries; green dots indicate other countries; toe = tons of oil equivalent.

around 570 million tons of oil equivalent (mtoe)—or more than all of South America's energy consumption in 2009.[4] The Russian Federation alone could save the equivalent of the energy use of the United Kingdom—or, with best available technology, the even-higher equivalent of France's energy consumption (Bashmakov et al. 2008; IEA 2011). Because most energy is still derived from fossil fuels, lowering energy use also helps keep greenhouse gas (GHG) emissions in check. The International Energy Agency (IEA) expects that energy efficiency could contribute up to half of the required emission reductions needed to limit temperature changes to presumably safe levels.

This chapter discusses an important empirical observation about energy intensities (the ratios of primary energy consumption to gross domestic product [GDP]) in ECA: although they are still high in many of the region's countries, they have much improved and are gradually moving toward Western European levels. This is significant for economic growth because lower energy intensities, through improved energy efficiency, raise the productivity of firms. Improved energy efficiency in ECA has led to a decoupling of economic growth and energy use. But because of rising GDP, energy consumption and GHG emissions have, at best, remained constant rather than falling, which would be necessary to keep climate change in check.

The chapter then briefly reviews the well-studied but largely unresolved issue of why seemingly profitable energy efficiency opportunities are often ignored; it subsequently summarizes the energy efficiency policies that have worked. Stuggins, Sharabaroff, and Semikolenova (2013) discuss household-level energy efficiency in more detail, and chapters 6–10 do so for specific sectors.

The chapter concludes with a discussion of fiscal and financing instruments. Environmental taxation (or ecotax) has become an important tool to encourage energy savings, while productive deployment of the resulting revenue can promote growth. Financing energy efficiency and investing in clean energy share many characteristics, so they are discussed together in this chapter. These financing instruments make it easier to obtain funding for projects that are perceived as risky or that have high initial costs with long payback periods.

Energy Intensities Still High but Converging

Energy intensities are still higher in much of ECA than in the European Union (EU)-15 countries, as shown in figure 2.3.[5] In 2009, the average ECA country used twice as much energy to produce a given amount of output as the average EU-10 country—0.29 kilograms of oil equivalent (kgoe) versus 0.14 kgoe per US$ purchasing power parity (PPP), GDP, respectively. The variation across countries is large. Uzbekistan has by far the highest energy intensities at 0.73 kgoe per US$, followed by Kazakhstan (0.49), Serbia (0.44), and Russia (0.42)—each of which is more than three times the EU-15 average. On the other hand, 16 countries in ECA do better than the worst performer in the EU-15: Finland. Albania, Croatia, and Turkey have the lowest energy intensities, all below the EU-15 average.

Between 1995 (when energy intensities peaked after the transition from Communism) and 2009, ECA countries lowered energy intensities by about 4 percent per year—to almost half their initial values. The EU-10, Turkey, and the Western Balkan countries[6] already had levels similar to those in the EU-15, as figure 2.4 illustrates. The most impressive reduction was achieved in the Southern

FIGURE 2.3

**Energy Intensities in Europe and Central Asia
Countries Relative to EU-15 Average, 2009**

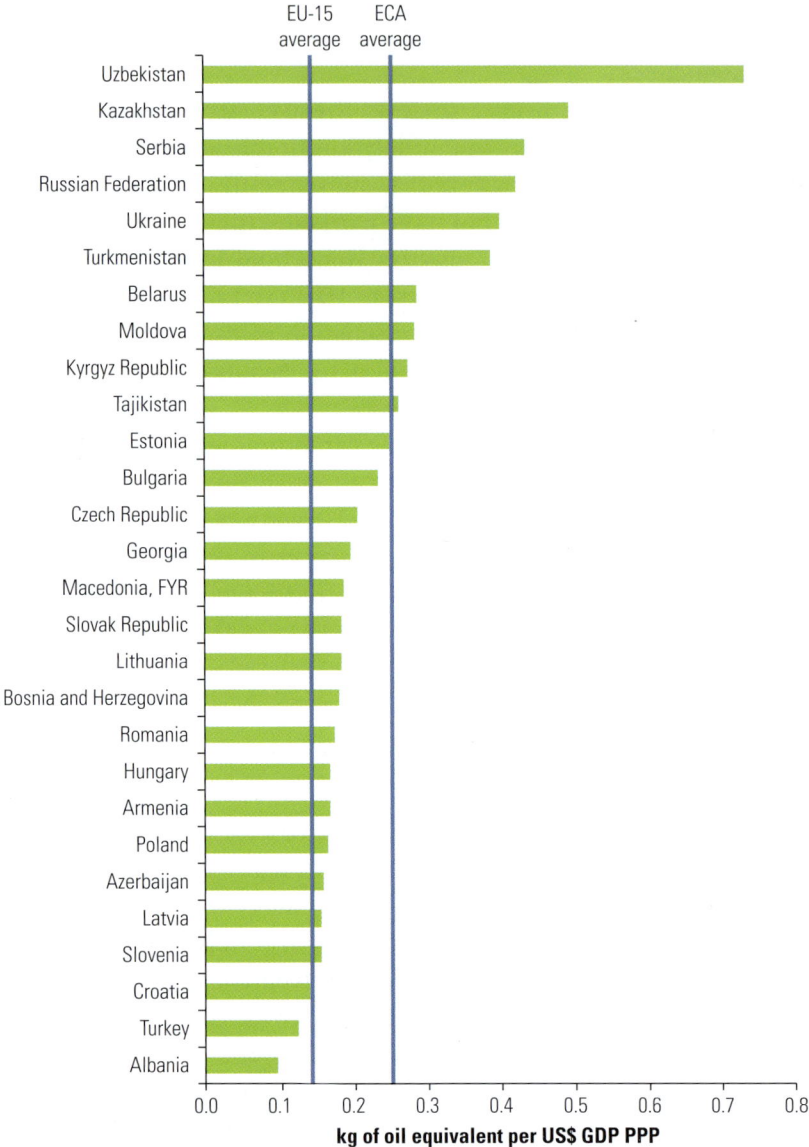

Source: IEA 2011.

Note: Energy intensity is the amount of energy used to produce US$1 of GDP. kgoe = kilogram of oil equivalent; PPP = purchasing power parity; EU = European Union.

Caucasus (Armenia, Azerbaijan, and Georgia), where Azerbaijan managed to reduce energy intensity by 12.5 percent per year—from 0.9 kgoe per US$ to 0.16 kgoe per US$—largely due to rapid GDP growth but also because of effective energy efficiency policies. Georgia improved its energy intensity by 8 percent per year, from 0.53 kgoe

FIGURE 2.4

Energy Intensities in Europe and Central Asia Relative to EU-15 Levels, 1990–2009

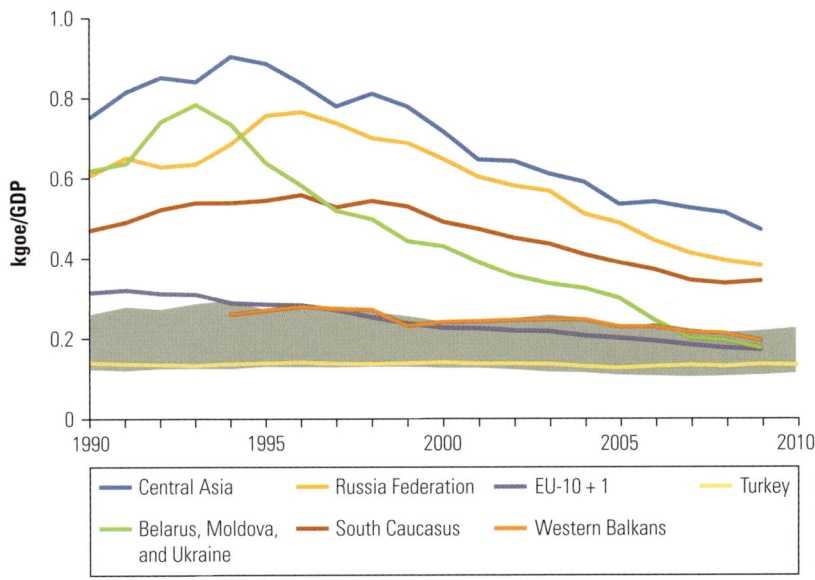

Sources: Calculations based on data from IEA 2011 and World Bank, *World Development Indicators.*
Note: Energy intensity is the amount of energy used to produce US$1 of GDP. The shaded area represents the minimum-maximum range of the EU-15. EU-10 + 1 includes the EU-10 countries plus Croatia; kgoe = kilograms of oil equivalent.

per US$ to 0.2 kgoe per US$. The Central Asian countries as well as a group including Belarus, Moldova, and Ukraine also showed rapid reduction, while Russia's energy intensity dropped by a relatively moderate 3.5 percent per year.

ECA countries with high initial energy intensity levels achieved faster improvements than those economies that already had lower energy intensity, as shown in figure 2.5, panel a. This is consistent with the general observation that high energy intensities in the region's countries reflected significant inefficiencies in energy use and not just a higher share of energy-intensive industries. All countries are gradually moving toward the energy intensity levels of the EU-15. The chart in figure 2.5, panel b, compares the distributions of countries' energy intensities in 1995 and 2009. Not only has the average intensity decreased, but the distribution is also narrowing so countries' energy intensities are becoming more similar.

Rapid drops in energy intensities in most ECA countries after 1990 are not a surprise, given the inefficient industrial structures the region inherited. But intrinsic energy efficiency gains (lower energy use to produce the same amount of a specific good) have been a more important driver of falling energy intensity than structural change

FIGURE 2.5

Convergence in Energy Intensity within Europe and Central Asia, 1995–2009

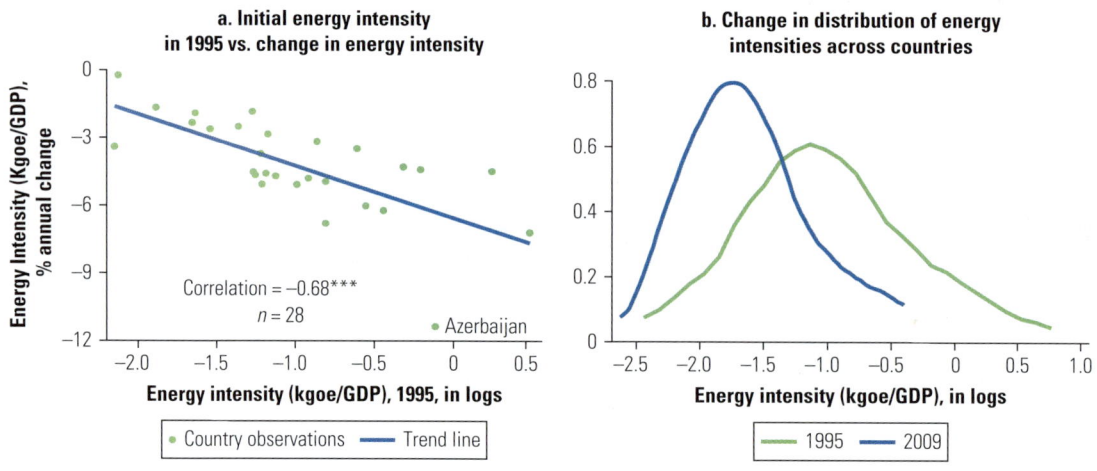

Source: Based on data from IEA 2011.
Note: Energy intensity is the amount of energy used to produce US$1 of GDP. kgoe = kilograms of oil equivalent.

(producing different goods that require lower energy inputs).[7] Chapter 7 discusses these trends for the production sector.

With Growth after Transition, Improvements in Energy Intensity

Figure 2.6 shows the same energy intensity trajectories as in figure 2.5 but in relation to per capita incomes. After 1990, economic output in Eastern European and Central Asian transition countries initially dropped for a number of years, in some countries quite significantly. Energy consumption also fell but not as much—because some basic energy services continued to be delivered even with lower income and production—which led to an initial worsening of energy intensities.

Subsequent declines in energy intensity and increases in per capita income occurred largely in tandem, though the trends differed considerably across ECA depending on the countries' varying starting points. In the Southern Caucasus region, the decline in energy intensity was accompanied by an acceleration in per capita economic growth (hence the curvature of the trace of points for that region in figure 2.6). In other regions, the comparisons of the two rates of change do not seem to show as discernible a change over time. The relationships look similar but less pronounced across Central Asia; a group including Belarus, Moldova, and Ukraine; and Russia,

FIGURE 2.6

Energy Intensities in Europe and Central Asia, by Country Group Relative to European Union, 1990–2009

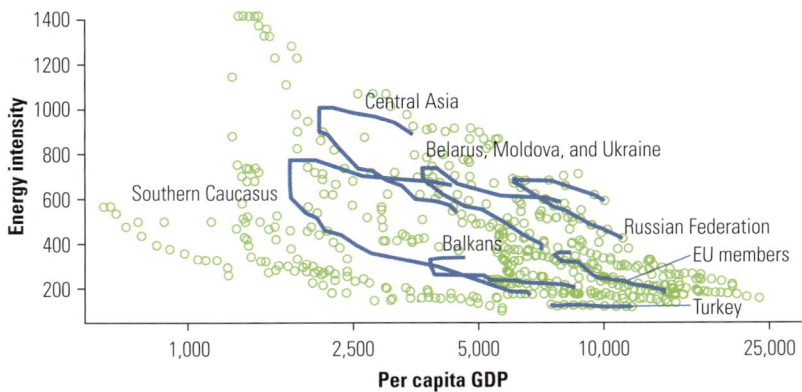

Source: Data from IEA 2011.

Note: Per capita GDP is indicated as PPP GDP in 2000 US$ (log scale). Energy intensity: kilograms of oil equivalent per US$1,000 PPP GDP. The blue lines show the average regional trajectories, starting at the higher energy intensity levels. Green circles indicate individual country values. "Caucasus" includes Armenia, Azerbaijan, and Georgia. "Balkans" include Albania, Bosnia and Herzegovina, Croatia, Kosovo, FYR Macedonia, Montenegro, Serbia and Slovenia.

although actual energy intensities and per capita incomes differ considerably among them. Much smaller declines in energy intensity occurred in the Western Balkans and especially in Turkey as per capita incomes rose.

As such, these patterns do not establish causality. However, they do highlight that one important potential channel for enhancing growth is increasing energy efficiency to overcome past distortions in energy use that have been economically wasteful. By lowering the use of a costly input, energy efficiency can increase productivity and households can use energy savings to buy other things. Thus the patterns in figure 2.6 again suggest how growth in countries with the highest energy intensity was, to a great extent, predicated on removal of preexisting distortions that also engendered serious energy inefficiency. In such circumstances, encouraging energy efficiency is a growth strategy with climate benefits (as box 2.1 further discusses).

Lower Energy Intensity Insufficient to Reduce Greenhouse Gas Emissions

Economic growth benefits from greater energy efficiency, but if dangerous climate change is to be avoided, climate action must also

BOX 2.1

Energy Efficiency, Energy Intensity, and Growth

A country's energy intensity tends to first rise as the economy shifts from agriculture to industries and then tends to fall as it moves toward higher-value manufacturing and services.[8] This pattern raises two questions: Can policies reduce the energy-intensity peak and induce a more rapid turnaround and reduction in energy intensity? And would speeding up this process also speed up growth?

The first question is easier to answer. Almost every country has policies to increase energy efficiency that either tax energy use or introduce regulations that force companies and households to lower energy inputs to production and consumption. Where large initial inefficiencies exist, the required investments have a high rate of return.

The second question is more difficult because isolating the causal effect of energy efficiency on growth is difficult. Growth is influenced by numerous other factors, and many are more significant than energy efficiency. Further, abundant domestic fossil energy endowments can both fuel economic growth through export earnings and encourage wasteful domestic energy use. Analysis is hampered by the paucity of data on energy efficiency proper and usually has to rely on a measure of energy intensity instead. Without controlling for other factors, higher-income countries in ECA generally have lower energy intensities, as figure B2.1.1, panel a, shows.

Countries that grew faster also reduced energy intensity the most over the past decade. This change of energy intensity is due to both GDP growth itself and genuine improvement in energy efficiency, with the relative contribution varying by country. Therefore, lower energy intensity does not necessarily imply falling total energy use.

FIGURE B2.1.1

Energy Intensity (2009) and Average Annual Change in Energy Intensity (2000–09) of Europe and Central Asia Countries

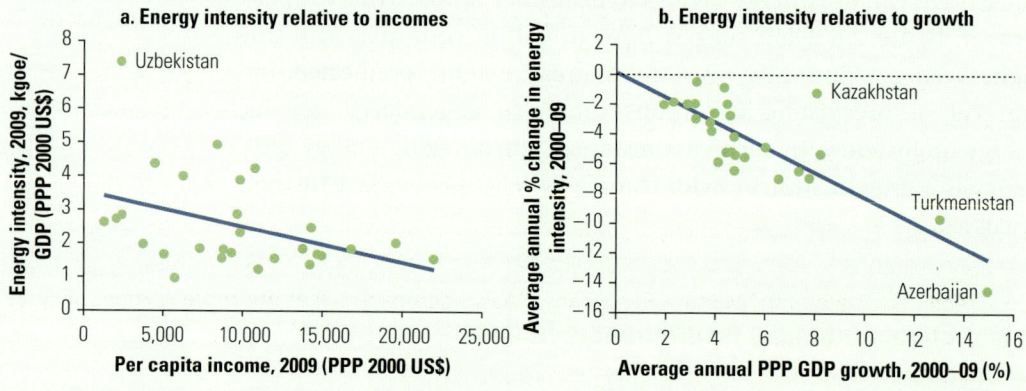

Source: IEA 2011.
Note: Energy intensity is the amount of energy used to produce US$1 of GDP; kgoe = kilograms of oil equivalent; PPP = purchasing power parity.

continued

BOX 2.1 *continued*

What might be the channels by which energy efficiency policies encourage growth?

- Higher energy costs will encourage firms to reduce the energy input in their production processes. Energy efficiency policies will stimulate investment and innovation, provided that an unrealized energy savings potential exists and taxes or regulatory costs are not excessive. If energy prices increase further—as is likely because of resource scarcity and rising energy exploration and production costs—energy-efficient firms will be more competitive internationally. This is particularly evident in sectors that are sensitive to energy prices.

- By taxing energy, governments raise revenue while encouraging energy efficiency. If there is a shortfall in public goods that affects companies, governments can spend revenues on growth-enhancing investments that lower companies' costs—such as education that increases the availability of skilled workers. Where government revenue is spent inefficiently, these benefits will not be realized and such policies could be growth reducing.

- Where nonwage labor costs like health care or social security are high, energy tax revenues can be used to lower them. This encourages companies to reduce their use of energy and to hire workers. Additional employment creates extra consumption, stimulating growth. Implemented well, such ecotax reforms yield a "double dividend," reducing environmental damages and lowering distortions or costs in the overall tax system (Cottrell 2011; Jones, Keen, and Strand 2012).

- For energy-importing countries, foreign exchange reserves used to buy energy abroad can instead be invested domestically.

- Higher energy costs or regulations also encourage households to use less energy. Even if there are initial investments, they will be recovered in the long-term savings that can result in growth-promoting consumption.

- Reduced output from polluting fossil-fuel power plants or industrial facilities due to improved energy efficiency lowers health impacts and associated costs to the economy (see chapter 5).

Energy efficiency can, but does not have to, increase growth and competitiveness. Some countries will grow even though they do not use energy efficiently. The United States and Canada are highly competitive and innovative despite consuming more than 60 percent and 110 percent more energy per unit of output, respectively, than the best performers in Western Europe. They still achieve relatively low energy intensity with very high energy consumption because output is also very high. But little of that output today comes from industries that directly compete with Western European or Asian companies that are more energy efficient. Most of the economic growth in the United States and Canada is in services and high-end manufacturing, where energy accounts for a small share of input costs.

significantly reduce the absolute volume of energy-related GHG emissions. There is a lively debate over whether energy efficiency will really be able to generate reductions in total energy consumption and emissions. This debate centers on the size, or indeed the existence, of the so-called energy rebound effect, or "Jevons' paradox," named after a 19th-century British economist, who wrote in 1865: "It is wholly a confusion of ideas to suppose that the economical use of fuel is equivalent to a diminished consumption. The very contrary is the truth" (Jevons 1866). In a narrow sense, the rebound effect implies that as energy efficiency improves, the cost of using energy-consuming goods and services goes down and people will use more of them. As fuel efficiency of cars goes up, people will drive their cars more because the cost per kilometer of travel goes down. A large literature on this topic suggests that this is indeed the case but that this increase is relatively modest, in the range of 10–20 percent (Madlener and Alcott 2009; Maxwell et al. 2011).[9]

In a broader sense, economywide effects can offset the gains from energy efficiency. Where energy efficiency increases productivity and accelerates growth, more economic activity occurs and wealthier households consume more goods. Rather than not owning a car or sharing one in a household, each adult family member will buy one. Energy use will grow accordingly. Even as there has been a decoupling of energy use from economic growth, total energy consumption and the emissions it causes have not come down in most ECA countries, whose citizens, at lower per capita incomes, still have a large pent-up demand for energy consumption, as shown in figure 2.7.[10] The Slovak Republic, Ukraine, and Uzbekistan are examples. Only a few EU-15 countries, such as Belgium and Germany, have been able to reduce emissions, even with slower drops in energy use. In these countries, an already high living standard—where most who want a car already own one—coincides with a smaller share of energy-intensive industries and effective energy efficiency *and* renewable energy policies.

Under a simple scenario for ECA, on the other hand—with economic growth at 4 percent and a continuation of past energy intensity trends moving toward a gradual convergence with the EU-15—total energy consumption would not drop at all, as figure 2.8 illustrates. Moreover, unless the region begins using greener energy sources, emissions would not drop, either.

Taken together, the charts in this section indicate that ECA countries are approaching the energy efficiency levels of their Western European neighbors—though at different speeds and with room to improve. These energy efficiency gains can support growth and raise

FIGURE 2.7

Trajectories of Economic Output, Energy Intensity, and Carbon Dioxide Emissions, Selected Eastern European and Central Asian and Western European Countries, 1990–2010

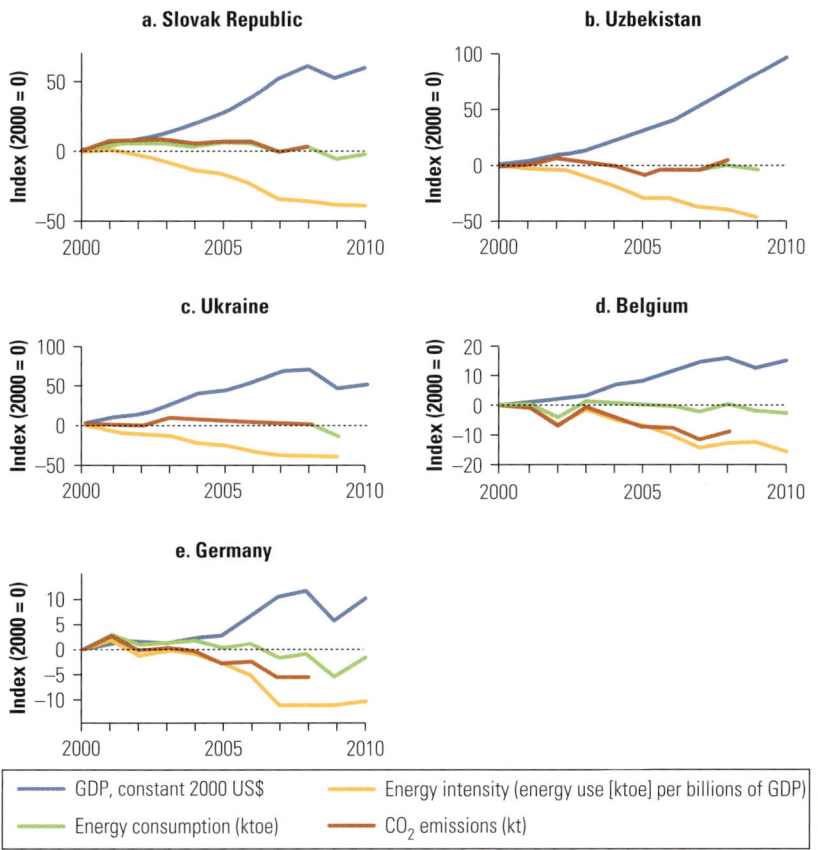

Source: IEA 2011.
Note: Energy intensity is the amount of energy used to produce a defined unit of GDP, in this case ktoe per US$ billions of GDP. Ktoe = kilotons of oil equivalent; CO_2 = carbon dioxide.

living standards, but they are insufficient to reduce overall energy consumption or to lower GHG emissions. At best, both energy use and emissions would remain flat for the foreseeable future.

For climate action, this means, as IEA scenarios suggest, that energy efficiency could indeed contribute about half of the emission reductions needed to achieve climate stabilization goals and will be a major component in national energy transition strategies. But this would mean that the ECA region will simply "tread water" in terms of emission growth. Therefore, energy efficiency should primarily be seen as a way to stabilize carbon emissions. Its role in reducing emissions will be indirect: by helping to achieve growth, energy efficiency

FIGURE 2.8

Projected Trajectories of Economic Output, Energy Intensity, and Carbon Dioxide Emissions in Europe and Central Asia Countries, 2000–30

2000 = 0

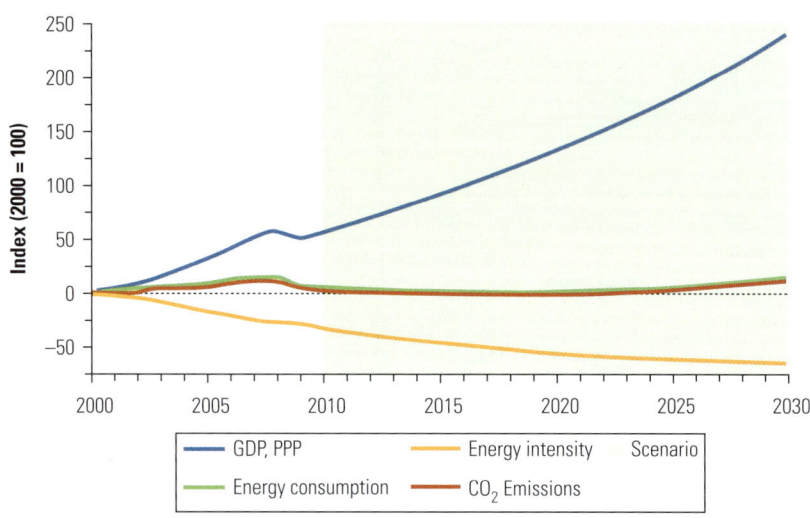

Source: IEA 2011.

Note: This scenario assumes 4 percent growth, and energy intensity convergence implies a continued decrease of 1.8 percent per year in EU-15 countries and 3.5 percent in ECA countries. This illustrative scenario differs from that in World Bank (2010b), referred to later in this report, which assumes a growth rate of 4.4 percent, energy intensity reduction of only 1.2 percent per year, and therefore an energy consumption growth of 3.1 percent. PPP = purchasing power parity; CO_2 = carbon dioxide. Energy intensity is the amount of energy use per unit of GDP.

also helps to generate the financial resources needed to fund the transition to a low-carbon energy future (see also Herring 2006).

Why Promoting Energy Efficiency Is Often Difficult

Improving energy efficiency appears deceptively simple. Economic and engineering studies—neatly summarized in marginal abatement cost curves—suggest that many energy efficiency investments have very attractive rates of return. They are expected to pay for themselves in a few years. But take-up in the real world, including in ECA, has been woefully slow aside from the relatively easy initial gains through economic restructuring and more market-oriented energy pricing.

Much has been written about why this is the case, but no clear consensus has emerged. There are a number of possible reasons:

- Engineering studies focus too much on up-front costs and ignore opportunity costs or unobserved factors such as the risk of using new equipment or lack of implementation capacity (see box 2.2).

BOX 2.2

MAC Curves Can Make the Energy Efficiency Challenge Look Easier than It Is

No report on climate change mitigation today is complete without showing a marginal abatement cost (MAC) curve. A MAC curve is a graphic representation of the relative economic costs of different mitigation technologies, many of which involve energy efficiency investments. The cost per unit of emission reduction is shown on the vertical axis, and the total amount of potential reductions on the horizontal axis. Mitigation technologies are arranged from lowest to highest costs, so the cheapest options appear on the left (typically with negative costs, meaning adopting them would actually save money), and the most expensive ones are toward the right. The basic idea is that any option with a unit cost below zero, or below a given carbon tax, can be profitably implemented. Most recently, MAC curves have been popularized by the global consulting firm McKinsey & Company. But electricity MAC curves were already used in the early 1980s in the aftermath of the oil crisis. For carbon emissions, they were first used in the early 1990s.

MAC curves are effective in summarizing a bewildering range of technical measures in an easily understood graphic. But they have some limitations (see, for example, Ekins, Kesicki, and Smith 2010):

- Basic economics would suggest that there should not be large unexploited opportunities at negative costs—that is, mitigation options that more than pay for themselves. If they are highly profitable, why have markets not seized these opportunities? One reason could be that MAC curve calculations ignore additional nontechnical costs, such as transaction costs or adjustment costs (for example, the employment impacts of switching from labor-intensive coal to imported gas for power generation). Another reason is that market and behavioral failures may prevent their adoption, as discussed elsewhere in this chapter.

- MAC curves sometimes underestimate not only costs but also the co-benefits of climate action. For example, excluding the health benefits of switching from coal to renewable power leads to overestimating the cost of solar and wind energy.

- MAC curves share the problem of many analytical tools in that they are a static representation of a dynamic process. Implementing one or more of the options will change the relative costs of others. For instance, large investments in renewable energy will affect the climate benefits from energy efficiency measures. In addition, assumptions that appear valid at the time of producing the MAC curve—such as the price of oil—may not hold in the future. So the shape of the curve is constantly changing, in large part as a direct response to policies. In addition to these general concerns, the utility of MAC curves is often difficult to evaluate when the data and assumptions that went into their construction are not published.

continued

BOX 2.2 *continued*

Abatement curves are useful as an initial ranking of mitigation options. But the MAC curve debate shows that reducing the task of climate change mitigation to a simple choice between competing technologies ignores some real-world complexity. Technology will play a crucial, perhaps dominant, role in addressing the climate problem. To properly deploy it also requires considering behavioral issues and market imperfections, being aware of the dynamic interactions between different policy choices, and paying attention to the inherent uncertainty in processes that will play out over many decades.

FIGURE B2.2.1

Marginal Abatement Cost (MAC) Curve of Climate Change Mitigation Technologies for Poland, 2030

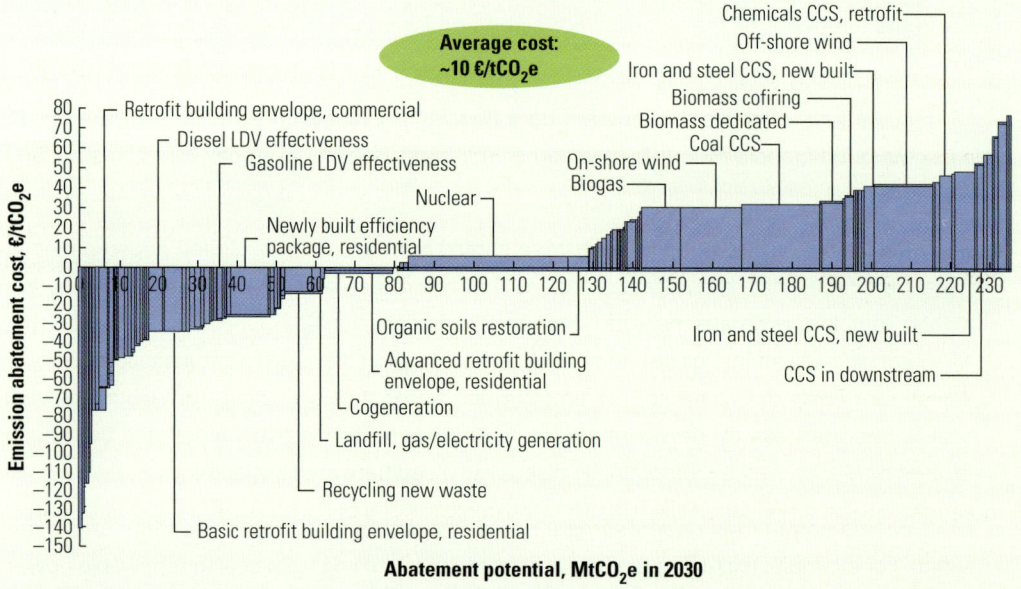

Abatement potential, MtCO₂e in 2030

Source: World Bank 2011.
Note: The bar width is the amount that emissions can be reduced against business-as-usual levels projected for 2030. €/tCO₂e = euros per ton of carbon dioxide equivalent; MtCO₂e = millions of tons of carbon dioxide equivalent; LDV = light duty vehicles; CCS = carbon capture and storage.

- Similarly, engineering estimates based on ideal conditions or motivated by commercial interests may not reflect real-world conditions. Predicted energy savings from building insulation, for instance, are often larger than those actually realized.

- Energy efficiency policies need to be tailored to numerous individual sectors and subsectors because energy is used in just about every economic and consumption activity (as described in chapters

BOX 2.3

Small Energy Savings Can Avoid Big Power Plant Investments

Improving energy efficiency in ECA will require numerous changes in technology and behavior. High-tech solutions will be required, but large gains can come from relatively simple actions by firms and home owners. One power company executive "fell in love with customer-communication technology, real-time price signals, and fantastic sensory capability in 1998. I have only now come to realize that what I really wish my customers would do [is] to use more caulking."

Because energy is used in all walks of life and all economic sectors, the energy efficiency challenge is highly diffuse. Relatively concentrated improvements in the industrial sector present large efficiency opportunities. The ECA region's iron and steel, cement, and pulp and paper industries could reduce energy consumption by half while maintaining or increasing output (see chapter 7). Supply-side efficiency projects have high rates of return. This is especially the case for efforts to reduce losses in transmission and distribution, which are more than 35 percent in countries like Albania, Kosovo, and Moldova. However, equally profitable are more diffuse targets, such as replacement of incandescent lighting with compact fluorescent or light-emitting diode (LED) lightbulbs. A modest US$5 million program to promote efficient lightbulbs in Poland saved more than 430 gigawatt-hours (GWh), equivalent to the total annual electricity consumption of 43,000 households. Overall, as discussed in the sectoral chapters (6–10), large energy savings will come from millions of individual firm and household-level investments and behavioral changes in practically all sectors of the economy.

A particularly impressive example from China illustrates the potential. First envisioned more than 90 years ago, China's Three Gorges Dam has an installed capacity of more than 20 gigawatts (GW), making it the largest power plant in the world. Its construction took 17 years; cost more than US$30 billion; and required 27 million cubic meters of concrete, 463,000 tons of steel, and the displacement of 102 million cubic meters of earth. The dam's social and environmental impacts are significant: 1.2 million people were relocated from the 1,084-square-kilometer reservoir area. The dam helps control flooding and facilitates shipping on the Yangtze River downstream, but it also causes sedimentation and erosion and might worsen water quality.

The Three Gorges Dam will produce about 85 terawatt-hours (TWh) of electricity per year. This is equivalent to the total electricity and heat output of countries such as Romania or Vietnam (IEA 2011). It is also equivalent to the annual amount of electricity that could be saved by 2020 if all of China's air conditioners and refrigerators complied with the country's 2005 appliance standards. By 2030 the equivalent of two Three Gorges Dams would be saved. The comparison is even more striking when expressed in monetary terms. The energy savings come at retail (consumer) price levels, which are typically twice the wholesale producer prices realized by the Three Gorges power plant operator.

continued

This example illustrates that energy efficiency should be considered a source of energy. Some call saved electricity "negawatts." Seemingly small energy savings at the household level add up to a point where they replace large investments in new power plants. A fast-growing economy such as China will require both new capacity and faster efficiency gains. But by investing in energy efficiency, construction of many new power plants can be avoided. A rule of thumb often quoted by energy economists is that a dollar invested in energy efficiency avoids two dollars spent on new capacity.

FIGURE B2.3.1

Projected Electricity Savings from Appliance Energy Efficiency Standards in China, 2015–30

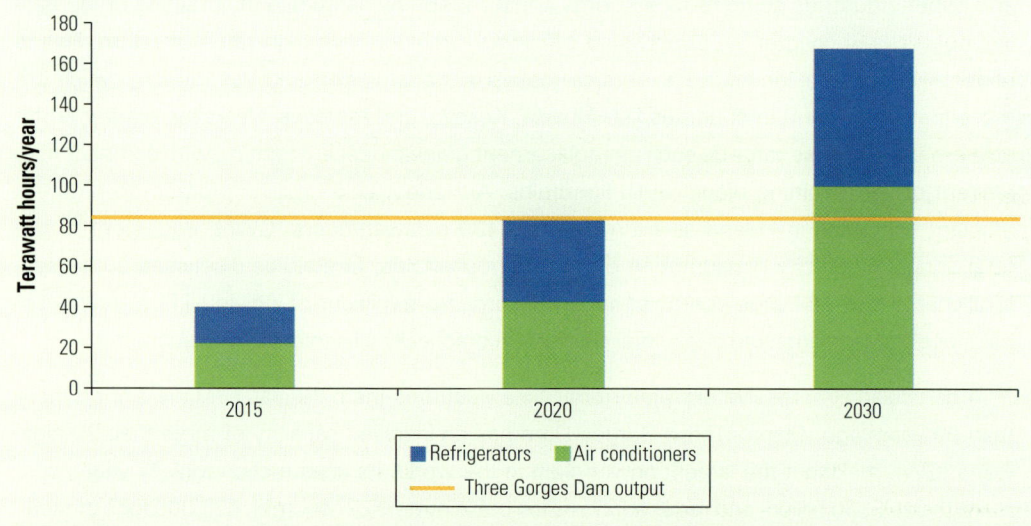

Sources: Rosenfeld and Poskanzer 2009; IEA 2007; Stone 2008; Embassy of China 2010; World Bank 2010b.
Note: Bars indicate electricity savings, in TWh per year, if all air conditioners and refrigerators meet existing standards. TWh = terawatt-hours.

6–10 of this report). Very large gains can come from many spread-out energy savings that individually are quite small (see box 2.3). However, policy coordination and effective implementation is a challenging task, especially in lower-income countries with lower institutional capacity.

- The most often-cited reason is that a number of failures in energy, capital, innovation, and information markets—as well as quirks in human behavior—inhibit the uptake of profitable energy efficiency investments, as shown in table 2.1 (Gillingham, Newell, and Palmer 2009).

TABLE 2.1

Market and Behavioral Failures Affecting Energy Efficiency and Low-Carbon Investments

Potential market failures	Potential policy options
Energy market failures	
Environmental externalities	Emissions pricing (tax, cap-and-trade)
Average-cost electricity pricing	Real-time pricing, market pricing
Energy security	Energy taxation, strategic reserves
Capital market failures	
Liquidity constraints	Financing, loan, energy services programs
Innovation market failures	
R&D spillovers	R&D tax credits, public funding
Learning-by-doing spillovers	Incentives for early market adoption
Information problems	
Lack of information, asymmetric information	Information programs
Principal-agent problems	Information programs
Learning by using	Information programs
Potential behavioral failures	**Potential policy options**
Prospect theory (preferences such as loss aversion)	Education, information, product standards
Bounded rationality (nonrational decision making)	Education, information, product standards
Heuristic decision making	Education, information, product standards

Source: Adapted from Gillingham, Newell, and Palmer 2009.
Note: R&D = research and development.

These market failures have two main consequences: One is that households and firms use more energy than is socially optimal, for instance by heating more than would be necessary with even small investments in reducing heat leaks around windows. The second is that they make inefficient investment decisions, focusing on up-front costs and discounting long-term energy savings when buying machines, cars, or appliances (Allcott and Greenstone 2012). A low cost of energy—below full-cost recovery levels in low-income countries and below the social cost in middle- and high-income countries—encourages excessive energy use because users respond to prices.

Direct and indirect subsidies to energy producers or consumers persist in several ECA countries and encourage the more liberal use of energy. In 2000, total underpricing of energy—the gap between market prices and prices charged—exceeded 10 percent of GDP in Albania, Azerbaijan, Georgia, the Kyrgyz Republic, Moldova, Serbia, and Tajikistan (Ebinger 2006). Subsidy reform is politically

BOX 2.4

Energy Price Reform: Lessons from a World Bank Review

The removal of energy subsidies is associated with substantial reductions of CO_2 emissions. In the transition countries, the World Bank has supported energy price reforms through technical assistance and policy advice as well as investment, and in most cases, the reforms have led to a decline in emission intensities. In Ukraine, for example, energy tariffs (electricity, gas, and coal) were increased by 25–50 percent during 2002–07, and CO_2 emissions dropped from 8.3 tons per dollar of GDP in 2002 to 5.9 tons in 2007.

Price reform should be accompanied by a poverty and social impact analysis (PSIA) and monitoring. Energy price reform poses social and political risks, but better-targeted and more effective social protection systems supported by strong analysis can mitigate these challenges. This process includes systematic analysis and monitoring of the distributional impact of energy price reform (see chapter 5). Experiences in both Ghana and Indonesia, where people were compensated when fuel prices increased, suggest that careful analysis and preparation can inform the design and implementation of price reforms and buffer their impacts.

Energy efficiency measures and a reduction in subsidies should be promoted to mitigate the burden of the low-carbon transition. Savings from reduced subsidy payments could fund efficiency investments, such as mass distribution of energy-efficient lightbulbs. However, few projects have employed such mechanisms. An exception is the China Heat Reform and Building Efficiency Project, which linked improved insulation with heat pricing. Strong collaboration across sectors is crucial for analyzing and managing the ripple effects of energy price reform. More broadly, climate change projects require cross-sectoral approaches that combine energy planning with water management, urban management, and social safety nets.

Source: World Bank 2010a.

difficult, but the experiences in a number of countries provide useful insights on how to approach it, as discussed in box 2.4 (see also Bacon, Ley, and Kojima 2010; and Vagliasindi 2012). New estimates for a selection of the region's countries, for which consistent data were available, suggest that several have greatly reduced power sector subsidies. Tajikistan's subsidies, for instance, dropped from 28 percent to less than 2 percent in 2009, as figure 2.9 shows. On the other hand, total energy sector subsidies (oil, gas, coal, and electricity) remain high in many countries. Consumption subsidies alone were US$39.2 billion in Russia in 2010, US$11.9 billion in Uzbekistan, US$7.7 billion in Ukraine, US$5 billion in Turkmenistan, US$4.3 billion in Kazakhstan, and US$0.8 billion in Azerbaijan.[11]

FIGURE 2.9

Progress on Power Sector Subsidy Reform, Selected Europe and Central Asia Countries, 2000 vs. 2009

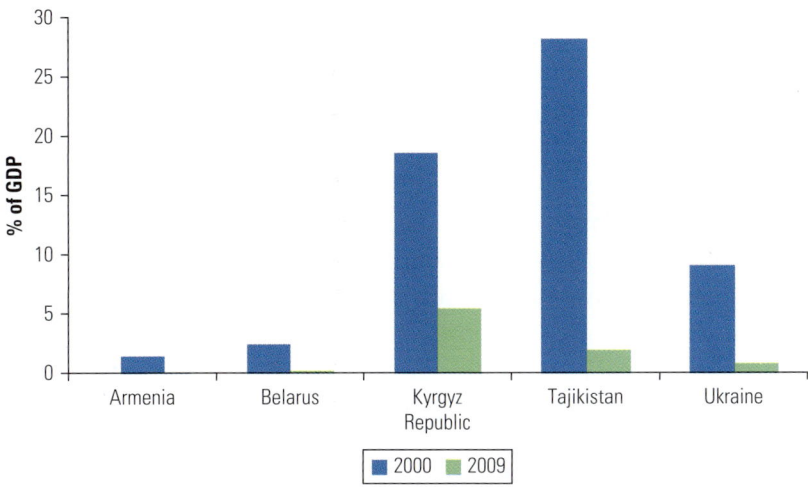

Source: Calculations replicating the methodology in Ebinger 2006.

Investment inefficiencies, on the other hand, mean that people do not sufficiently invest in energy-efficient goods. The most effective policies to address this problem are (a) information programs that make consumers and firms aware of energy-efficient options and their long-term savings, or (b) programs that encourage disclosure of energy use for durable goods or homes (encouraging buyers and renters to make energy-saving choices). Other instruments, such as subsidies to purchase energy-efficient goods or performance or fuel-efficiency standards, may sometimes be necessary but tend to be more costly. One study in the United States suggests that avoiding one ton of CO_2 using the Corporate Average Fuel Economy (CAFE) standard costs US$222, while achieving the same with a gas tax costs US$92 (Jacobsen 2010).

Policies that Have Worked

There is a case for active government programs that help overcome these market failures or inefficiencies. Energy efficiency policies must have two objectives: they should address investment barriers holding back the use of energy-efficient goods, and they should reduce the energy-consuming use of those goods. One without the other could mean that, for instance, people buy the right number of fuel-efficient cars but use them too much. Information programs and energy taxes are the preferred instruments. However, because problems are difficult to manage or politically sensitive, countries typically use a mix of instruments—*sticks* in the form of regulations and standards and *carrots* in the form of tax breaks and subsidies, augmented by investments in information provision, research and development (R&D), and energy efficiency improvements in the public sector.

Although the set of instruments to promote energy efficiency is quite standard—prices, regulations, and investments—their design and implementation is sector-specific. How they support efficiency improvements across all the facets of the economy is discussed in the sectoral chapters of this report (chapters 6–10). But some general points can be made.[12] The first is that the sheer complexity of the energy efficiency challenge can give rise to a bewildering array of programs. A U.S. Government Accountability Office review identified 11 federal agencies involved in 94 green building initiatives.[13] The complexity of the challenge suggests an important role for a national energy efficiency strategy coordinated by a dedicated agency. Some countries in ECA have taken action. Russia, for instance, created an Energy Agency in 2009 tasked with promoting a 40 percent reduction in energy intensity by 2020.

The second point is that given the broad range of energy efficiency levels in the ECA region and its neighborhood, there is much scope for learning from each other. Western European countries that have significantly lowered their energy intensities, such as Ireland or Sweden, provide many lessons to countries with medium energy intensities in the EU-10 that are catching up to their peers. Their experience can inform policies in the Western Balkan countries and EU partnership countries, which outperform Central Asia.[14] Better mechanisms to transfer and adapt these lessons within the region could have high payoffs.

Third, while most energy efficiency programs have been used widely for a long time, there is still scope for disseminating policy innovations. One relatively new instrument is market-based trading of energy savings certificates, also known as "white certificates." The principle is similar to trading of emission allowances or of green certificates for production of renewable energy. The following section reviews environmental tax instruments that, in contrast to regulations such as efficiency standards, raise revenue that can be reinvested. These instruments have been introduced in some ECA countries, but their use could be expanded. This discussion is followed by an overview of instruments that help overcome financing constraints. Some of the major issues are unique to energy efficiency, but some also relate to financing the low-carbon energy investments further discussed in chapter 3.

Environmental Taxes[15]

Tax instruments that change consumption through prices are often more efficient than regulations in achieving environmental goals. They also raise revenue, which sometimes blurs the line between environmental taxation and general revenue generation. Turkey has the highest gasoline prices in the world, which largely explains why the country ranks highest in the OECD in share of total revenue from environmentally related taxation. Turkey's fuel taxes probably exceed an efficient level that reflects the cost of environmental externalities (Heine, Norregaard, and Parry 2012). Although these taxes reduced the growth in vehicle ownership and fuel consumption, they were in fact introduced because these transaction-based taxes are easier to collect than income taxes.

Environmental taxes are still unfamiliar in many countries. In the simplest case, environmental taxes would just flow into general revenue to fund unspecified investments or tax cuts elsewhere. But to increase acceptance, they are often introduced as a revenue-neutral "ecotax reform." Switzerland reimburses revenue from a CO_2 tax uniformly to all citizens through the universal health insurance system, largely for ease of implementation. Other countries earmark environmental taxes to achieve other policy objectives. In the simplest case, this earmarking shifts the tax burden to activities that should be discouraged from those that should be encouraged ("taxing the bads and not the goods"). So a tax on energy replaces a distortionary tax on wages, for instance. If the policy goal is environmental, an energy tax is usually a proxy for taxing pollution or carbon emissions that are much harder to measure and monitor.

A successful tax shift generates a "double dividend," reducing damage to the environment and encouraging higher output and employment. Two problems may arise, however:

• Net employment will increase only if the new environmental taxes do not have an employment-reducing effect themselves as they make energy-intensive products more expensive. This requires that the tax rates be set carefully and that firms or households have the ability to reduce energy consumption by investing in energy efficiency. A tax reform that yields a "strong" double dividend—achieving environmental gains and reducing the over-all distortionary costs of taxation—can be a no-regrets strategy, even if it brings fewer environmental benefits from changes in energy consumption than expected. However, because such reforms can distort hiring decisions in energy-using sectors, a double dividend is not guaranteed. Tax policy design matters a lot.

- If ecotax reform is very successful, the tax base will shrink as firms and households reduce energy consumption. Employment goals will get harder to finance. In principle, tax design could include an escalation factor that raises the tax in proportion to the energy savings achieved in the previous period (Von Weizsäcker et al. 2009). This provides a dynamic incentive for further energy efficiency improvements and helps keep a check on rebound effects.

Environmental or ecological tax reform has been credited with large energy savings and employment gains in a number of countries. It raised GDP and employment by up to 0.5 percent in EU countries that introduced such reforms: Denmark, Finland, Germany, the Netherlands, Slovenia, Sweden, and the United Kingdom (Andersen and Ekins 2009). In ECA, the EU-10 countries are subject to environmental taxes, including a minimum tax on energy. Some countries have gone beyond that. Slovenia introduced a CO_2 tax as early as 1997. By 2009 its environmental tax revenue represented 3.6 percent of GDP. The Czech Republic and Estonia have implemented environmental taxes to comply with EU directives that include a tax-shifting element, as shown in figure 2.10. Estonia introduced air pollution

FIGURE 2.10

Gasoline and Electricity Taxes, Relative to EU Energy Tax Directive Minimums, in Estonia and the Czech Republic, 2005–11

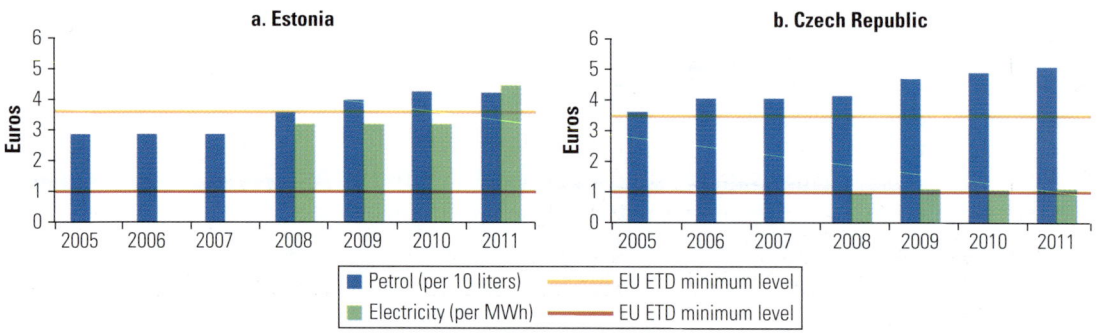

Source: European Commission; see Cottrell 2011.
Note: EU = European Union; ETD = energy tax directive; MWh = megawatt-hour.

charges in 2006 that were expanded to CO_2 charges for electricity companies and later also included transport fuels and natural gas and electricity. Initially revenue went to an Environmental Investment Centre and to the general budget. More recently, revenue has been used to reduce labor and income taxes. The Czech Republic's fiscally neutral environmental tax reform followed the German model. Taxes on fuel and electricity lowered social security contributions by 1 percent for employees and 1.5 percent for employers.

The Czech Republic and Estonia use a high share of fossil fuels in electricity generation and have high emissions as a result. Estonia has the highest per capita emissions in ECA because it relies on oil shale. The relatively large revenue raised by tax reforms has therefore been largely due to the high energy and emission intensity rather than tax rates (see figures 2.10 and 2.11). Though Estonia's tax rates are increasing, they remain comparatively low, and their environmental benefits were comparatively low as well. For instance, emissions in Estonia fell by about 2 percent by 2012 over 1997 levels, compared with Finland's 7 percent between 1990 and 2005 as a result of energy and carbon taxes.

EU accession countries will need to revise tax systems to conform to EU directives. A review of environmental instruments in Commonwealth of Independent States countries[16] also suggests there is potential for efficiency gains from environmental tax reform (Cottrell 2011). The economically most efficient approach would consist of pollution taxes on CO_2 as well as other pollutants that reflect their

FIGURE 2.11

Overall Energy Tax Rates in the Czech Republic and Estonia Compared with Selected EU Countries, 2000–09

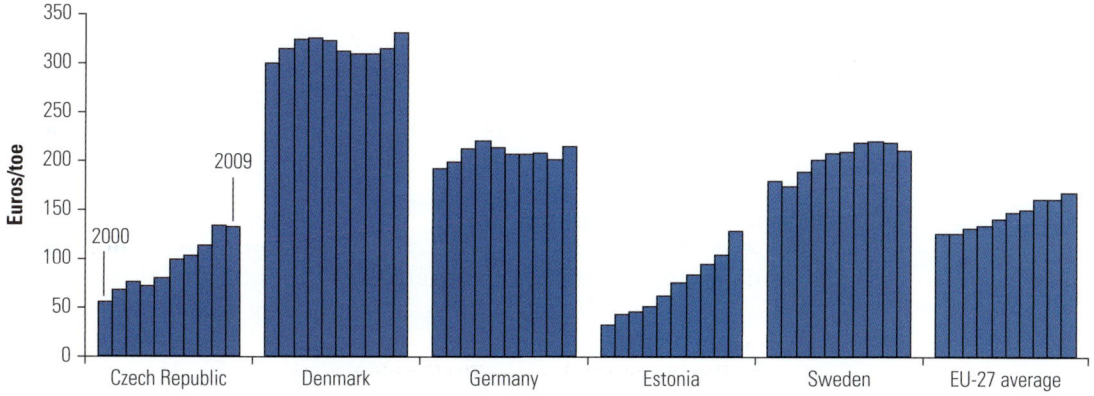

Sources: Eurostat online database (http://epp.eurostat.ec.europa.eu); see Cottrell 2011.
Note: EU = European Union; toe = tons of oil equivalent.

total environmental cost. Ideally these should be levied upstream to basic energy sources—for example, imported oil or natural gas—with a credit for downstream capture, once operational, through carbon capture and storage or other abatement. One difficulty is measuring the size of these costs. (Chapters 5 and 8 discuss this in the context of health and transport.)

Where environmental levies need to be linked to offsetting tax cuts to increase acceptability or where environmental tax reform could reduce distortions in the existing tax code, a few general principles are useful: The tax policy needs to be transparent with a clear definition of environmental targets because this should determine tax rates. A clear focus on a limited set of important pollutants, including GHGs, helps clearly define the tax base. Policy analysis can predict the effects of policy changes, including negative impacts on employment or social outcomes.

Financing the Energy Transition[17]

The scale and scope of needed investments to significantly reduce emissions from the energy sector vastly exceeds governments' ability to raise and redirect financial resources. A key challenge for ECA countries is therefore to mobilize private investment in energy efficiency and clean-energy, especially because private gains from these categories of energy investments often equal or exceed public benefits significantly.

To do so, governments have two main tasks: First, there must be sufficient long-term financial incentives for private investors to be engaged in energy efficiency and clean-energy markets. Credible and predictable climate policies such as cost-reflective energy pricing align incentives and motivate private participation. It is policy uncertainty that deters investors more than anything. Second, investments in energy efficiency and clean energy pose specific financing barriers. They require high initial investments but generate long-term savings that could be attractive to financial investors.

Barriers to Investment

The most important barriers to energy efficiency and clean-energy investments are the following (Taylor et al. 2008; World Bank 2010a; World Bank 2012):

- Many energy efficiency projects have high financial rates of return and short payback periods, often between one and five years.

However, many banks are unfamiliar with energy efficiency financing. Benefits in the form of calculated cost savings streams appear to be less tangible. Many banks either do not understand that the savings flow could back a loan, do not know how to appraise that flow, or simply overestimate the risks of these loans.

- There is a lack of long-term financing for renewable energy. The capital-intensive nature of most renewable technologies results in a higher demand for capital, a longer payback period, and therefore greater exposure to market and regulatory risks. Uncertainties regarding the performance of new renewable technologies and the variable availability of natural resources (such as hydrology and geology) further increase the risks associated with investments in renewables.

- There are also more general barriers that affect all types of lending. For example, households or small and medium-size companies are generally considered less creditworthy and face credit constraints regardless of the type of investment. When dealing with the residential sector, projects tend to be small scale and dispersed. In this case, transaction costs can prove daunting to banks.

Financing Instruments

Innovative financing instruments can address these barriers to make private investment feasible. These instruments have to be tailored to the specific capital market barriers, market segments, and local context. The next paragraphs discuss the most important categories of publicly sponsored financing instruments, and table 2.2 summarizes their main features.

Dedicated Credit Line through Local Financial Institutions

Under this instrument, governments, multilateral development banks, and donors provide concessional loans to participating banks in developing countries, who in turn on-lend to beneficiaries at either concessional or market rates. In the case of renewable energy financing, the loans are with longer maturity than those available in the local market.

This instrument can be used to demonstrate the financial viability of energy efficiency and renewable projects and to generate interest among local commercial banks to enter the sector. For example, the success of the Turkey Renewable Energy Project led local banks to launch long-term financing for renewable energy projects. When

TABLE 2.2

Market Segments for and Barriers to Energy Efficiency and Clean-Energy Financing Instruments

Financing Instrument	Market Segments	Market Barriers	Examples
Dedicated credit line	• Local banks' traditional clients: large and medium enterprises	• To reduce the perceived high risks of EE and RE lending by demonstrating the viability of EE and RE projects and overcoming banks' unfamiliarity with these projects • To provide long-term financing for RE	• Bulgaria Residential Energy Efficiency Credit Line • Turkey Renewable Energy Project
Partial credit or risk guarantees	• Less-creditworthy borrowers: SMEs and first-time ESCOs • RE technologies with resource or technology risks	• To overcome the liquidity constraints of less-creditworthy borrowers • To mitigate the private sector's risks of financing RE technologies	• Hungary Energy Efficiency Co-financing Program • Geothermal Fund in Eastern Europe
Equity and mezzanine finance	• Start-up ESCOs • SME developers of RE projects and early-stage RE technologies • More-developed financial markets	• To address limited access to equity fund for SMEs • To mitigate private creditors' risks	• European Investment Bank • Global Energy Efficiency Renewable Energy Fund (GEEREF)
ESCO financing and equipment leasing	• SMEs, residential sectors, and more-developed financial markets	• To overcome high up-front costs and SMEs' and households' liquidity constraints • To bundle similar projects and reduce transaction costs	• Poland GEF Energy Efficiency Project • Argentina Renewable Energy in the Rural Market
Utility financing	• Utility customers	• To overcome high up-front costs and households' liquidity constraints • To reduce transaction costs	• Brazil end-use energy efficiency program • US EE/DSM Fund in California

Source: Adapted from World Bank 2012.

Note: EE = energy efficiency. RE = renewable energy. ESCO = energy service company. SMEs = small and medium enterprises. GEF = Global Environment Facility. EE/DSM = Energy Efficiency and Demand Side Management.

combined with technical assistance, such loans act like training wheels to help local banks get familiar with appraisal and the structure of energy efficiency loans. For instance, the China Energy Efficiency Financing Program was designed specifically to build capacity of local banks.

Partial Risk Guarantees

A risk guarantee scheme can offer individual project guarantees or portfolio guarantees. This instrument can reduce the perceived repayment risks associated with energy efficiency projects (as in the Hungary Energy Efficiency Co-financing Program) or cover the specific technology risks such as insuring partial drilling costs for unsuccessful geothermal exploration wells (for example, the Geothermal Fund in Eastern Europe). The guarantees could be useful for less-creditworthy borrowers (such as those with short track records) in underdeveloped financial markets (World Bank 2010a).

Loan guarantees can also be combined with technical assistance, as in the China Utility-Based Energy Efficiency Finance Program, to overcome banks' unfamiliarity with energy efficiency lending.

Equity Funds and Mezzanine Finance

Multilateral development banks (the International Finance Corporation and the Global Environmental Facility) and donors as well as, recently, governments in a few Asian countries (India and Thailand) have provided specialized equity funds targeting renewable energy and energy efficiency. Mezzanine finance—an unsecured loan at a higher interest rate—is a higher-risk debt in terms of its priority of payment or in liquidation.

Equity funds and mezzanine finance provided by the public sector could, by their example, ease some of the concerns that private creditors may have regarding extending credit to higher-risk renewable energy or energy efficiency projects. In addition, private lenders typically demand that borrowers take a certain equity ratio in risky investment. But small and medium-size developers have only limited equity funds to make the essential contribution. Public equity funds can be used to fill the gap. Both instruments tend to be used in relatively mature financial markets.

Energy Service Company Financing and Equipment Leasing

Energy service companies (ESCOs) provide both finance and technical know-how to their clients through performance contracting. In some cases (such as in Croatia and Poland), ESCOs arrange for project financing and provide a wide range of energy efficiency services, from auditing to project implementation. Customers bear no financial obligation other than to pay a percentage of the energy savings to the ESCO over a specified period. In other cases, ESCOs simply become equipment leasing companies that install energy efficiency equipment and collect monthly payments over the lease period. In the case of renewable energy, a consumer receives renewable electricity services through an ESCO, generally at equal or lower cost than conventional service. The ESCO owns the facility (for example, solar panels), obtains available public incentives, and is responsible for maintenance and repair over the life of the service contract (such as in projects in Argentina and California).

The ESCO model allows for the repayment of the capital cost to be treated as operating expense, therefore helping end-users defray the high up-front costs. It can also overcome barriers of perceived high risks and is effective in aggregating small-scale projects to reduce transaction costs. Today these instruments are used in more-developed financial markets, where contracts are easily enforced.

Utility Financing

This instrument targets the household sector to overcome high up-front capital investments and consumer liquidity constraints. Under such an arrangement, the utility provides or arranges for investment financing and collects the loan repayments through the customer's utility bill. Such an arrangement reduces both the transaction costs of recovering the loan repayments and the risk of default. Public funds are sometimes used to buy down the interest rate to further increase loan affordability.

Strengthening Sustainable Energy Finance

In summary, there is usually not a single, universally applicable support mechanism or policy for financing low-carbon investment. Most countries rely on a mix of financing instruments addressing various market barriers. In deciding how to best leverage public support to mobilize private funds, some general rules are important.

First, program design requires a firm understanding of the specific market barriers that restrict low-carbon financing. For instance, dedicated credit lines that demonstrate the viability of firm-level energy efficiency lending will not be effective if the bank's concern is in fact with the general creditworthiness of the firm (World Bank 2010a). Likewise, partial credit guarantees that help reduce banks' perceived risks in energy efficiency projects will be inefficient if the high transaction costs are the real concern. Often there will be several barriers. As in other public policy areas, solving multiple problems usually requires multiple instruments.

Second, market readiness is an important factor determining the effectiveness of financing instruments. In this regard, government can play a role in introducing regulatory reforms and investing in education, R&D, and information dissemination to actively shape the clean-energy markets and nudge private behavior. In Hungary, the regulatory changes and promotion of energy efficiency in the housing sector created a big push for banks to serve this particular market segment. In the Czech Energy Efficiency Project, regulatory change and EU subsidies for renewable energy gave an important boost for investment.

Notes

1. Stuggins, Sharabaroff, and Semikolenova (2013) provide a comprehensive discussion of energy efficiency policies on which this section partly draws.

2. See "A Brief History of Con Edison" (http://www.coned.com/history/electricity.asp) and *Edison, His Life and Inventions* by Frank Lewis Dyer and Thomas Commerford Martin (http://www.gutenberg.org/ebooks/820).

3. A country's energy prices are, of course, not determined solely by its reserves, especially because coal, gas, and oil are globally traded, so prices reflect overall supply-demand balances. Nevertheless, many countries with large fossil energy reserves also maintain policies that keep domestic energy prices low and thereby encourage higher energy use—notably lower taxes or even energy consumption subsidies.

4. Based on International Energy Agency (IEA) data and purchasing power parity (PPP) GDP. ECA's kilogram of oil equivalent (kgoe) per US$ of PPP GDP in 2009 was 0.29 compared with OECD's 0.16. Rather than 1,333 mtoe, consumption would be 763 mtoe with OECD efficiencies. The hypothetical savings would be much larger when GDP is measured at exchange rates. South American countries consumed 485 mtoe in 2009.

5. The EU-15 countries include Austria, Belgium, Denmark, Finland, France, Germany, Greece, Ireland, Italy, Luxembourg, the Netherlands, Portugal, Spain, Sweden, and the United Kingdom.

6. The EU-10 countries include Bulgaria, the Czech Republic, Estonia, Hungary, Latvia, Lithuania, Poland, Romania, the Slovak Republic, and Slovenia. The Western Balkan countries include Albania, Bosnia and Herzegovina, Croatia, Kosovo, FYR Macedonia, Montenegro, Serbia, and Slovenia.

7. EBRD (2011) presents a detailed decomposition analysis.

8. In reality, most countries have crossed the peak long ago, and energy intensities have generally been converging during the period for which good data are available.

9. The rebound effect could be slightly higher for car travel (for example, see Hymel, Small, and Van Dender 2010).

10. The income elasticity of demand for energy tends to be higher in lower- and middle-income than in high-income countries (for example, regarding household demand for electricity: 0.4 for Turkey versus 0.2 for Norway; see Halicioglu 2007 and Nesbakken 1999).

11. Data from the IEA web page: http://www.worldenergyoutlook.org/resources/energysubsidies/.

12. These points have been adapted from the very comprehensive review of Stuggins, Sharabaroff, and Semikolenova (2013). See also Sarkar and Singh (2010).

13. See the table, "Green Building: List of Initiatives" at http://www.gao.gov/assets/590/588818.pdf#page=349.

14. EU eastern partnership countries include Belarus, Moldova, Ukraine, and the South Caucasus countries of Armenia, Azerbaijan, and Georgia.

15. This section draws on Cottrell (2011); Fullerton, Leicester, and Smith (2008); and Heine, Norregaard, and Parry (2012).
16. The Commonwealth of Independent States includes Armenia, Azerbaijan, Belarus, Georgia, Kazakhstan, the Kyrgyz Republic, Moldova, Russia, Tajikistan, Turkmenistan, Ukraine, and Uzbekistan.
17. Because energy efficiency and clean or renewable energy investments share many characteristics, they are discussed together in this section.

References

Allcott, Hunt, and Michael Greenstone. 2012. "Is There an Energy Efficiency Gap?" *Journal of Economic Perspectives* 26 (1): 3–28.

Andersen, Mikael Skou, and Paul Ekins, eds. 2009. *Carbon-Energy Taxation: Lessons from Europe.* Oxford, U.K.: Oxford University Press.

Bacon, Robert, Eduardo Ley, and Masami Kojima. 2010. "Subsidies in the Energy Sector: Measurement, Impact, and Design." Background paper for the World Bank Group Energy Strategy, World Bank, Washington, DC.

Bashmakov, Igor, Konstantin Borisov, Maxim Dzedzichek, Inna Gritsevich, and Alexei Lunin. 2008. "Resource of Energy Efficiency in Russia: Scale, Costs, and Benefits." Center for Energy Efficiency report developed for the World Bank, Moscow.

BP (British Petroleum). 2011. "Statistical Review of World Energy." BP, London. www.bp.com/statisticalreview.

Cottrell, Jacqueline. 2011. "Ecological Tax Reform in Europe and Central Asia." Background paper, Europe and Central Asia Region, World Bank, Washington, DC.

Ebinger, Jane. 2006. "Measuring Financial Performance in Infrastructure: An Application to Europe and Central Asia." Policy Research Working Paper 3992, World Bank, Washington, DC.

EBRD (European Bank for Reconstruction and Development). 2011. "The Low Carbon Transition." Special Report on Climate Change, EBRD, London.

Ekins, Paul, Fabian Kesicki, and Andrew Z. P. Smith. 2010. "Marginal Abatement Cost Curves: A Call for Caution." Report to Greenpeace UK, UCL Energy Institute, University College London.

Embassy of China. 2010. "Some Facts about the Three Gorges Project." http://www.china-embassy.org/eng/zt/sxgc/t36512.htm. Accessed December 22, 2011.

European Commission. n.d. Eurostat online database. http://epp.eurostat.ec.europa.eu.

Fullerton, Don, Andrew Leicester, and Stephen Smith. 2008. "Environmental Taxes." In *The Mirrlees Review: Reforming the Tax System for the 21st Century.* London: The Institute for Fiscal Studies.

Gillingham, Kenneth, Richard G. Newell, and Karen Palmer. 2009. "Energy Efficiency Economics and Policy." Working Paper 15031, National Bureau of Economic Research, Cambridge, MA.

Halicioglu, F. 2007. "Residential Electricity Demand Dynamics in Turkey." *Energy Economics* 29 (2): 199–210.

Heine, Dirk, John Norregaard, and Ian W. H. Parry. 2012. "Environmental Tax Reform: Principles from Theory and Practice to Date." Working Paper WP/12/180, Fiscal Affairs Department, International Monetary Fund, Washington, DC.

Herring, Horace. 2006. "Energy Efficiency—A Critical View." *Energy* 31 (1): 10–20.

Hymel, Kent M., Kenneth A. Small, and Kurt Van Dender. 2010. "Induced Demand and Rebound Effects in Road Transport." *Transportation Research Part B: Methodological* 44 (10): 1220–41.

IEA (International Energy Agency). 2007. *World Energy Outlook 2007: China and India Insights*. Special Report. Paris: Organisation for Economic Co-operation and Development and IEA.

———. 2011. *World Energy Outlook 2010*. Paris: Organisation for Economic Co-operation and Development and IEA.

Jacobsen, Mark. 2010. "Evaluating U.S. Fuel Economy Standards in a Model with Producer and Household Heterogeneity." University of California at San Diego. http://econ.ucsd.edu/~m3jacobs/Jacobsen_CAFE.pdf.

Jevons, W. Stanley. 1866. *The Coal Question*. London: Macmillan.

Jones, Benjamin, Michael Keen, and Jon Strand. 2012. "Fiscal Implications of Climate Change." Policy Research Working Paper 5956, World Bank, Washington, DC.

Madlener, Reinhard, and Barry Alcott. 2009. "Energy Rebound and Economic Growth: A Review of the Main Issues and Research Needs." *Energy* 34 (3): 370–76.

Maxwell, Dorothy, Paula Owen, Laure McAndrew, Kurt Muehmel, and Alexander Neubauer. 2011. "Addressing the Rebound Effect." Report for the European Commission DG Environment, Brussels.

Nesbakken, Runa. 1999. "Price Sensitivity of Residential Energy Consumption in Norway." *Energy Economics* 21 (6): 493–515.

Neumayer, Eric. 2002. "Can Natural Factors Explain Any Cross-Country Differences in Carbon Dioxide Emissions?" *Energy Policy* 30 (1): 7–12.

Rosenfeld, Arthur H., and Deborah Poskanzer. 2009. "A Graph Is Worth a Thousand Gigawatt-Hours: How California Came to Lead the United States in Energy Efficiency." *Innovations: Technology, Governance, Globalization* 4 (4): 57–79.

Sarkar, Ashok, and Jas Singh. 2010. "Financing Energy Efficiency in Developing Countries—Lessons Learned and Remaining Challenges." *Energy Policy* 38 (10): 5560–71.

Stone, Richard. 2008. "Three Gorges Dam: Into the Unknown." *Science* 321 (5889): 628–32.

Stuggins, Gary, Alexander Sharabaroff, and Yadviga Semikolenova. 2013. "Lessons Learned from Energy Efficiency Success Cases." Europe and Central Asia Region, World Bank, Washington, DC.

Taylor, Robert P., Chandrasekar Govindarajalu, Jeremy Levin, Anke S. Meyer, and William A. Ward. 2008. *Financing Energy Efficiency: Lessons from Brazil, China, India, and Beyond*. Washington, DC: World Bank.

Vagliasindi, Maria. 2012. *Implementing Energy Subsidy Reforms: Evidence from Developing Countries*. Washington, DC: World Bank.

Von Weizsäcker, Ernst Ulrich, Karlson Hargroves, Michael H. Smith, Cheryl Desha, and Peter Stasinopoulos. 2009. *Factor Five: Transforming the Global Economy through 80% Improvements in Resource Productivity*. London: Earthscan.

World Bank. 2010a. *Climate Change and the World Bank Group—Phase II: The Challenge of Low-Carbon Growth*. Independent Evaluations Group (IEG) Study Series. Washington, DC: World Bank.

———. 2010b. *Lights Out? The Outlook for Energy in Eastern Europe and the Former Soviet Union*. Washington, DC: World Bank.

———. 2011. "Transition to a Low-Emissions Economy in Poland." Low-carbon growth study series, Europe and Central Asia Region, World Bank, Washington, DC.

———. 2012. "Maximizing Leverage of Public Funds to Unlock Commercial Financing for Clean Energy in East Asia." East Asia and Pacific Region, World Bank, Washington, DC.

———. Various years. *World Development Indicators*. Washington, DC: World Bank.

CHAPTER 3

Cleaner Energy

Main Messages

- To ensure a reliable energy supply, by 2030 Europe and Central Asia will need to invest in up to 778 gigawatts (GW) of new power generation capacity (additions and replacement of outdated plants), US$522 billion in transmission and distribution, and additional resources in an expansion of natural gas networks. These large investments present a chance to transition to a more reliable and more sustainable energy system.

- Planning long-lived energy sector infrastructure faces three types of uncertainty: regulatory uncertainty (such as the scope and scale of future environmental regulation), technological uncertainty (relative price shifts between different power generation options), and climate uncertainty (changes in water availability for hydropower and thermal generation).

- The principles for dealing with these uncertainties are predictability of climate action policies, flexibility in the pace and composition of energy sector investments, and reliability of supply through greater diversity and a larger share of climate-resilient generation techniques.

- These principles call for even greater efforts to improve energy efficiency to avoid or delay investments as well as a gradual move to a more decentralized, diversified energy mix with a rising share of renewables.

Few areas of the Soviet economy were considered as important as the power sector. Following Lenin's famous dictum—"Communism is Soviet power plus electrification of the whole country"—planners ventured to expand electrification to production in all branches of the economy. This was assumed to yield vast increases in resource and labor productivity, thus facilitating the transition from socialism to communism (Michel and Klain 1964). However, despite large investments (initially in hydropower, later in thermal and nuclear power), the last decades of the Soviet Union saw frequent power shortages. These were felt especially among households because supply preference was given to industry.

The reliability of supply improved during transition, but the power sector across the Europe and Central Asia (ECA) region still faces major challenges over the next several decades. These include the needs to secure electricity supply in the face of increasing demand, refurbish a deteriorating capital stock, and reduce greenhouse gas (GHG) emissions. One recent scenario expects demand for electricity to increase by 3.1 percent per year over the next two decades.[1] Much of the generating capacity is quite old, particularly thermal capacity, so many of the existing power plants need to be retired or rehabilitated.

The scale of required investments is enormous. The World Bank scenario suggests that additions and rehabilitations will need to add capacity from about 84 gigawatts (GW) in the 2006–10 period to nearly 233 GW in 2026–30 to meet demand (World Bank 2010). These massive capital investments will not only secure reliable electricity supply, but will also provide an opportunity to significantly reduce the carbon intensity of power generation—one of the largest contributors to GHG emissions, accounting for 21 percent of total emissions in ECA in 2008. International Energy Agency (IEA) scenarios consistent with warming below 2 degrees Celsius (°C) foresee a continuing role for fossil fuels (in the future, using carbon capture and storage [CCS] technologies) but a gradual shift to greater contributions from nuclear, hydro, and renewable energy sources (IEA 2011). This shift would allow emissions to fall even with an expansion of generation capacity.

This transition to lower-emission electricity generation will not proceed according to a regional blueprint. It will depend on national-level decisions by planners, regulators, utilities, and investors. Many of the technical and economic issues that guide this process have been discussed in the World Bank's *Lights Out?* report (World Bank 2010). Chapter 6 of this volume discusses options for low-emission power production. This chapter takes a step back and discusses some general concepts that could serve as a guide for energy sector planning. It focuses on three types of uncertainty in making decisions about long-lived infrastructure investments against a backdrop of constantly changing economic, technological, and natural environments:

- *Regulatory uncertainty*. Although a global climate agreement currently looks unlikely, some form of a carbon constraint will likely be introduced, sooner or later, in all countries.

- *Technological uncertainty*. Constant changes in resource costs and massive investment in energy research and development imply that relative prices between different energy and power generation technologies will continue to shift frequently.

- *Climate uncertainty*. Climate change itself could affect hydro, as well as thermal, nuclear, and some solar power production, all of which rely on abundant water for cooling.

This chapter suggests that decision makers can reduce these uncertainties in ways that also provide major climate benefits. *Predictability* of policies, such as those addressing carbon emissions, gives investors a clear framework for long-term investment planning. *Flexibility* means moving away from a supply expansion paradigm in power planning to one favoring both supply- and demand-side efficiency as well as smaller, decentralized generation. Efficiency and decentralization help avoid or delay larger and infrequent or lumpy investments when technological change could quickly alter the relative costs of power generation options. Finally, *reliability* of supply can be improved by designing power systems that are resilient to changes in hydrology and temperature due to climate change.

Supplying Reliable, Sustainable, and Affordable Power

During evening rush hour on August 20, 2010, a failure of antiquated equipment at two substations near St. Petersburg left the

Russian Federation's second-largest city paralyzed. Half of the city was plunged into darkness. Traffic lights were off on the city's central avenue, causing traffic jams. Commuter and long-distance trains got stuck, forcing many passengers on the world's deepest subway to walk through tunnels and up escalators to exit. Other basic services including water, mobile, and Internet communications were subsequently disrupted. The power outage lasted for more than an hour and cost the city more than US$3 million.

Electricity disruptions such as the one in St. Petersburg have been increasing again in parts of ECA during the past few years. The reason is that the power sector has suffered from the deterioration of an aging infrastructure that dates back to the planned-economy era. There is a vast backlog of existing capacity that needs upgrading and modernizing, and too few new facilities were added over the past two decades. The installed capacity in ECA in 2009 totaled an estimated 519 GW, compared with 443 GW in 1991. Most of the increase was in Russia and Turkey, where generation capacity increased by 12 GW and 28 GW, respectively. In the rest of the region, the capacity increase has been almost negligible, as figure 3.1 shows.

FIGURE 3.1

Total Installed Electricity Capacity in Selected Regions, 1991–2009

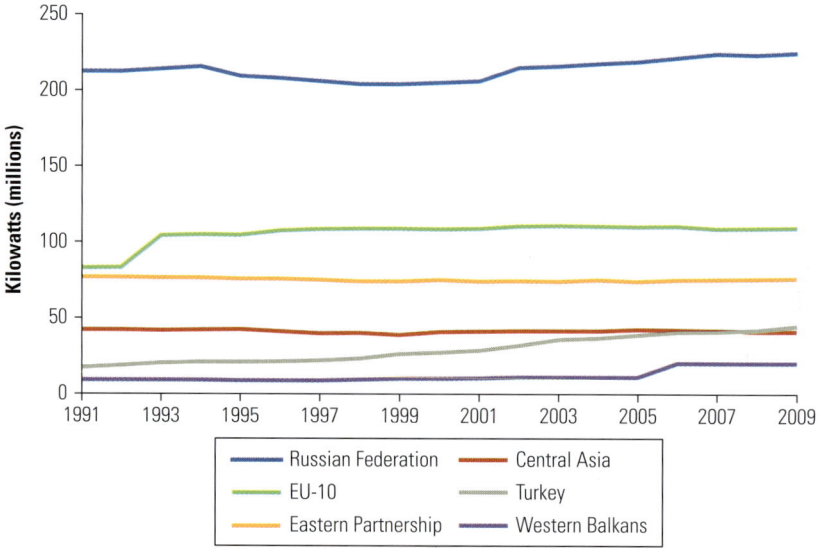

Source: U.S. EIA n.d.

Note: EU = European Union; Central Asia = Kazakhstan, the Kyrgyz Republic, Tajikistan, Turkmenistan, and Uzbekistan; Western Balkans = Albania, Bosnia and Herzegovina, Croatia, Kosovo, FYR Macedonia, Montenegro, Slovenia, and Serbia; EU-10 = Bulgaria, the Czech Republic, Estonia, Hungary, Latvia, Lithuania, Poland, Romania, the Slovak Republic, and Slovenia; Eastern Partnership = Armenia, Azerbaijan, Belarus, Georgia, Moldova, and Ukraine.

The problem of underinvestment was not immediately felt in the 1990s. After the collapse of the Soviet Union, electricity demand dropped markedly with lower industrial output, particularly the output of heavy industry. Although electricity output also shrank, generation still exceeded demand. However, from the mid-1990s on, with rising incomes and industrial growth recovering, electricity demand quickly increased. By 2008, the legacy of abundant electricity infrastructure had disappeared. Energy importers were experiencing shortages leading to periodic brownouts and blackouts. However, currently even energy exporters face problems. In Russia, a large share of power stations will reach the end of their planned life spans in a matter of years and will need replacement—even without an increase in demand (Ketting 2008).

Meanwhile, the whole ECA region could see electricity consumption rising at about 3 percent annually for the next two decades. Over 20 years, that translates into demand growth of over 90 percent from 2007 levels. Taking together additions, rehabilitations, and retirements, the required new generation capacity is equivalent to almost 560 new standard-size coal-fired plants, as figure 3.2 indicates. This could cost US$970 billion (in 2008 dollars) by 2030, with another

FIGURE 3.2

Projected Capacity Additions, Rehabilitations, and Retirements in the Electricity Infrastructure of Europe and Central Asia, 2006–30

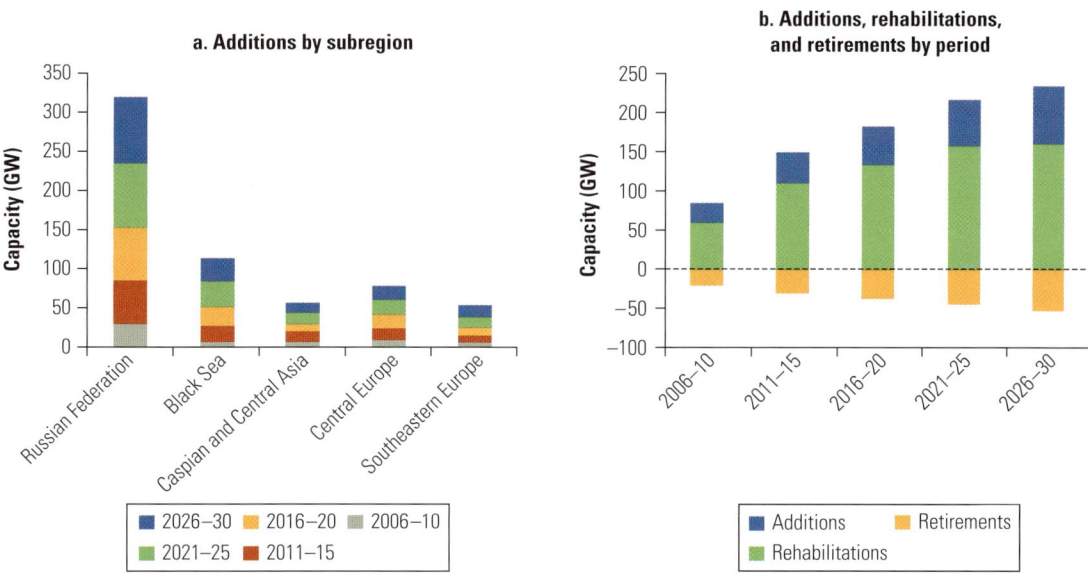

Source: World Bank 2010.
Note: GW = gigawatts.

US$520 billion needed for investment into the transmission and distribution network (World Bank 2010).

These massive investments will stretch the financial and implementation capacity in many ECA countries. But with the challenge also comes an opportunity to create a more reliable and sustainable power supply infrastructure. This development will involve a transition from carbon-intensive fuels such as coal to natural gas and to a much larger share of renewables. The Commonwealth of Independent States (CIS) countries already rely heavily on natural gas, which gradually increased from 39 percent to 42 percent of total electricity generation from 1990 to 2008, as shown in figure 3.3, panel a.[2] Coal is the CIS region's second most important fuel, at around 20 percent. Nuclear power and hydropower each provided about 17 percent by 2008. The biggest change has been with oil, whose share dropped from almost 15 percent to just 1.6 percent over the past two decades. The share of non-hydro renewable energy sources remains negligible.

Among the European Union (EU)-10 and Western Balkan countries, coal is the dominant fuel for electricity, at around 60 percent on average, as shown in figure 3.3, panels b and c.[3] For the EU-10 countries, nuclear power represents around 21 percent of total generation, followed by hydropower and natural gas at about 8 percent. Non-hydro renewables have increased from almost zero in 1990 to 2 percent in 2008.

Hydropower is the second-largest source of electricity (after coal and peat power) in the Western Balkan countries, fluctuating between 30 percent and 40 percent in recent years. Gas is around 3–4 percent, while oil has declined from 8 percent to 3 percent. The share of non-hydro renewables is 0.8 percent.

In Turkey, gas significantly increased, from 18 percent to 50 percent from 1990 through 2008, as shown in figure 3.3, panel d. Meanwhile, coal's share—once at 35 percent—has declined to about 29 percent. Hydropower is next at 16 percent, having declined from almost 40 percent 20 years ago. Non-hydro renewables is at 0.6 percent.

Overall, ECA is heavily dependent on fossil fuels and nuclear power, with hydropower the leading source of renewable energy. From a very low base, non-hydro renewables—mainly solar, wind, and biomass—have increased rapidly, especially in Turkey and the EU-member countries. The ECA region (shown as "Non-OECD Europe and Eurasia" in figures 3.4 and 3.5) does not have high carbon intensity relative to other world regions because of the region's relatively large share of lower-emission energy sources including gas, hydro, and nuclear, as figure 3.5 illustrates.

However, while overall carbon intensities have remained flat, individual country experience varies. In Azerbaijan, as natural gas almost

FIGURE 3.3

Electricity Generation Sources, by Fuel Type, Europe and Central Asia Subregions, 1990–2008

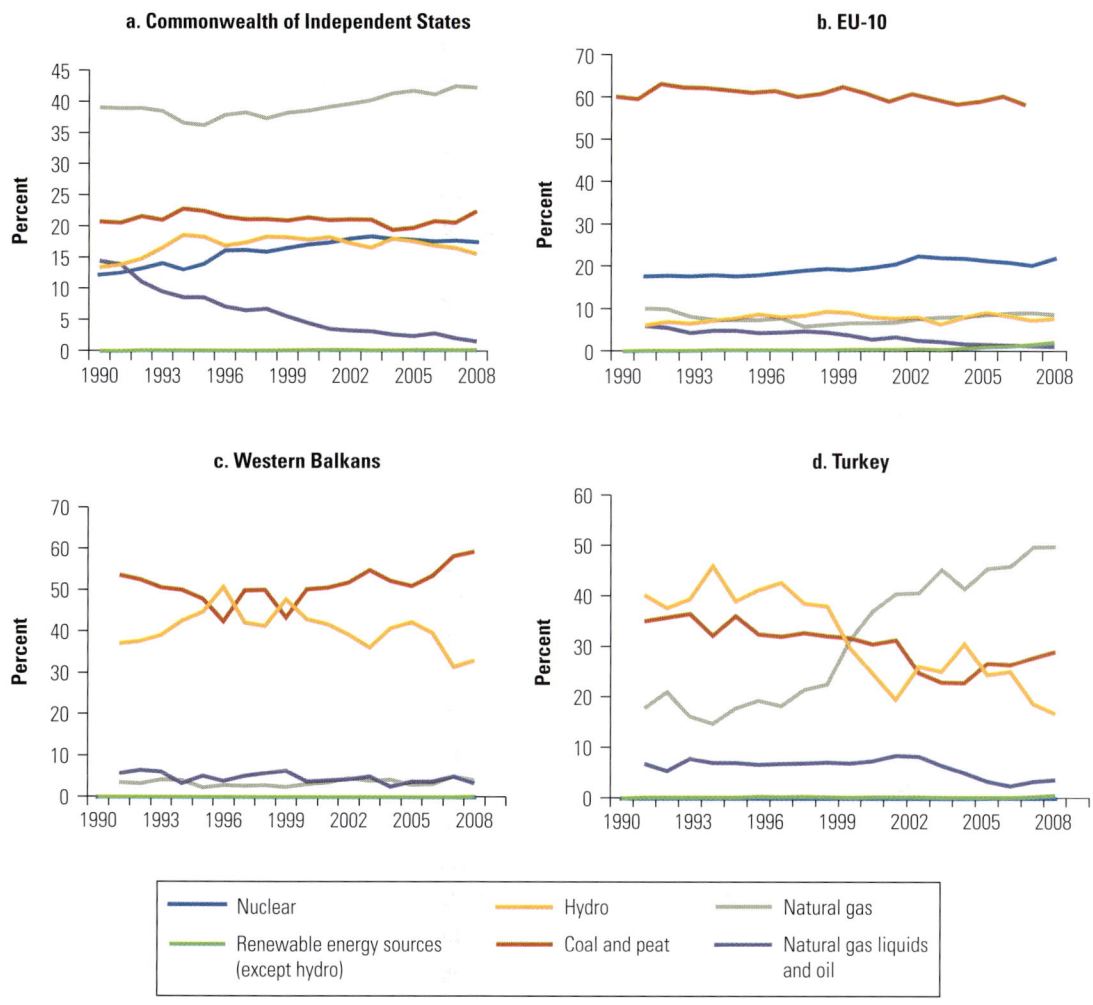

a. Commonwealth of Independent States

b. EU-10

c. Western Balkans

d. Turkey

Legend:
- Nuclear
- Renewable energy sources (except hydro)
- Hydro
- Coal and peat
- Natural gas
- Natural gas liquids and oil

Source: IEA n.d.
Note: Commonwealth of Independent States = Azerbaijan, Armenia, Belarus, Georgia, Kazakhstan, the Kyrgyz Republic, Moldova, the Russian Federation, Tajikistan, Turkmenistan, Uzbekistan, and Ukraine. EU-10 = Bulgaria, the Czech Republic, Estonia, Hungary, Latvia, Lithuania, Poland, Romania, the Slovak Republic, and Slovenia. Western Balkans = Albania, Bosnia and Herzegovina, Croatia, the former Yugoslav Republic of Macedonia, Kosovo, Montenegro, Serbia, and Slovenia.

completely replaced oil for electricity generation (from almost nothing to 82 percent), the country's carbon intensity fell from 864 grams (g) per kilowatt-hour (kWh) to 443 g per kWh from 1992 to 2009. Bosnia and Herzegovina and Turkmenistan, on the other hand, have some of the most carbon-intensive power sectors in the world at 776 g per kWh and 789 g per kWh, respectively. Estonia and Turkmenistan have also significantly increased their carbon intensity in the past two decades.

FIGURE 3.4

Carbon Intensity of Electricity and Heat Generation, by World Region, 1992–2009

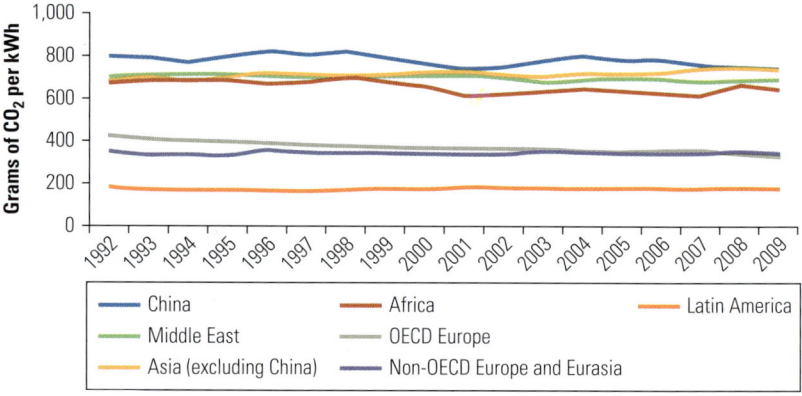

Source: IEA 2011.
Note: IEA's "Non-OECD Europe and Eurasia" region includes all ECA countries except the Czech Republic, Hungary, Poland, the Slovak Republic, and Turkey. OECD = Organisation for Economic Co-operation and Development; CO_2 = carbon dioxide; kWh = kilowatt-hour.

FIGURE 3.5

Electricity Generation Sources in World Regions, by Fuel Type, 2009

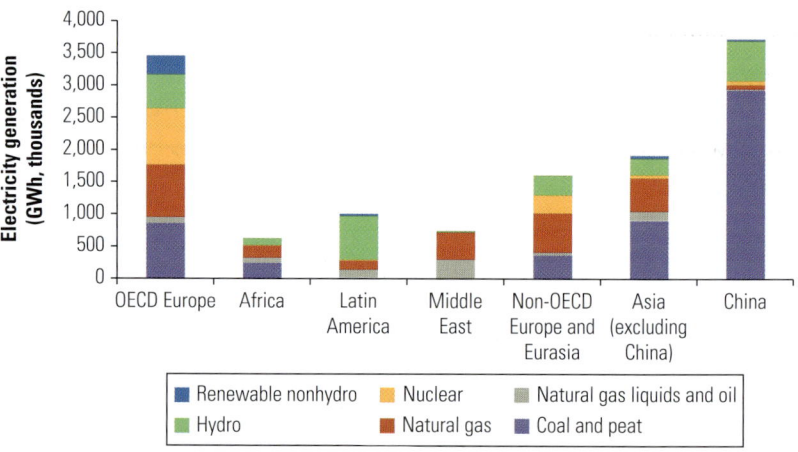

Source: IEA n.d.
Note: The International Energy Agency's "Non-OECD Europe and Eurasia" region includes all ECA countries except the Czech Republic, Hungary, Poland, the Slovak Republic, and Turkey. OECD = Organisation for Economic Co-operation and Development; Gwh = gigawatts per hour.

When carbon intensities remain stable in the power sector, expansion of generation capacity raises total emissions. In 2008, the region's total emissions were up to 2,678 million tons from 2,427 million tons in 2000. But stabilizing atmospheric GHGs at 450 parts per million (ppm) would mean that ECA's power sector will have to reduce carbon dioxide (CO_2) emissions by 76 percent by 2050—which would require a drop in CO_2 intensity to 67g per kWh, according to the IEA 450 ppm scenario (IEA 2011). To achieve this goal, the IEA scenario assumes that renewables will account for 48 percent of power generation, nuclear will provide 24 percent, and 17 percent of the plants will be equipped with CCS technologies.

Uncertainties: Regulation, Technology, and Climate

Energy infrastructure such as electric power facilities is a long-term business. Consumers may change mobile phones every 2 years, and traditional centralized planning works in 5-year increments, but a power plant built today may still be operating 60 or 70 years from now. It is also an expensive big-ticket item. A new coal plant may cost as much as US$2 billion. With such a long planning horizon and the enormous amount of capital at stake, planning and decision making must take into account the large uncertainties in future regulations, technology, and climatic conditions.

Regulatory Uncertainty

Just a few years ago, the United States was thought to be in a new coal rush. By 2006, 150 new coal-burning plants were on drawing boards across the nation. Things have changed considerably in the past few years. By 2010, power companies dropped or delayed plans to build 80 coal units while announcing that they would retire 48 aging, inefficient ones. One reason is a price drop in natural gas, which also requires lower capital investments in plant construction. But environmental issues, particularly CO_2 emissions, were the most cited reason for cancellations and delays (*Washington Post* 2011).[4] The fate of the long-planned Glades County coal facility in Florida is one example. In 2007, the state utility commission rejected the proposal to build what would have been the nation's largest coal-burning power plant—a 1,960-megawatt (MW) ultra super critical coal plant. In its decision to reject the proposal, the commission said that "the plant was cost-effective in fewer than half the scenarios examined"— a major reason being uncertainty about the future cost of curbing CO_2 emissions.

The rapid shift away from coal in the United States shows how concerns about future carbon emission penalties affect planning in the power industry. A price on carbon, either through a cap-and-trade system or a tax, can profoundly alter the comparative economics of different power generation technologies. With a price on carbon emissions, the cost differential between fossil-fuel plants and low-carbon alternatives shrinks and sometimes disappears. The expected cost of carbon compliance will be highest for conventional coal-based power generation.

Coal is the most carbon-intensive fuel. A traditional coal-fired power plant produces roughly one ton of CO_2 emissions for every one megawatt-hour (MWh) of electricity. Projected prices of one ton of CO_2 range widely, from US$8 to US$364, depending on the timing and stringency of carbon control (ICCG 2010). That translates into added costs of anywhere between 13 percent and 587 percent of the original price of coal power. The IEA shows that the cost of coal-based electricity is far more sensitive to changes in CO_2 costs than the cost of electricity generated by natural gas or coal with CCS, as shown in figure 3.6.

Natural gas is considered a cleaner and more environmentally attractive fuel than coal. Although natural gas is low in carbon, it is not carbon free: generating one MWh of gas-fired electricity releases about half a ton of CO_2, half as much as coal.[5] GHG emissions from venting and leakage during the production and transport processes raise the carbon intensity of gas plants (as further discussed in chapter 6). Therefore, natural gas can help to reduce emissions substantially in

FIGURE 3.6

Projected Sensitivity of the Costs of Fossil Fuel–Based Generation to CO_2 Pricing under Two Discount Rate Scenarios

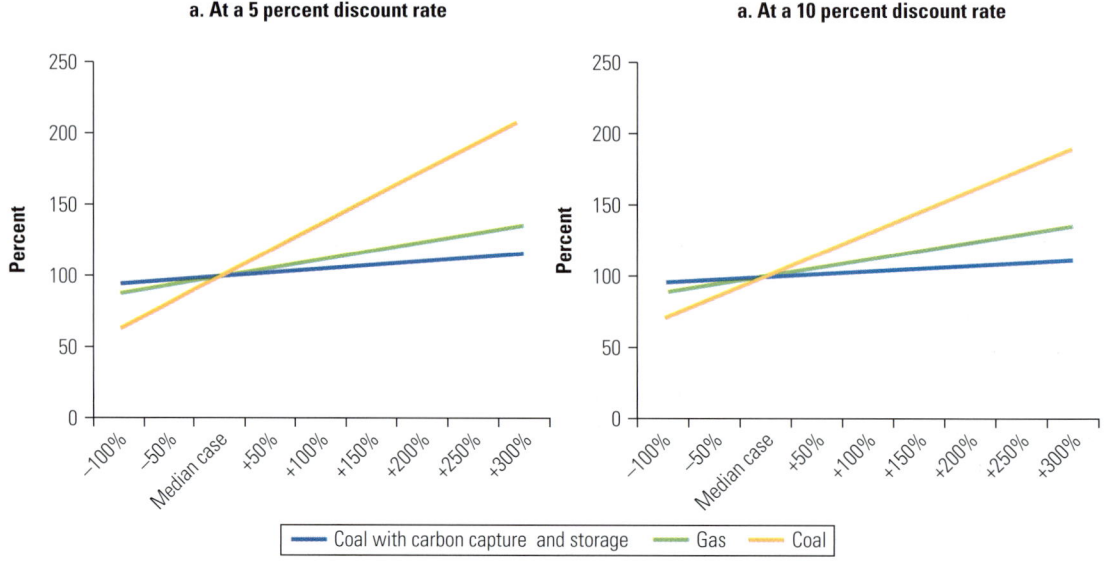

a. At a 5 percent discount rate

a. At a 10 percent discount rate

Coal with carbon capture and storage — Gas — Coal

Source: IEA 2010.

Note: Unabated coal generation is most sensitive to future carbon prices. A +/– 50 percent variation in carbon costs translates into a +/– 18 percent, +/– 6 percent, or a +/– 3 percent change in the total LCOE of coal-fired plants, gas-fired plants, and coal plants equipped with CCS technology, respectively. LCOE = levelized cost of electricity; CCS = carbon capture and storage.

the short and medium term, but even gas could be under pressure in the long term when carbon prices are sufficiently high—unless CCS becomes feasible for natural gas as well as coal-fired generation. In the United Kingdom, for instance, onshore wind is projected to match the most-efficient gas plants as the least cost-generating option in 2017, given a carbon price of US$26 per ton (Mott Macdonald 2010).

It is not clear for most ECA countries when or how carbon emissions will be priced in the absence of a successor to the Kyoto Protocol.[6] Regulatory uncertainty differs significantly among the countries in the region. The EU-member countries have made the clearest commitment, pledging to reduce carbon emissions to at least 20 percent below 1990 levels by 2020. An EU-wide emission trading scheme has already put a mandatory and declining cap on carbon emissions. Although carbon prices traded on the EU market have been volatile and difficult to predict, they nonetheless send price signals, allowing climate considerations to influence long-term technology choices.

The EU candidates and prospective members are not yet subject to EU regulations on carbon mitigation. But EU aspiration sets these countries on a path toward a carbon-conscious future. Turkey, for example, is exploring a domestic cap-and-trade market for carbon emissions under the World Bank-led Partnership for Market Readiness. Although none of these countries has yet adopted mandatory mitigation targets, they have incentives to explore imposing a carbon constraint. The CIS countries, in contrast—especially energy exporters rich in fossil fuels—currently have the least incentive in the region to significantly reduce carbon emissions. Investors in these countries could face the most extreme scenarios if binding climate policies are eventually adopted.

Technological Uncertainty

Another feature of power sector investment is that the future costs of technologies, especially emerging low-carbon technologies, are uncertain. For conventional fossil-fuel power plants, a large portion of the risk lies in the fuel costs. Figure 3.7 shows large price swings in all major fossil fuels. Relatively recently, 10 years ago, the natural gas price rose in Europe with increased use of gas in power generation, but the price dropped in the United States as unconventional sources such as shale gas entered the market.

There are no fuel costs for most renewable energy, so the variable costs are low, and while initial capital investments are substantial, they are predictable at the time of investment. Intermittent power availability and grid integration of many decentralized generation

FIGURE 3.7
Volatility in Fossil-Fuel Energy Prices, 1997–2012

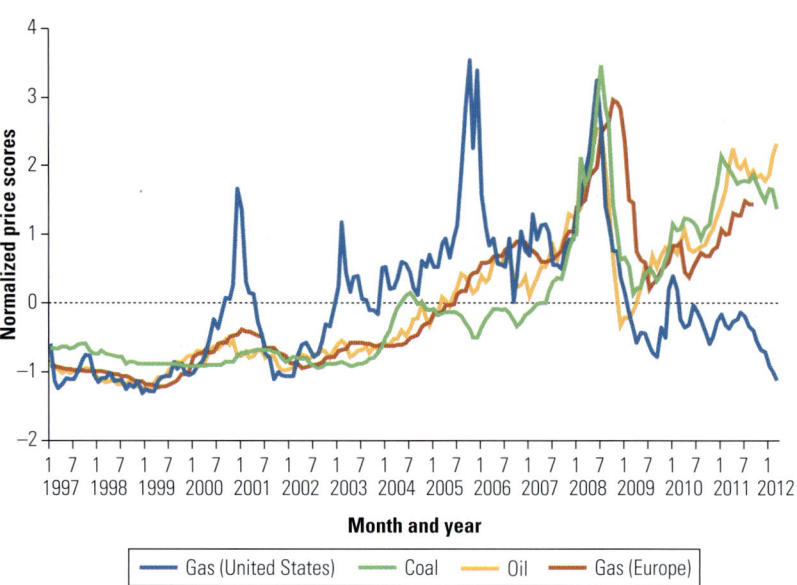

Sources: U.S. EIA n.d.; World Bank n.d.; www.indexmundi.com.

Note: Normalized price scores of (a) Europe Brent Oil Spot Price FOB (dollars per barrel); (b) Natural gas, Europe, $ per mmbtu (million British thermal units); (c) Henry Hub Gulf Coast Natural Gas Spot Price; and (d) Australian thermal coal FOB Newcastle/Port Kembla.

units create additional obstacles that increase costs. The challenge for power system planning is that the future technology costs are difficult to predict. Just as coal has been sidelined in some countries for both regulatory and market reasons, continued innovation and possible technological breakthroughs could shift relative prices in favor of renewable energy within the next few decades—well within the life span of large power infrastructure.

Among renewable energy technologies, onshore wind, biomass combustion, and geothermal are relatively well developed. They have been widely used and can often compete with conventional energy in highly suitable areas, with access to the grid and relatively high energy prices. Emerging renewable generation technologies include offshore wind, several types of bioenergy, and solar photovoltaic. These technologies are proven technologically but will require substantial cost reduction for large-scale, unsubsidized use. In a third group are the technologies still mostly in the research and development (R&D) phase, including concentrated solar power, ocean energy, CCS (coal and gas), and more advanced forms of bioenergy such as fuel from algae. Chapter 6 discusses low-emission electricity generation in more detail.

Whether and when renewables and other clean technologies (such as coal with CCS) become economically viable depends on a number of variables. The more widespread the use of renewable technologies, the cheaper they will become, so the pace at which they are adopted affects their price. A larger share of renewables will require large investments in hardware (transmission capacity) and software (grid management) for integrating a very large share of dispersed electricity generating units. This shift toward renewables will affect the profitability of conventional power plants that may run less frequently. Government policies have a large influence on these dynamics. A shift in the support system for renewables in Germany, for instance, encouraged a far faster expansion of renewable energy generation than almost all previous scenarios or forecasts had predicted (as shown in figure 3.8; see also figure 1.4 in chapter 1). Although this process is policy induced, deployment will come from the private sector. In the United States, clean technology represented 16 percent of total venture capital investment, rising from $317 million in 2001 to US$3.7 billion in 2010.[7]

Technology uncertainty affects energy planning in all countries, whether they have a strong domestic R&D capacity or not. Larger and higher-income ECA countries have the resources to create both supply push (in the form of R&D) and demand pull (in terms of subsidies

FIGURE 3.8

Renewable Energy Generation Trends in Germany, Predicted vs. Actual, 1995–2020 (as of 2009)

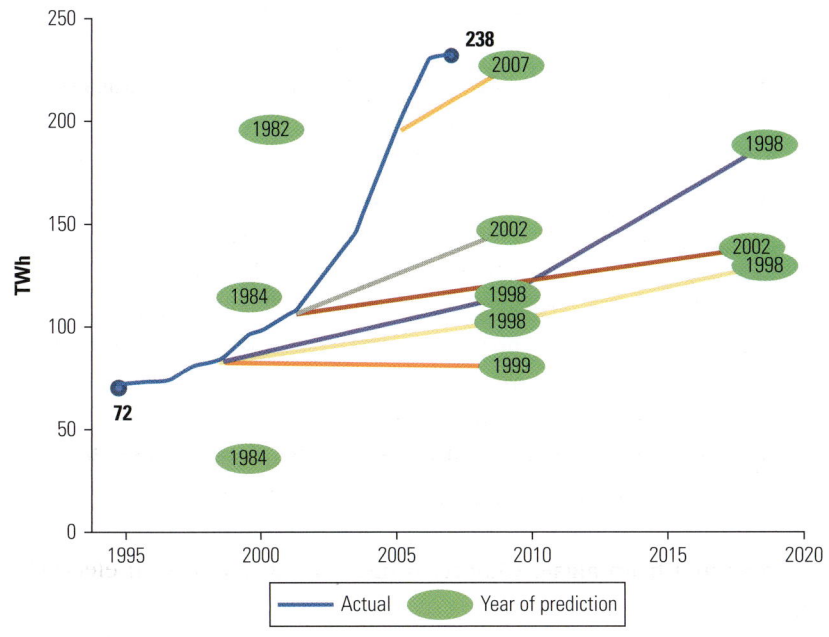

Source: Pieprzyk and Rojas Hilje 2009.
Note: TWh = terawatt-hours.

for large-scale deployment) of low-carbon technologies. They therefore influence the direction and the speed of technology development. Smaller countries do not. Although learning may occur at the local level, technology uncertainty is largely beyond the control of these countries. Either way, in the absence of trade barriers protecting incumbent power systems, all countries will be subject to global trends in the relative prices of energy supply technologies.

Climate Uncertainty

The last and probably least discussed source of uncertainty is the difficulty of predicting climate change and its impact on the power supply. This is where climate change mitigation and adaptation most clearly overlap. Climate change could affect thermal generation such as in fossil-fuel and nuclear power plants that require a constant supply of cooling water to discharge the surplus heat while at the same time producing steam, which in turn drives a turbine and produces electricity, (table 3.1).

When high water temperatures and low water levels occur simultaneously, thermal power supply could be severely disrupted. This is what happened during the 2003 summer heat wave in Europe, the hottest on record since at least 1540. France, where over 70 percent of the electricity comes from nuclear power, was hit especially hard. Electricity demand reached peak levels, but cooling-water shortages forced about a third of the nuclear power stations to shut down. Millions of people across France suffered through an extended power shortage.

Climate change will also affect hydropower production in regions subject to changing patterns of precipitation or snowmelt.

TABLE 3.1
Water Withdrawn and Consumed for Power Plant Cooling
gallons of water required per MWh of electricity produced

Fuel Source	Once-through		Wet-recirculating with cooling tower		Dry-cooling	
	Withdrawal	Consumption	Withdrawal	Consumption	Withdrawal	Consumption
Coal	20,000–50,000	300	500–600	480	0	0
Natural gas combined cycle	7,500–20,000	100	230	180	0	0
Nuclear	25,000–60,000	400	800–1,100	720	—	—
Solar thermal (through)	—	—	600–850	—	—	—
Solar	0	0	0	0	0	0
Wind	0	0	0	0	0	0

Source: U.S. GAO 2009.
Note: — = no data available; MWh = megawatt-hour.

Hydroelectric generation is very sensitive to changes in water supply. Every 1 percent decrease in precipitation results in a 2–3 percent drop in stream flow, and every 1 percent decrease in stream flow in a river basin may result in a 3 percent drop in power generation (ORNL 2007).

Such magnifying sensitivities—occurring because water flows through multiple power plants in a river basin—underscore the vulnerability of a power system dominated by hydro resources. In 2001, a severe drought triggered the worst energy crisis in Brazil's history. The hydroelectric power plants that had generated 94 percent of Brazil's electricity supply before the crisis were left with reservoirs 70 percent empty. The crisis led to electricity rationing for nine months with cutbacks in use ranging from 15 percent to 20 percent. Six years later, during the winter of 2007–08, record-low hydrologic conditions occurred in the Kyrgyz Republic and Tajikistan, where 90 percent of the electricity is generated by water. Acute electricity shortages had major social impacts as people needed heat and electricity to cope with the winter cold.

The best current projections for changes in hydrological conditions suggest that the northern and eastern parts of ECA will see increases, while the southern and western parts of the region will see decreasing water availability in the long run, as map 3.1 illustrates. The area around the Mediterranean, for instance, could see decreases of hydropower potential of 20–50 percent (IPCC 2011). However, the intrayear changes in water availability are hard to predict.

Currently, nuclear power provides about 17 percent of the electricity supply of ECA, while hydropower provides about 16 percent. Climate change-related extreme weather events such as heat waves and water shortages are expected to increase. These events could cause more frequent disruptions in the use of nuclear and hydropower plants. Planning and design of new power infrastructure thus needs to take into account the likelihood of the impact—especially because many projects built today could still be in place in 50 or more years.

Climate change could also affect non-hydro renewable energy sources. For example, changing cloud cover affects solar energy resources although current models predict only minor changes in sunshine intensity (IPCC 2011). Changes may be more significant for wind energy resources and for temperature and precipitation patterns that affect biomass production. The limited research to date suggests that ECA may benefit from increased solar radiance, especially around the Mediterranean (Ebinger et al. 2008). A moderate temperature rise will increase timber supply and flatten or reduce prices, therefore promoting biomass production in the region (Kirilenko and

MAP 3.1

Projected Global Changes in Water Availability, 1980–99 vs. 2090–99

large-scale changes in annual runoff (water availability, in percent)

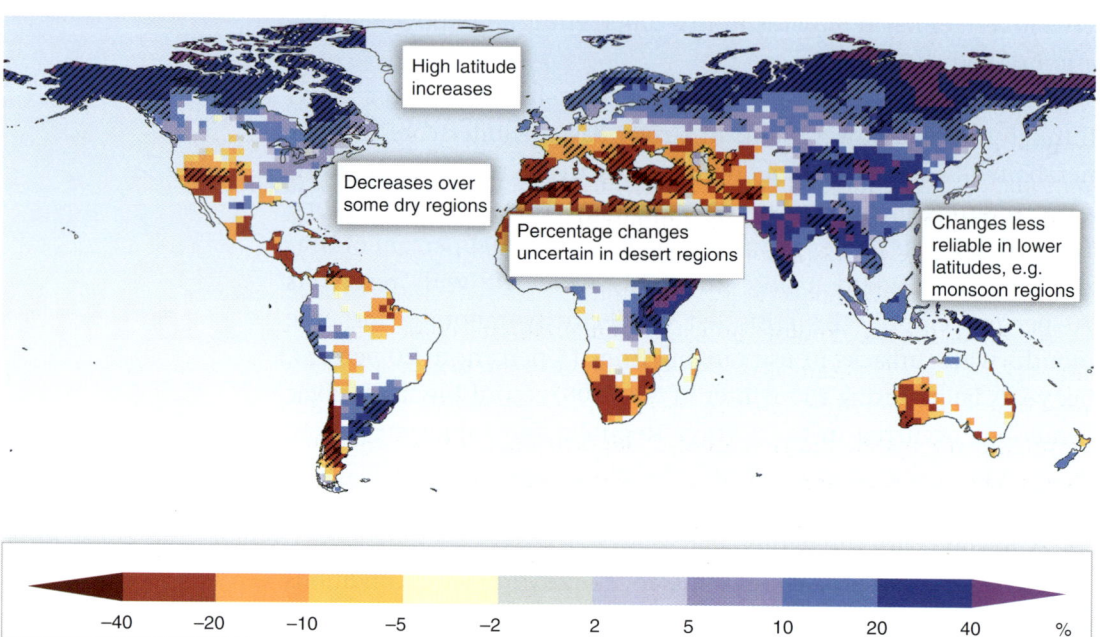

Source: IPCC 2011.

Note: Values represent the median of 12 climate model projections using the Special Report on Emissions Scenarios (SRES) A1B scenario of the Intergovernmental Panel on Climate Change (IPCC). (The A1B scenario describes a balanced technological emphasis across all energy sources—both fossil intensive and non-fossil intensive.) White areas are where fewer than 66 percent of the 12 models agree on the sign of change, and hatched areas are where more than 90 percent of models agree on the sign of change.

Sedjo 2007). Wind strength will rise, particularly in northern parts of ECA, but there will also be increased variability. However, the overall impact will likely be small (Pryor and Barthelmie 2010).

Dealing with Uncertainty in Power Sector Planning

Power sector planning has always had to deal with uncertainty, especially because many projects are very large, putting enormous amounts of capital at stake. Climate change policies and potential impacts further complicate power investment decisions. Given the long-lived nature of power infrastructure, uncertainty can lead to decisions that ultimately impose unnecessary costs on the economy. Locking in carbon-intensive technology, for example, could result in facilities that soon end up obsolete if they subsequently prove to be inadequate to meet future carbon regulations. At the same time, power planners must keep up with electricity demand and cannot always adopt a wait-and-see attitude. There is no simple solution to dealing with this uncertainty, but emphasizing predictability, flexibility, and reliability can help improve the security of affordable power supply while also supporting climate change mitigation goals.

Predictable Policies

The likelihood of a new global climate treaty appears small at the moment, but many countries, including most in ECA, have made national climate change mitigation commitments following the 2009 United Nations Climate Change Conference, commonly known as the Copenhagen Summit. EU-member states and accession countries are obligated to further emission reductions.

For all countries, credible and predictable climate change mitigation policies will facilitate long-term planning in the power sector for two main reasons. First, resolving regulatory uncertainty will help companies to better assess risks and opportunities and to optimize their investment decisions. Given the limited funds available for public investment, private capital will be crucial to address the large investment needs in the power sector. But when rules and policies are vague, private financing will be either delayed or come at higher costs. A recent pan-European survey on the power sector finds that regulatory and policy changes are perceived as the most significant risk for investments in electricity generation capacity (DNV 2009). In the United Kingdom, perceptions of greater risk are estimated to result in a capital cost that is 3 percentage points higher for low-carbon technologies (Redpoint Energy 2010). IEA analysis shows that the risk premium of climate change uncertainty can add 40 percent to construction costs of the plant for power investors and 10 percent of price surcharges for the electricity end users (IEA 2007).

Providing a stable regulatory environment is therefore important for attracting private financing, especially for capital-intensive low-carbon alternatives. For example, in 2005 and 2007, the Turkish government passed two laws favoring renewable investments. Investment levels were noticeably low in those years as firms held back investments and waited for clearer rules, as indicated in figure 3.9. Once it was announced that the new law would provide a 10-year purchase agreement coupled with a guaranteed price of 5–5.5 € cents per kWh for renewable electricity, the private sector became engaged. A guaranteed purchase agreement and a floor price significantly reduced uncertainties of investment in renewable power. In 2006, investment more than doubled. In 2008, investment further tripled relative to 2006.[8]

The second argument for predictable policies is that acting now will allow time for modest initial steps, followed by a gradual phase-in of

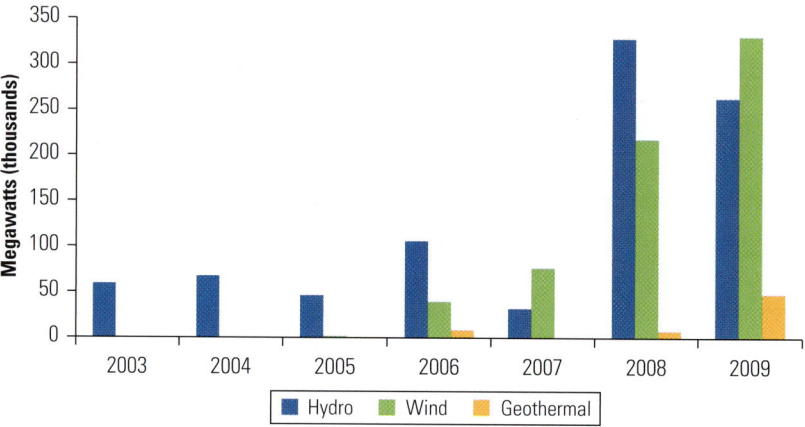

FIGURE 3.9

New Renewable Energy Capacity in Turkey, 2003–09

Source: World Bank 2009.

more stringent requirements. A smoother transition will allow economic systems to gradually adjust to changing energy prices and technologies rather than waiting until it becomes necessary to implement drastic reductions in a shorter time frame (Grubb, Chapuis, and Ha-Duong 1995; Grubb 1997). The overall adjustment costs will be lower—sometimes by a factor of six or more—when policies are announced early, defined clearly, and implemented consistently (Atkeson 1997). This is the basic insight of the Nobel Prize–winning work by the economists Finn E. Kydland and Edward C. Prescott on the time consistency of economic policy.

For the power sector specifically, gradual and predictable phase-in of climate policies minimizes the adjustment costs. To change the carbon intensity of power generation, it is often necessary to change the fuel and equipment used. When the reduction targets are known with enough certainty and far enough in advance, companies can match the cycle of capital stock turnover with the ramping up of mitigation efforts. They can avoid the costs of stranded assets or investments that have become obsolete. In countries where domestic climate policies are not foreseeable in the near future, other strategies may be needed to overcome investment hurdles imposed by regulatory uncertainty. One alternative is for the government to provide revenue guarantees (power purchase agreements, for example) to mitigate investment risks. Such an approach will require the government to take on some of the risk itself. The uncertainty associated with the scope and continuity of the guarantee policies may also make investors less interested in long-term projects.

Flexible Systems

Technology uncertainty complicates investment decisions in areas where technological change occurs quickly. When there is a risk to make a wrong choice, such as selecting a power-generation technology that later turns out more expensive than alternatives, there is value in being able to delay the decision. In other words, there is an opportunity cost to investing rather than waiting for new information.[9] This insight represents a general investment philosophy and is also the basic conclusion derived from the "putty-clay" investment model, so called because capital in monetary form can be turned into any type of investment, but once capital has been used to buy equipment, it hardens like clay and is no longer malleable (Albrecht and Hart 1983).

Keeping options open suggests a two-pronged strategy for ECA's power sector. The first is to encourage efficiency, so there is a close relationship between energy efficiency (as discussed in chapter 2) and cleaner energy. The safest investment is the one that does not need to be made. So if more energy can be produced with existing equipment (supply-side efficiency) or if households and firms can be encouraged to reduce their energy use (demand-side efficiency), the risks of misallocating investments are smaller. Second, where new capacity needs to be added or existing equipment replaced, investors could favor smaller, incremental investments in response to price and technology changes. Such a flexible expansion will favor a model of decentralized generation.

Supply-Side Efficiency

Generating, transmitting, and distributing electricity more efficiently delays or avoids the need for large-scale investment in new capacity. It will also avoid carbon emissions from the power sector. The scope for efficiency gains in ECA's power generation is large. The average efficiency covering all generation technologies in the region is 43 percent, as shown in figure 3.10—slightly better than the world and the Organisation for Economic Co-operation and Development (OECD) averages, which are both at around 40 percent. However, for thermal generation, the average efficiency of ECA plants is 30 percent compared with the world average of 35 percent and the OECD average of 41 percent.

There are also substantial differences among countries. Tajikistan has the lowest thermal efficiency, at 15 percent, but also the highest overall efficiency owing to a large share of hydroelectric generation. Turkey has the highest thermal efficiency, at 43 percent, as well as a very high overall generation efficiency level at 46 percent. The

FIGURE 3.10

Average Efficiency of Power Generation in Global Regions and Selected Europe and Central Asia Countries, 2008

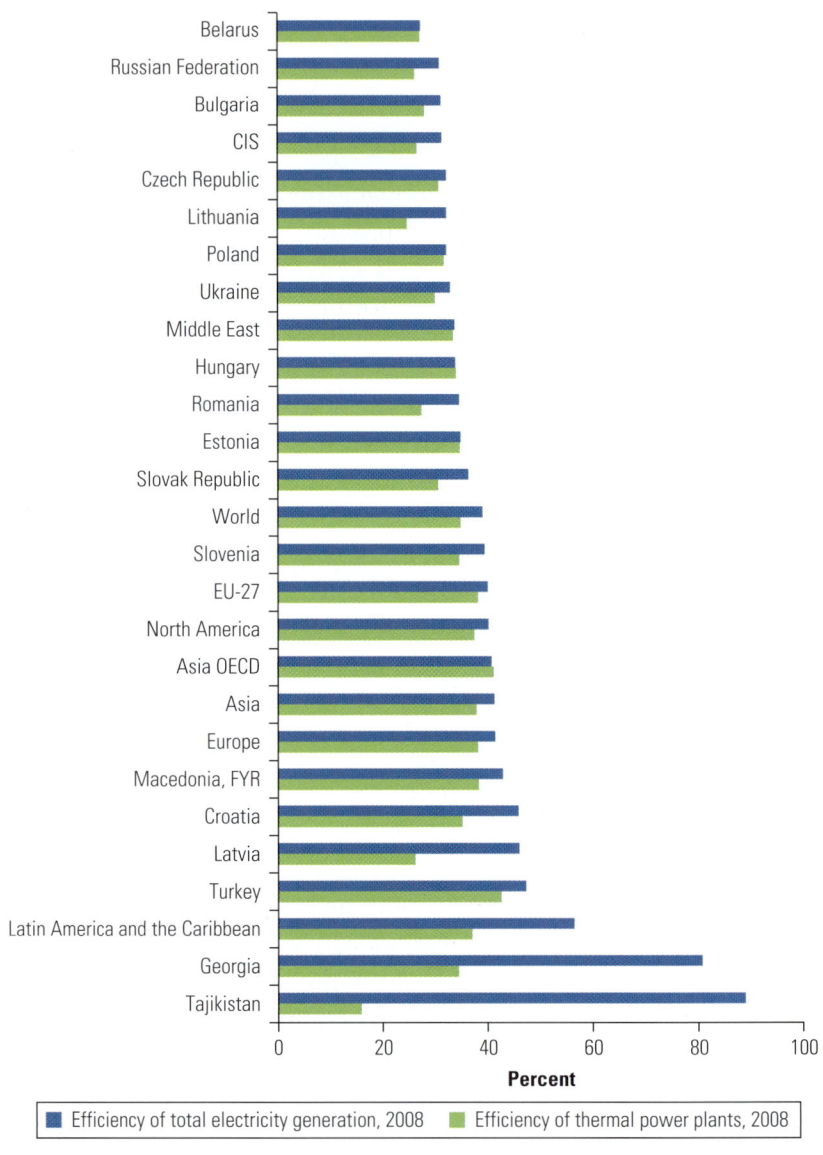

Source: Enerdata (http://wec-indicators.enerdata.eu/power-generation-efficiency.html).
Note: Data for remaining ECA countries are not available. CIS = Commonwealth of Independent States; EU = European Union; OECD = Organisation for Economic Co-operation and Development.

CIS countries as a whole performed poorly on both accounts. Their average thermal efficiency is 26 percent, and average overall efficiency is 31 percent.

The main reason for the low thermal efficiency is the large number of plants built during the 1960s and 1970s that use outmoded technologies and are still operating. For example, many gas-fired plants in ECA are traditional steam-cycle units or gas turbines in open cycles. Compared with the latest generation of combined-cycle gas turbine units capable of 52–60 percent efficiency, these old units can at best deliver around 40 percent. A further problem is that many of these assets have also deteriorated significantly because of insufficient maintenance since the 1990s. As a consequence, they operate at below the designed capacity.

Rehabilitation of existing infrastructure is a cost-effective way to improve efficiency, bring in additional generation, and reduce emissions. This typically involves refurbishing or upgrading generating facilities, such as gas turbines and boilers, and putting in place improved operating and maintenance practices. Rehabilitation and refurbishment can significantly increase the operational life and capacity of a plant at a fraction of its replacement cost. It is also faster than replacement, avoiding many steps in the planning, site selection, and construction cycle.

In Poland, Romania, and the Western Balkan countries, extensive restoration and upgrade programs are in progress in preparation for meeting EU environmental standards. Generation efficiency has been encouraged through funding by multilateral organizations such as the World Bank and the European Bank for Reconstruction and Development (EBRD), often supported by the Kyoto Protocol's Clean Development Mechanism, including projects in Albania, Azerbaijan, Kazakhstan, Russia, and Ukraine. Given the carbon intensity of thermal power generation, these programs can result in large emission reductions. The rehabilitation of the largest thermal power station, AzDRES in Azerbaijan, will increase plant efficiency by 14 percent, leading to an increase in output by 20 percent and a reduction in CO_2 emissions of 2.2 million tons a year (EBRD 2009).

There is also great potential to improve efficiency of electricity transmission and distribution (T&D). T&D losses represent the single biggest "use" in any electricity system. In OECD countries, they consume 6–8 percent of electricity generated. In ECA, the average T&D loss has decreased slightly, from 13 percent in 2000 to 11 percent in 2009. There is also large variation between countries, as figure 3.11 shows. In 2009, the Slovak Republic had the region's lowest T&D losses, at 3 percent. In contrast, T&D losses in Moldova reached almost 40 percent, six times higher than the EU average. At a national average retail price of electricity at US$0.1 per kWh, those losses cost the Moldavan economy US$360 million in 2009.

T&D losses come in two forms: commercial and technical. Commercial losses are caused by electricity theft or malfunctioning meters. Enforcing payment discipline is the key to reducing commercial losses. To encourage lower technical losses, improving regulation

FIGURE 3.11

Electricity Transmission and Distribution Losses in Europe and Central Asia, by Country, Relative to World and Selected Regional Averages, 2009

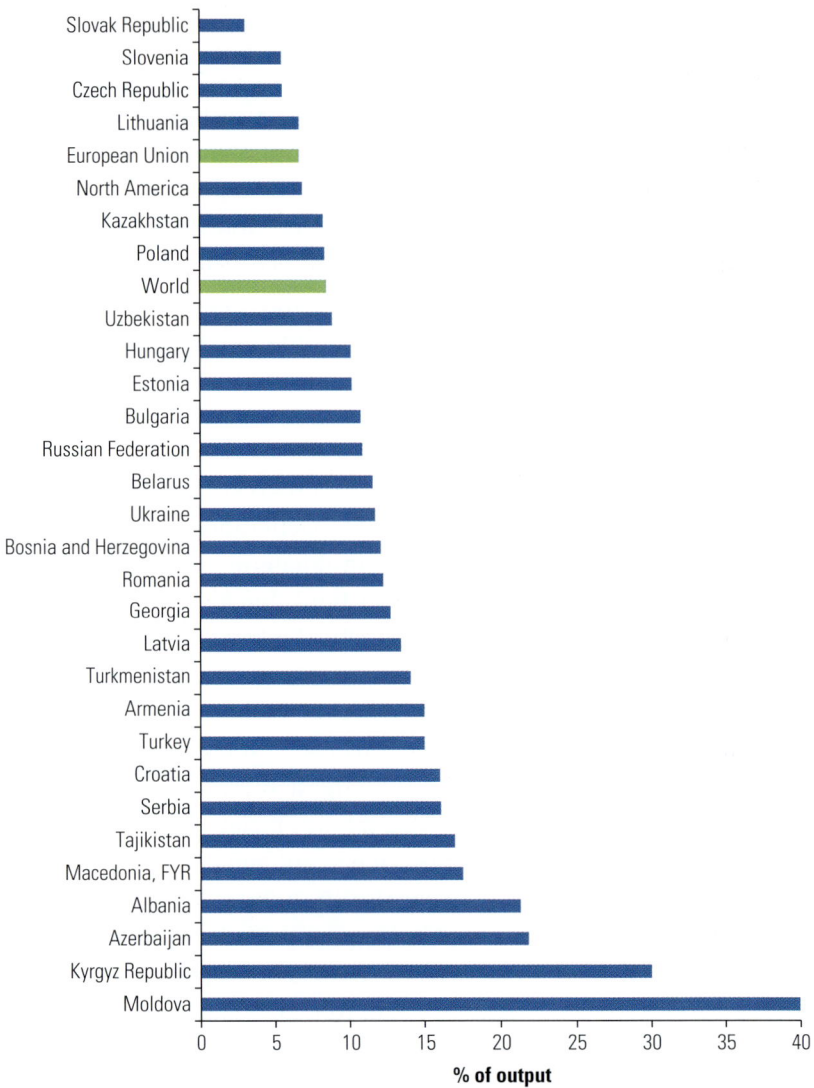

Source: World Bank, *World Development Indicators*, 2012.

is critical. Currently in most of the ECA countries, the cost of network losses can be entirely passed on to the customer. This tariff system produces a strong disincentive for investing in network efficiency because the network operator does not cover the losses. In some EU countries, maximum values are set for the amount of network losses that can be passed on. This practice forces network operators to prevent losses from increasing, but it does not yet stimulate them to reduce losses. An effective regulatory approach to reduce network losses has been used in Estonia and the United Kingdom. Every year, Estonia lowers the maximum network loss that can be passed on by 1 percent of the total load. In the United Kingdom, the distribution network operator is charged £48 per MWh for losses that exceed a certain target rate (Leonardo Energy 2010).

Rehabilitation and improved maintenance further reduce technical losses. In Albania, the rehabilitation work carried out on the T&D system has significantly reduced electricity losses from 31.5 percent in 2008 to 17 percent in 2010. Optimizing grid operation can also reduce technical losses. For example, moving from a bilateral or intraday market (which is common in ECA) to real-time balancing could improve grid reliability as well as transmission efficiency. This is because when transmission flows can be adjusted within a short time frame, power plants can change their output to avoid congestion and redundancy. As a result, the electricity system is operated more efficiently.

More flexible market mechanisms and more responsive infrastructure and grid management constitute one aspect of so-called smart grids, whose development is being accelerated to enable integration of a growing share of intermittent renewable energy. In ECA, Poland has become the first country to invest in smart grid technology. In 2011, with World Bank support, Poland started a US$187 million investment program to build a regional pilot smart grid. The purpose is to boost energy efficiency and accommodate a greater share of renewable energy. About 65 percent of the funding is to be spent on smart grid infrastructure, while the remaining 35 percent will be invested in renewable energy and low-energy lighting. Smart grid technology promises improved efficiency throughout the energy system. The IEA analysis estimates that smart grids could help to achieve global net annual emissions reductions of 0.7–2.1 gigatons (Gt) of CO_2 by 2050 (IEA 2011).

Demand-Side Efficiency

Demand-side efficiency encompasses more general energy efficiency issues, as further discussed in chapters 7–10. The perspective is usually from the energy consumer who should be induced by public policies (such as prices or regulations) to change behavior or make more energy-efficient purchasing decisions. However, utilities also have an important role in increasing demand-side efficiency. With regulations that ensure that profits do not depend solely on the amount of energy sold, utilities have a greater incentive to promote energy efficiency among their customers (see box 3.1).

Reducing, not Just Producing, Megawatts

California pioneered incentives that reward utilities for conserving rather than just selling energy. The state decoupled profits from energy sales in 1982 and later allowed utilities to earn additional profits if energy-saving targets were met or exceeded. Under the California program, utilities get to keep as profit about 12 percent of the value of the energy they help their customers save. In 2008, the San Francisco-based Pacific Gas & Electric Company earned US$41.9 million in profit this way.

In neighboring Nevada, regulators treat expenditures on energy efficiency programs, once savings are proven, essentially the same way they treat expenditures on a new power plant: both are capital outlays that the utility can recoup through regulated prices. In fact, the regulations add a small premium on investments in energy efficiency so that energy efficiency outlays earn the highest rate of return of any investments a Nevada utility can make. This regulatory approach moves power generation companies toward "a business model in which *reducing* megawatts is treated the same way from an investment point of view as *producing* megawatts," says James Rogers, CEO of Duke Energy, a U.S. energy utility.

The beneficial side effect for utility companies is that these programs avoid risky investments in new generation capacity, saving capital while also reducing carbon emissions. Between the early beginnings of California's energy efficiency programs in 1974 and 1998, no large power plant was built in California—a staggering 24 years in a state with one of the world's largest economies. Rising demand in a state with a rapidly growing population was entirely covered by energy efficiency gains (the largest contributor), small-scale independent producers, cogeneration, renewables, and imports.

Sources: Fox-Penner 2010; Rosenfeld and Poskanzer 2009.

Regulation gives utilities the incentives to help customers reduce electricity use. The customer side of smart grid technologies provides the means to do so. An advanced smart grid system will have three components: advanced metering that allows real-time communication between utilities and customer; dynamic tariffs that react to supply and demand constraints; and automated response technology that allows customers' electricity-using systems to react automatically to changes in prices.

Advanced metering infrastructure enables remote reading of meters by sending information directly back to the utility, which allows it to monitor demand patterns in real time. At the same time, it provides homeowners with information about how much electricity they are using at any given moment so they can adjust their load accordingly to save money. Dynamic, or real-time, pricing adjusts tariffs in response to current utility production costs. During peak hours—for instance, during midday in hot areas with high air conditioning use—prices per kWh will increase. Customers therefore have the incentive to shift energy-intensive tasks to times when tariffs are low, which in turn reduces the peak capacity utilities need to maintain. The next step is to equip electricity-using appliances or machines to react independently to changes in prices. For instance, temperature settings on air conditioners could be raised, or equipment doing tasks that are not time sensitive could be turned off when prices peak. Similarly, large customers can give utilities the permission to turn off nonessential equipment to avoid system shortages.

The penetration rate of smart meters in ECA has so far been quite low, at below 1 percent in 2009, as shown in table 3.2. Most of these meters do not use the advanced features described above. In ECA countries where electricity theft has been a serious issue, smart meters have been installed mainly to curb commercial losses. The new meters are theft proof and will allow the utility to detect commercial losses.

Small-Scale and Decentralized Generation

Energy efficiency and efforts to reduce demand may not alone be enough to bridge the demand-supply gap. In cases where new capacity is needed, one solution to avoid the risk of being locked in is to invest in smaller-scale units and deploy a range of different generation technologies for electricity and heat. In doing so, investors also obtain

TABLE 3.2

Penetration Rate of Electricity Smart Meters in Selected Europe and Central Asia Countries, 2009
percentage

Country	Household	Commercial	Commercial or Industry	Commercial or Industry	Total
	Low voltage			Medium voltage	
Albania	0	0	0	0	0
Bosnia and Herzegovina	0.2	0.2	11.8	38	0.2
Croatia	0	7.1	100	100	0.75
Georgia	—	—	—	100	—
Kosovo	0.33	0.91	0.42	100	0.42
Macedonia, FYR	0	0	0	0	0
Montenegro	—	—	—	100	—
Serbia	0.9	2.8	31	100	1
Turkey	0	0	2	—	0.02

Source: Energy Community Regulatory Board 2010.
Note: — = not available.

useful operational experience for future large-scale deployment of new technologies. Distributed generation relies on smaller, modular generators such as wind installations and solar photovoltaic installations, interconnected through a low-voltage distribution system that can function either in concert with, or independent of, the larger grid. Because renewable energy sources are by nature small and widely spread out, distributed generation facilitates the use of different types of renewable resources. For example, distributed generation units could burn landfill gases near landfills or other locally available biomass resources and use them to generate electricity on site. In some scenarios, because distributed generation accounts for most of all generation, it removes the need for traditional larger plants (see box 3.2).

BOX 3.2

The Future of Base Load Power

The backbone of traditional power systems consists of large generation plants that run continuously, that usually use coal or nuclear fuel, and that are slow to start and shut down but whose operating costs are relatively low. These plants produce the bulk of what is required for a country's energy needs. Power plants that can react faster, such as natural gas plants, smooth out variable or peak demand. However, scenarios in some countries with ambitious goals for renewable energy, such as Denmark or Germany, see future energy systems relying less and less on base load plants. Those countries want to phase out coal-fired power plants, and some have also decided to shut down nuclear plants. With a large share of intermittent and distributed power sources, supply would be adjusted flexibly with the lowest-cost combination of currently available sources. In fact, large base load plants become too inflexible to fit into such a system. Simulation scenarios suggest that even large national power systems could be run entirely with renewable energy sources without traditional base load power, provided that some critical technical problems can be resolved and the economics of renewables continue to improve. The main challenge is to develop better transmission infrastructure and the "software" of market-based power system integration and management.

These transformations will probably take many decades to unfold because they require sophisticated grid management and a larger capacity to store electricity. However, their impacts could be substantial. As Stanford University researcher Roy Amara noted, "We tend to overestimate the effect of a technology in the short run and underestimate the effect in the long run." The analogy to computing is instructive: In the 1970s, few could have imagined a future without powerful mainframe computers. Instead, the Internet today is fueled by millions of small servers, home and office computers, and personal digital devices that provide flexible, inexpensive, and easily scalable computing services where they are needed. The power system of the future may look similar.

Source: World Bank; Nitsch et al. 2012.

The most common distributed generation is a combined heat and power (CHP) system, also known as cogeneration. A CHP system generates electricity by burning fuel and then captures heat from the combustion process to produce steam or hot water. Alternatively, electricity is generated as a by-product of heat production. CHP relies heavily on on-site distributed generation, thus avoiding energy losses during heat transmission and storage. Compared with separate generation of heat and electricity, CHP generation may result in energy conservation, varying from 10 percent to 30 percent depending on the size of the cogeneration units (Voorspools and D'haeseleer 2002). Many ECA countries have in fact successfully expanded the use of CHP. Hungary, Latvia, the Slovak Republic, and Poland are leaders in CHP-based power generation in the region, as figure 3.12 shows.

Two additional advantages of distributed generation are increased reliability (because any individual plant is less critical in the overall supply) and smaller investment requirements that encourage experimentation and faster rollout of promising technologies. This experimentation and speed also encourages learning and development of locally appropriate solutions because simply adopting internationally developed technology is often risky (Neuhoff 2005; IEA 2003). On the other hand, a more decentralized power system with a large number of generators will be more difficult to manage. Western European countries that aggressively increase their share of renewable energy are currently developing the tools and operational procedures for efficient management of complex power systems.

Regulation that rewards power generators' efficiency efforts, smart grids, more decentralized generation, and more complex grid management will change the profile of electric utilities. In the future, utilities may operate the power grid and its control systems but may not actually own or sell the power delivered by the grid. Their mission will be to transmit and deliver electricity with high reliability at prices set by regulator-approved market mechanisms. Unlike traditional utilities—which may have incentives to block the entry of renewable generation if it takes market share from their conventional power plants—future utilities will accommodate any sources if they increase reliability and reduce costs. Power markets in some countries are moving in this direction.

Reliable Networks

Diversification is a risk-hedging strategy in the face of uncertainty, including the uncertainty brought on by climate change. There are two basic measures for diversification: fuel diversification and supply-source diversification. Fuel diversification can be achieved through the development of alternative generation methods such as solar and

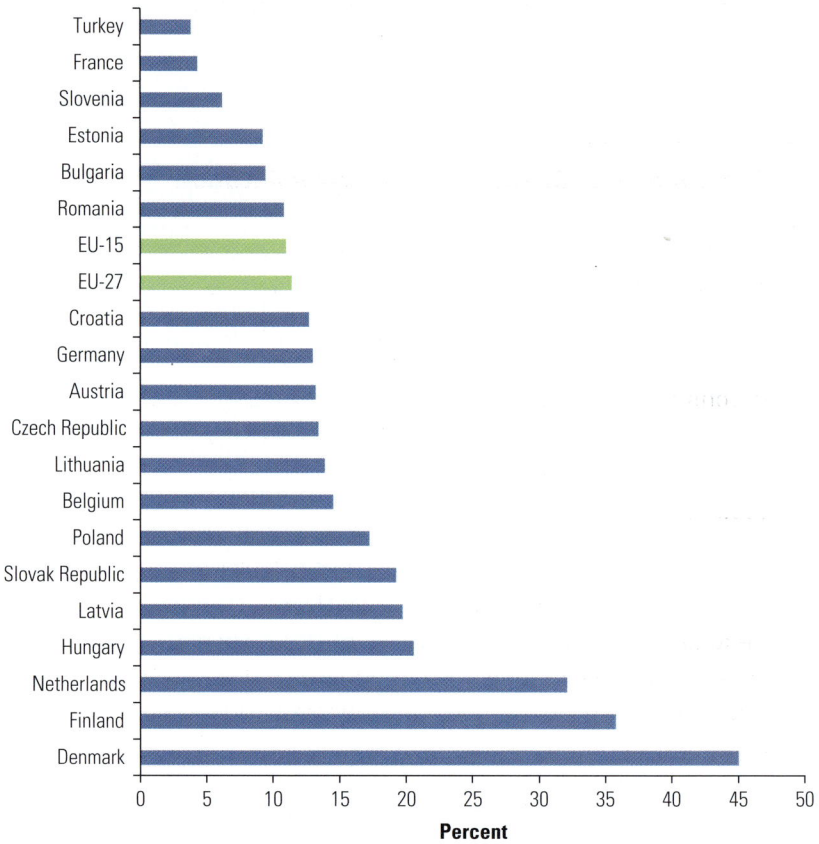

FIGURE 3.12

CHP Use as a Share of National Electricity Generation, Selected Europe and Asia Countries Relative to Selected European Countries, 2009

Source: IEA n.d.
Note: Data for other Europe and Central Asia countries not available. CHP = combined heat and power system; EU = European Union.

wind. Following the energy crisis in 2001, for example, the Brazilian government created a program to foster alternative sources of electric power, promoting wind and biomass as the primary alternatives to lessen the country's dependence on hydropower. Wind is considered a good hedge against low rainfall because, in Brazil, wind energy's greatest potential is during the dry season. According to the IEA, wind energy capacity in Brazil has since grown from 22MW in 2003 to 602MW in 2009, with a goal to reach 10GW by 2020. Wind generation increased by 200 percent during the 2003–09 period.

Regional energy cooperation through trade and power swaps is another way to diversify supply and increase energy security when countries share complementary power sources and different supply and demand patterns (see also box 3.3). When the heat wave crippled France's nuclear power generation, much of the shortfall was met by imports from the United Kingdom, which, since 1986, has been linked to the French power grid by a 45-kilometer subsea power

---

BOX 3.3

The Benefits of Energy Portfolio Diversification

Energy security—reliable access to energy at affordable prices—is a frequently stated policy goal. It is often conflated with energy independence, but the two concepts are different. Few countries today can rely entirely on domestic energy resources. For those that cannot, the more integrated their energy systems are with those of their neighbors and the more their energy imports are spread across different exporters and energy sources, the more secure will be their overall supply and the less affected they will be by price swings in any one energy source. So rather than energy independence, it is energy diversification that contributes most to energy security.

Energy diversification has several aspects: (a) *variety*—the number of options in the energy mix (such as coal, oil, gas, solar, wind, and hydro); (b) *balance*—the contribution each source can make to the energy mix; and (c) *disparity*—the degree of independence of price and other characteristics between alternative sources (such as between fossil fuels and renewables) (Stirling 1994, 2011; Grubb, Butler, and Twomey 2006). Each of these conditions is necessary but individually insufficient because each contributes differently to the energy security goal.

For ECA countries that rely heavily on (often imported) fossil fuels, a larger share of renewables diversifies the energy mix. What would be the diversification benefits if all of the region's countries reached a target of 20 percent renewable energy by 2020? A few countries (Albania, Estonia, Georgia, the Kyrgyz Republic, Latvia, Romania, and Tajikistan) already use more than 20 percent emission-free energy. Others are between 2 percent and 20 percent short of the goal, with Azerbaijan, Kazakhstan, Turkmenistan, Ukraine, and Uzbekistan having the largest gaps.

A simple scenario replaces the most carbon-intensive fuels with renewables that have the highest potential in the country, as shown in figure B3.3.1. The measure of benefit is borrowed from financial portfolio theory. By allocating funds across many uncorrelated assets (stocks, bonds, and so on), an investor trades off some potential return for lower risk. Applied to energy portfolios, the shift to renewables lowers *portfolio returns*—the amount of electricity generated per US$0.01 of investment. However, it also reduces volatility risk as measured in *standard deviations* (a simple measure of variability of investment returns) because a less-correlated portfolio of energy sources, with a higher share having no fuel costs, reduces the impact of price volatility (as previously shown in figure 3.7).

continued

BOX 3.3 *continued*

Figure B3.3.2 shows that many ECA countries could achieve large risk reductions at fairly modest costs. Uzbekistan, for instance, reduces volatility by about one-third for a drop in portfolio returns of about 10 percent. This is a significant cost, but lower risk also has an economic value because volatility reduces overall economic growth by raising inflation and unemployment and by depressing the value of financial and other assets (Awerbuch and Sauter 2006).[a]

FIGURE B3.3.1

Implied Fuel Switch with 20 Percent Renewable Target in Europe and Central Asia Countries, 2020

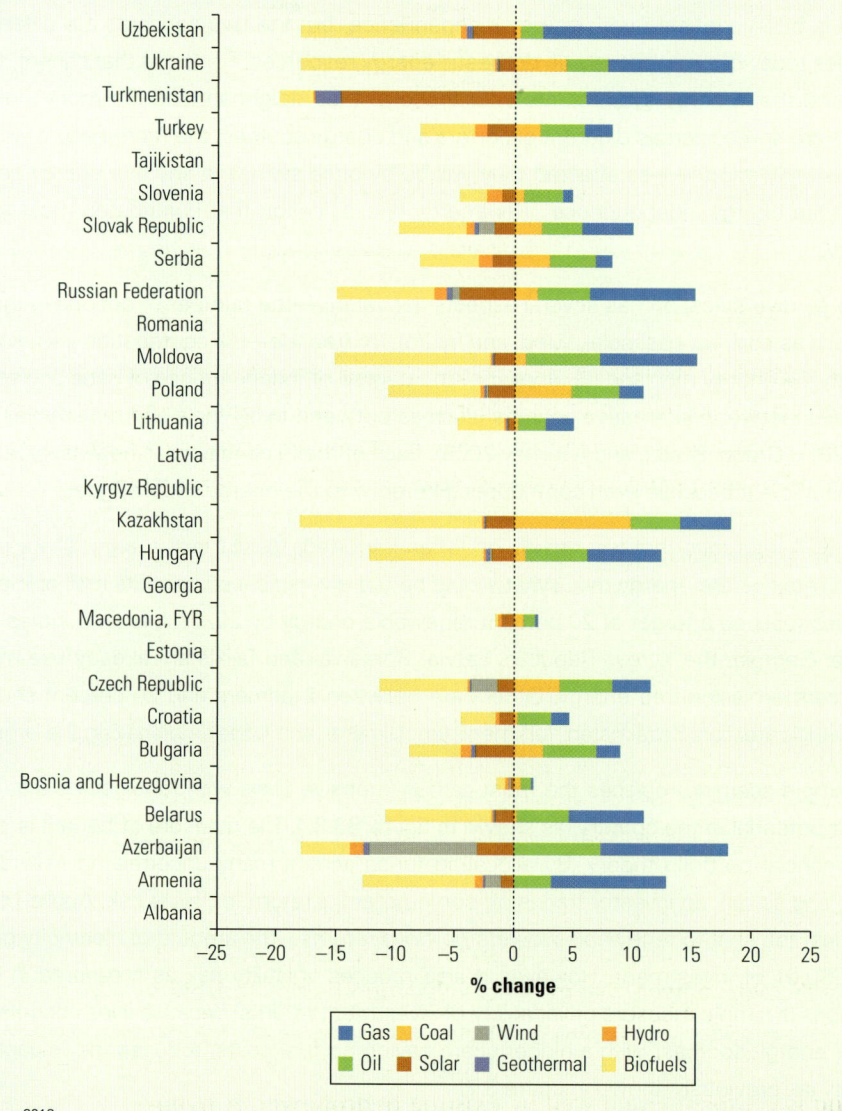

Source: Meisner 2012.

continued

BOX 3.3 *continued*

FIGURE B3.3.2

Risk-Return Profile of Fuel Mix Changes under the 20 Percent Renewable Target, Europe and Central Asia Countries, 2020

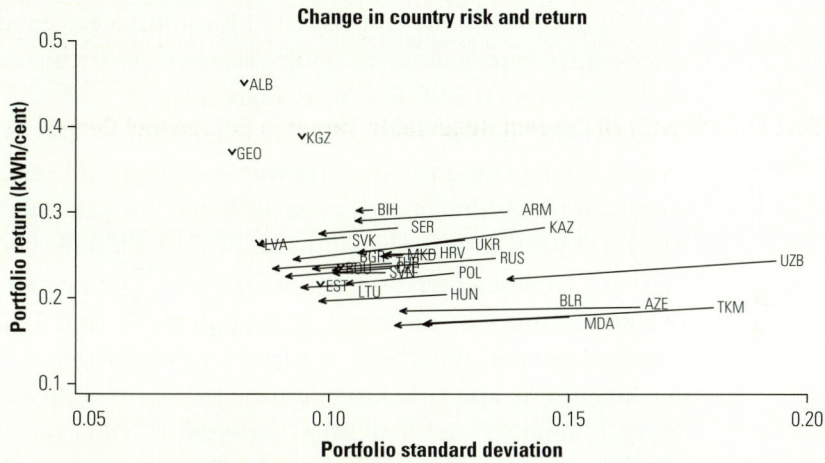

Change in country risk and return

Source: Contributed by Craig Meisner; see Meisner 2012.
Note: kWh = kilowatt-hour.

Fuel importers, such as Belarus, Moldova, and Ukraine, see additional benefits from improved current account balances. According to the scenario shown here, Armenia, Belarus, Kazakhstan, Moldova, Russia, Turkmenistan, Ukraine, and Uzbekistan could all achieve "savings" of more than 1 percent of gross domestic product (GDP). The total annual savings for the ECA region amount to US$68 billion. These benefits could offset some of the additional costs of deploying renewable energy sources.

a. For instance, the oil-GDP effect is estimated to lead to losses on the order of 0.5 percent of GDP for a 10 percent oil price increase.

cable. Where a large share of renewable energy is part of the fuel mix, supply diversification helps manage the intermittency of renewable energy sources. In Denmark, for example, interconnection to neighboring power systems supports a 44 percent share of renewable energy. Denmark exports wind-generated electricity to Norway when demand is low and imports Norwegian hydropower during peak periods. Norway's hydropower system effectively smooths the variability of wind power along with cheap energy storage in the form of water reservoirs. Quebec and the U.S. Pacific Northwest are other regions where wind is complemented well by existing hydropower. A hypothetical Europe-wide power grid analysis has shown that wind energy penetration levels could reach as high as 70 percent. With the cost of the extra transmission lines at around 10 percent of the turbine cost, this could yield power at around present-day prices (Murray n.d.).

Exploiting this portfolio effect of diversification depends importantly on the cross-border transmission capacity and coordinated market rules and regulation. In ECA, the Southeast European (SEE) countries are taking the lead in developing an integrated power market. Under the Athens Process, 10 SEE countries have entered into a cooperative agreement to develop a regional electricity market and to integrate it into the EU electricity market.[10]

However, in the rest of ECA, even though the potential for regional cooperation is high, power exchange still remains largely confined to within countries. For instance, the Central Asian countries had been historically connected by a bulk power transmission network and depended on one another for energy and water. The water resources in upstream Kyrgyz Republic and Tajikistan provided abundant and cheap electricity to downstream Kazakhstan, Turkmenistan, and Uzbekistan during the summer. In exchange, the Kyrgyz Republic and Tajikistan imported thermal power from the downstream countries in winter when hydropower conditions were poor. However, since the breakup of the Soviet Union, the regional power exchange has collapsed, falling from 25,413 GWh in 1990 to 3,714 GWh in 2008.

Intraregional power trade would be a cost-effective option to address winter energy shortages in the Kyrgyz Republic and Tajikistan. The World Bank, together with the Asian Development Bank, has been facilitating policy dialogue among the Central Asian countries to overcome obstacles to regional cooperation. Another World Bank-funded study has analyzed the prospect of regional power cooperation in the South Caucasus region where Armenia, Azerbaijan, and Georgia could export power to Turkey (World Bank 2012). The potential trade could reach between US$1 billion and US$1.7 billion annually in the next 5–10 years.

Notes

1. World Bank (2010), p. 10, assumes an annual gross domestic product (GDP) growth rate of 4.4 percent and a 3.1 percent increase in energy consumption.
2. The Commonwealth of Independent States includes Armenia, Azerbaijan, Belarus, Georgia, Kazakhstan, the Kyrgyz Republic, Moldova, the Russian Federation, Tajikistan, Turkmenistan, Ukraine, and Uzbekistan.
3. The EU-10 countries include Bulgaria, the Czech Republic, Estonia, Hungary, Latvia, Lithuania, Poland, Romania, the Slovak Republic, and Slovenia. The Western Balkan countries include Albania, Bosnia and

Herzegovina, Croatia, the former Yugoslav Republic of Macedonia, Kosovo, Montenegro, Serbia, and Slovenia.

4. A more complete discussion of coal-fired plants planned and canceled can be found in NETL (2011).

5. Values from the 2006 Intergovernmental Panel on Climate Change guidelines for National Greenhouse Gas Inventories, Vol. 2, Chapter 2, "Stationary Combustion," p. 2.16. http://www.ipcc-nggip.iges.or.jp /public/2006gl/index.html.

6. The Kyoto Protocol is an international agreement, linked to the United Nations Framework Convention on Climate Change, that set legally binding obligations on 37 industrialized countries and the European community to reduce their GHG emissions. It was initially adopted on December 11, 1997, and entered into force on February 16, 2005. See http://unfccc.int/kyoto_protocol/items/2830.php.

7. Historical trend data for the United States come from Thomson Reuters, as published by PriceWaterhouseCoopers on its Money Tree Report website: https://www.pwcmoneytree.com/MTPublic/ns/nav .jsp?page=notice&iden=B.

8. Many other factors also contributed to the uptake of private investment in renewable power generation, such as the establishment of a competitive wholesale market and the development of a necessary legal and regulatory basis for exploiting renewable resources. The Project Implementation Completion Report of the Renewable Energy Project in Turkey (World Bank 2009) provides a detailed discussion.

9. This is an insight from real options theory, which implies that every investment competes for funds not only against alternative investments but also against itself in the future. See McDonald and Siegel 1986; Dixit and Pindyck 1994.

10. The signatories to the Athens Memorandum are Albania, Bosnia and Herzegovina, Bulgaria, Croatia, Kosovo, the former Yugoslav Republic of Macedonia, Montenegro, Romania, Serbia, and Turkey.

References

Albrecht, James W., and Albert G. Hart. 1983. "A Putty-Clay Model of Demand Uncertainty and Investment." *Scandinavian Journal of Economics* 85 (3): 393–402.

Atkeson, A. 1997. "Models of Energy Use: Putty-Putty vs. Putty-Clay." Staff report, Federal Reserve Bank of Minneapolis, Minneapolis, MN.

Awerbuch, S., and R. Sauter. 2006. "Exploiting the Oil-GDP Effect to Support Renewables Deployment." *Energy Policy* 34 (17): 2805–19.

Dixit, A. K., and R. S. Pindyck. 1994. *Investment under Uncertainty*. Princeton, NJ: Princeton University Press.

DNV (Det Norske Veritas). 2009. "Cleaner Energy Survey: Risks and Opportunities Facing the European Power Sector." Report, Det Norske Veritas, Oslo.

Ebinger, Jane, Bjorn Hamso, Franz Gerner, Antonio Lim, and Ana Plecas. 2008. "How Resilient Is the Energy Sector to Climate Change?" Background paper, Europe and Central Asia Region, World Bank, Washington, DC.

EBRD (European Bank for Reconstruction and Development). 2009. "Modernization of Power Sector Evaluation." Evaluation report, EBRD, London.

Energy Community Regulatory Board. 2010. "A Review of Smart Meters Rollout for Electricity in the Energy Community." Report R10-CWG-13-06, Energy Community Regulatory Board, Vienna, Austria.

Fox-Penner, P. 2010. *Smart Power: Climate Change, the Smart Grid, and the Future of Electric Utilities*. Washington, DC: Island Press

Grubb, M. 1997. "Technologies, Energy Systems, and the Timing of CO_2 Emissions Abatement: An Overview of Economic Issues." *Energy Policy* 25 (2): 159–72.

Grubb, M., L. Butler, and P. Twomey. 2006. "Diversity and Security in UK Electricity Generation: The Influence of Low-Carbon Objectives." *Energy Policy* 34 (18): 4050–62.

Grubb, M., T. Chapuis, and M. H. Ha-Duong. 1995. "The Economics of Changing Course: Implications of Adaptability and Inertia for Optimal Climate Policy." *Energy Policy* 23 (4/5): 417–31.

ICCG (International Center for Climate Governance). 2010. "Economic Models for the Long-Term Carbon Price Evaluations." ICCG, Venice, Italy.

IEA (International Energy Agency). 2003. *Creating Markets for Energy Technologies*. Paris: Organisation for Economic Co-operation and Development.

———. 2007. *Recent Analysis into Indicators for Industrial Energy Efficiency and CO_2 Emissions*. Paris: Organisation for Economic Co-operation and Development.

———. 2010. "Projected Costs of Generating Electricity." Joint report by the IEA and the Organisation for Economic Co-operation and Development (OECD) Nuclear Energy Agency, Paris.

———. 2011. *World Energy Outlook 2010*. Paris: IEA.

———. n.d. World Energy Statistics and Balances (database). IEA, Paris. http://www.iea.org/stats/index.asp.

IPCC (Intergovernmental Panel on Climate Change). 2011. "Special Report on Renewable Energy Sources and Climate Change Mitigation." Research summary, IPCC, Abu Dhabi.

Ketting, J. 2008. "The End of the Russian Electricity Sector and the Beginning of a New One." *European Energy Review*, March/April.

Kirilenko, A. P., and R. A. Sedjo. 2007. "Climate Change Impacts on Forestry." *Proceedings of the National Academy of Sciences* 104 (50): 19697–702.

Leonardo Energy. 2010. "Treatment of Losses by Network Operators." Public consultation paper, Leonardo Energy, Brussels.

McDonald, R., and D. Siegel. 1986. "The Value of Waiting to Invest." *Quarterly Journal of Economics* 101 (4): 707–28.

Meisner, Craig. 2012. "Energy Security and the Benefits of Fuel Mix Diversification." Background paper, Europe and Central Asia Region, World Bank, Washington, DC.

Michel, Aloys A., and Stephen A. Klain. 1964. "Current Problems of the Soviet Electric Power Industry." *Economic Geography* 40 (3): 206–20.

Mott MacDonald. 2010. "UK Electricity Generation Cost Update." Report for the UK Department of Energy & Climate Change, Mott MacDonald, Brighton, UK.

Murray, James. n.d. "Wind Holds Key to European Super Grid." Article published on the EurActiv.de website, "renewable UK." http://www.euractiv.de/fileadmin/images/czisch_bwea-1.pdf.

NETL (National Energy Technology Laboratory). 2011. "Tracking New Coal-Fired Power Plants." NETL, Pittsburgh, PA.

Neuhoff, K. 2005. "Large-Scale Deployment of Renewables for Electricity Generation." *Oxford Review of Economic Policy* 21 (1): 88–110.

Nitsch, Joachim, Thomas Pregger, Yvonne Scholz, Tobias Naegler, Dominik Heide, Diego Luca de Tena, Franz Trieb, Kristina Nienhaus, Norman Gerhardt, Tobias Trost, Amany von Oehsen, Rainer Schwinn, Carsten Pape, Henning Hahn, Manuel Wickert, Michael Sterner, and Bernd Wenzel. 2012. "Long-Term Scenarios and Strategies for the Deployment of Renewable Energies in Germany in View of European and Global Developments." Final report, Deutsches Zentrum für Luft- und Raumfahrt (DLR), Fraunhofer Institut für Windenergie und Energiesystemtechnik (IWES), Ingenieurbüro für neue Energien (IFNE), Stuttgart Kassel, and Teltow Germany.

ORNL (Oak Ridge National Laboratory). 2007. "Effects of Climate Change on Electricity Production and Use in the United States." Final Review Draft, U.S. Climate Change Scientific Program, Synthesis and Assessment Product 4.5, ORNL, Oak Ridge, TN.

Pieprzyk, Björn, and Paula Rojas Hilje. 2009. "Erneuerbare Energien - Vorhersage und Wirklichkeit." (Renewable Energy: Prediction and Reality). Agentur für Erneuerbare Energien, Berlin.

Pryor, S. C., and R. J. Barthelmie. 2010. "Climate Change impacts on Wind Energy: A Review." *Renewable and Sustainable Energy Reviews* 14 (1): 430–37.

Redpoint Energy. 2010. "Electricity Market Reform: Analysis of Policy Options." Report in association with Trilemma UK for UK Department of Energy & Climate Change. http://www.decc.gov.uk/assets/decc/Consultations/emr/1043-emr-analysis-policy-options.pdf.

Rosenfeld, Arthur H., and Deborah Poskanzer. 2009. "A Graph Is Worth a Thousand Gigawatt-Hours: How California Came to Lead the United States in Energy Efficiency." *Innovations: Technology, Governance, Globalization* 4 (4): 57–79.

Stirling, A. 1994. "Diversity and Ignorance in Electricity Supply Investment: Addressing the Solution Rather than the Problem." *Energy Policy* 22 (3): 195–216.

———. 2011. "The Diversification Dimension of Energy Security." In *The Routledge Handbook of Energy Security*, ed. B. Sovacool, 146–75. New York: Routledge.

U.S. EIA (Energy Information Administration). n.d. International Energy Statistics (database). EIA, U.S. Department of Energy, Washington, DC. http://www.eia.gov/cfapps/ipdbproject/IEDIndex3.cfm.

U.S. GAO (General Accounting Office). 2009. "Energy-Water Nexus: Improvements to Federal Water Use Data Would Increase Understanding of Trends in Power Plant Water Use." Report to the Chairman, Committee on Science and Technology, U.S. House of Representatives, GAO, Washington, DC.

Voorspools, K., and W. D'haeseleer. 2002. "The Evaluation of Small Cogeneration for Residential Heating." *International Journal of Energy Research* 26(13): 1175–90.

Washington Post. 2011. "Coal's Burnout." January 2. http://www.washingtonpost.com/wp-dyn/content/article/2010/12/31/AR2010123104110.html.

World Bank. 2009. "Project Implementation Completion Report of the Renewable Energy Project in Turkey." World Bank, Washington, DC.

———. 2010. *Lights Out? The Outlook for Energy in Eastern Europe and the Former Soviet Union*. Washington, DC: World Bank.

———. 2012. "Stock Taking: Regional Power Trade in Southern Caucasus." Unpublished manuscript, Energy Team, Europe and Central Asia Region, World Bank, Washington, DC.

———. Various years. *World Development Indicators*. Washington, DC: World Bank.

———. n.d. Global Economic Monitor (database). World Bank, Washington, DC. http://data.worldbank.org/data-catalog/global-economic-monitor and http://data.worldbank.org/data-catalog/commodity-price-data.

Emission Trends in the Europe and Central Asia Region

ECA's heat-trapping emissions are falling relative to output, but per capita levels have risen to an average of 6.9 tons in 2011. With more ambitious policies, they could fall back to around 6 tons of CO2 per capita by 2035. Action is most pressing for the five largest emitters (in order of emission volumes): the Russian Federation, Poland, Ukraine, Turkey, and Kazakhstan, which collectively account for 77 percent of the region's total.

The reasons for ECA's large contribution to global greenhouse gas (GHG) emissions—relative to the size of its population and economy—have been well documented (see, for example, Fay, Block, and Ebinger 2010; World Bank 2010). Several countries in the region have abundant coal, gas, and oil resources which, when combined with low energy prices, encourage wasteful energy consumption. There has been progress, but there is still a persistent legacy of inefficient energy use in private buildings, industry, and transport as well as a backlog of modernization of the public capital stock (public buildings, street lighting, water treatment plants, and so forth), leading to high energy use in power production, water supply, and public transportation. The following paragraphs illustrate the scale of the region's emissions. This summary mostly focuses on CO_2 emissions that come largely from fossil-fuel combustion, but it also discusses other GHGs emitted from noncombustion activities.

ECA's share of global emissions is larger than its share of GDP or population. The region accounts for 7 percent of world population, 7.3 percent of global GDP (purchasing power parity [PPP]-based), 11.2 percent of all GHG emissions (5.4 billion of 48.4 billion tons of CO_2 equivalent), and 13.1 percent of CO_2 emissions from fuel combustion (3.8 billion of 29 billion tons). Among World Bank regions, it ranks second in total emissions, behind East Asia and Pacific, although combined emissions from high-income economies are still far larger, as shown in figure S2.1.

Per capita CO_2 emissions from fossil-fuel burning in ECA (6.9 tons of CO_2) are the highest among World Bank regions but are still relatively low compared with the OECD average (which is pulled up by very high emissions in North America) and slightly below those of the EU-15. There are notable exceptions at the country level: the national average emissions vary from less than half a ton in Tajikistan to more than 12 tons in Kazakhstan (see figures S2.2 and S2.3).

Per capita emissions are likely to rise considerably without effective climate action. Figure S2.4 shows future per capita CO_2 emissions based on three scenarios developed by the IEA (2011):

• The first assumes that current energy and climate change policies will prevail in the region over the next 25 years.
• The second assumes that already-announced new policies are effectively implemented.

FIGURE S2.1
Shares of Global CO$_2$ Emissions among World Bank Regions, 2009

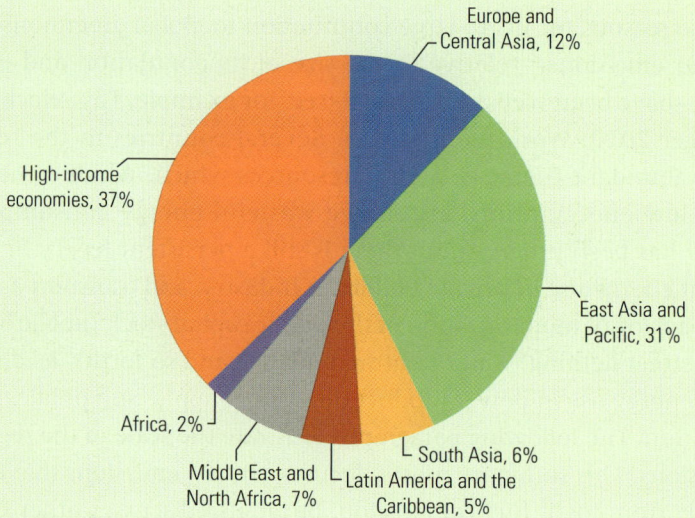

Source: IEA 2011.
Note: CO$_2$ = carbon dioxide.

FIGURE S2.2

Per Capita CO$_2$ Emissions among World Bank Regions, 1970–2010

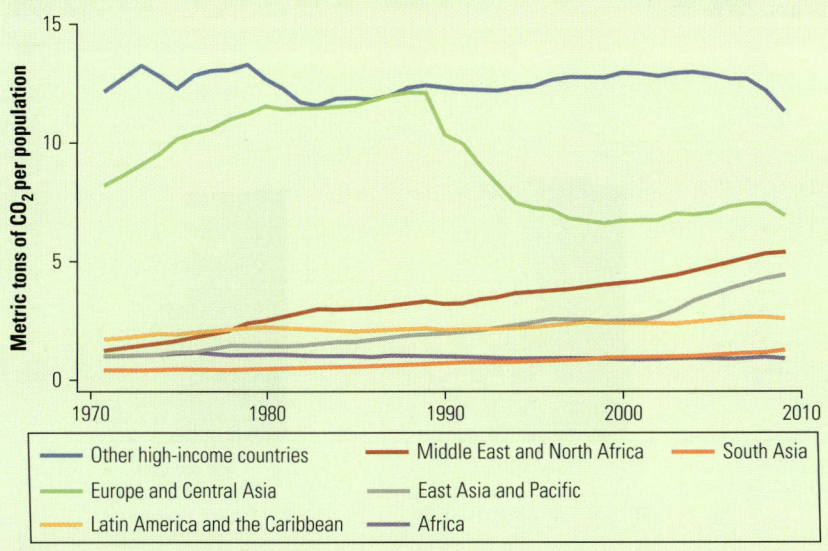

Source: Calculations based on IEA 2011.
Note: CO$_2$ = carbon dioxide.

FIGURE S2.3

Per Capita CO$_2$ Emissions from Fuel Combustion, Europe and Central Asia Countries Relative to OECD, EU, and World Averages, 2009

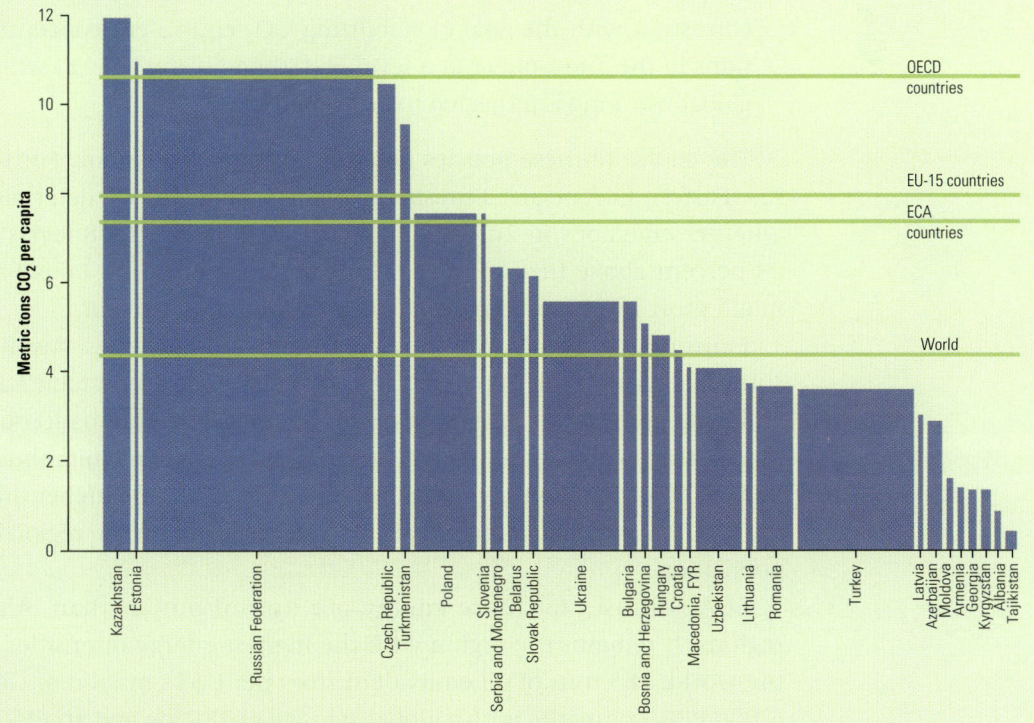

Source: Calculations based on data from IEA 2011.
Note: Width of bars is proportional to population. CO$_2$ = carbon dioxide. OECD = Organisation for Economic Co-operation and Development. EU = European Union.

FIGURE S2.4

Projected Per Capita CO$_2$ Emissions from Fuel Combustion in the Europe and Eurasia Region, under Three Policy Scenarios, 2008–35

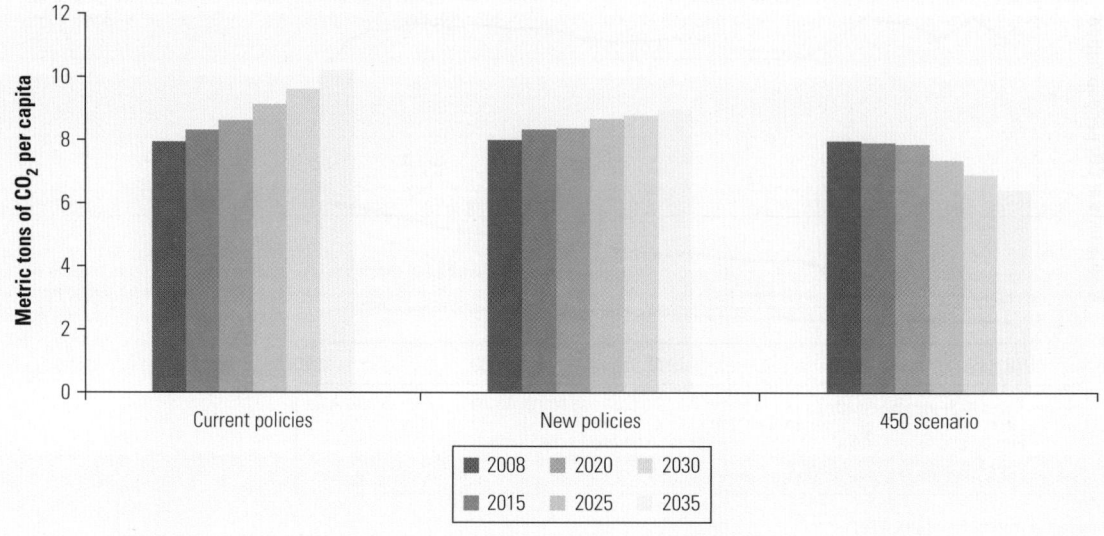

Source: IEA scenarios from IEA 2011; UNPD n.d.
Note: IEA's Europe and Eurasia region includes all ECA countries except the Czech Republic, Hungary, Poland, the Slovak Republic, and Turkey. The 450 Scenario indicates policies consistent with the goal of stabilizing CO$_2$ equivalent concentrations in the atmosphere at a level of 450 ppm. CO$_2$ = carbon dioxide.

- The third assumes policies will be adopted in the region that are consistent with the goal of stabilizing CO$_2$ equivalent concentrations in the atmosphere at a level of 450 ppm—the level at which global warming is projected to be limited to 2°C.

The choice of these policies determines future emissions. For the IEA Eastern Europe and Eurasia region as a whole, it will determine whether emissions in 2035 will be 23 percent below 2008 levels or 24 percent above them. For the IEA Caspian subregion,[1] emissions might grow by either 58 percent or by only about 9 percent.

Emissions are highly concentrated among ECA countries. Just five countries account for 77 percent of emissions (as shown in table S2.1 and figure S2.3). The 15 countries with the smallest emissions account for less than 6 percent of the region's total. Russia alone emits almost half of the region's emissions. Its per capita emissions are higher than the OECD average, and it is by far the largest country in the region by population.

ECA still uses far more energy per unit of output than other regions. It remains the region with the highest energy intensities in the world: 285 tons of oil equivalent (toe) per US$1 million of GDP (2000 PPP) compared with a global average of 189 toe and an EU-15 average of 133. The region's share of global total primary energy supply is 11 percent.

TABLE S2.1

Five Largest CO$_2$ Emitters from Fuel Burning in Europe and Central Asia, 2009

Country	Per capita emissions (tons of CO$_2$)	Population (millions)	Total emissions (tons of CO$_2$, millions)
Russian Federation	10.8	141.9	1,532.6
Poland	7.5	38.2	286.8
Ukraine	5.6	46.0	256.4
Turkey	3.6	71.9	256.3
Kazakhstan	12.0	15.9	190.0

Source: IEA 2011.
Note: CO$_2$ = carbon dioxide.

Emission intensities—emissions per unit of GDP—are consequently also higher than in other regions. Uzbekistan emits 1.7 kilograms (kg) of CO$_2$ per US$1 of GDP (2000 PPP), compared with a world average of 0.5 kg per US$1. Kazakhstan, Russia, Serbia, and Uzbekistan also have emission intensities above 1 kg per US$1. These levels are not primarily due to the region's carbon intensity of power generation (the amount of CO$_2$ per unit of energy produced), which has steadily decreased because of the increased share of natural gas in the fuel mix. (It is now approximately the same as the world average and only 17 percent higher than in the EU-15, as figure S2.6 shows.) However, the ECA region uses considerably more energy to produce a given output. Emission intensities have fallen steadily over the past four decades, with the notable exception of the immediate transition period, when GDP fell far more than fuel combustion (see figures S2.5 and S2.6). So the trend is in the right direction, and the challenge is to maintain and accelerate it. Emission intensities also vary greatly among ECA subregions, with Central Asia having the highest values (see figure S2.7).

In addition to burning fossil fuels, noncombustion activities also release significant amounts of CO$_2$, methane, and other GHGs. Venting and flaring is the major source of methane and noncombustion CO$_2$. Russia is the world's leading gas-flaring country (responsible for 26 percent of flaring in 2010), and Kazakhstan is ranked seventh (responsible for 3 percent). Agricultural and forestry practices emit the largest amount of nitrous oxide in addition to CO$_2$ and methane. Manufacturing of aluminum and cement and other industrial products releases CO$_2$ and industrial gases, such as hydrofluorocarbons (HFCs) and perfluorocarbons (PFCs). In 2008, fuel combustion constituted 64 percent of total GHG emissions, among which natural gas contributed 26 percent, followed by coal at 24 percent and oil at 14 percent. Noncombustion-related GHGs totaled 1.9 billion tons of CO$_2$ equivalent, or 36 percent of ECA's overall emissions, including 15 percent from venting and flaring; 6 percent each from agriculture,

FIGURE S2.5

CO$_2$ Emission Intensities in Europe and Central Asia Relative to Other Regions, 1970–2010

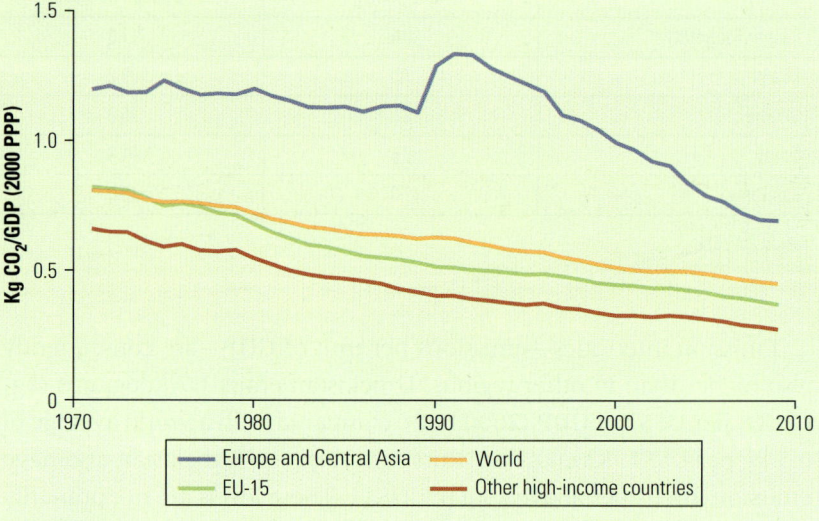

Source: IEA 2011.
Note: GDP using purchasing power parities (PPP) in 2000 U.S. dollars. CO$_2$ = carbon dioxide; kg = kilogram.

FIGURE S2.6

CO$_2$ Emission per Unit of Energy Use Relative to Other Regions, 1970–2010

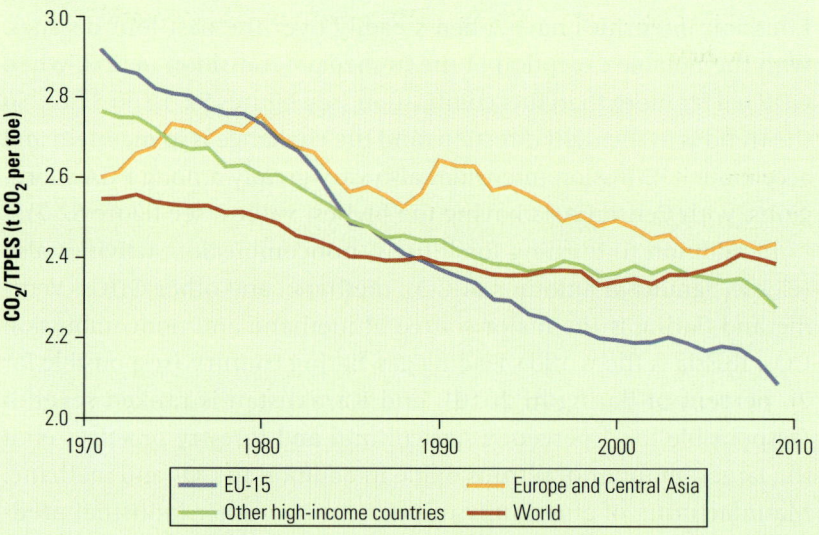

Source: Calculations based on IEA 2011.
Note: CO$_2$ = carbon dioxide; TPES = total primary energy supply; toe = tons of oil equivalent.

FIGURE S2.7

CO$_2$ Emission Intensities in Europe and Central Asia, by Subregion, 1991–2009

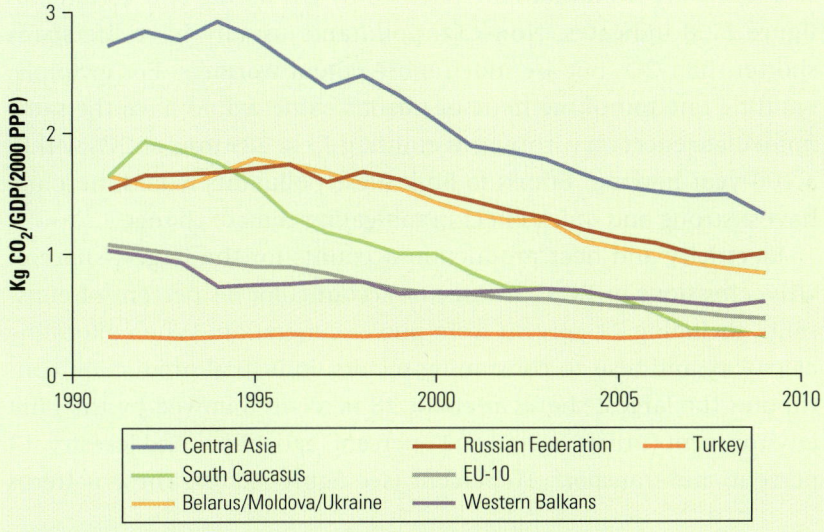

Source: IEA 2011.
Note: GDP using purchasing power parities is in 2000 U.S. dollars. CO$_2$ = carbon dioxide; EU = European Union; South Caucasus = Armenia, Azerbaijan, and Georgia; Western Balkans = Albania, Bosnia and Herzegovina, Croatia, Kosovo, FYR Macedonia, Montenegro, Serbia, and Slovenia; EU-10 = Bulgaria, the Czech Republic, Estonia, Hungary, Latvia, Lithuania, Poland, Romania, the Slovak Republic, and Slovenia.

FIGURE S2.8

Per Capita Emissions of Non-CO$_2$ Greenhouse Gases, by World Region, 2008

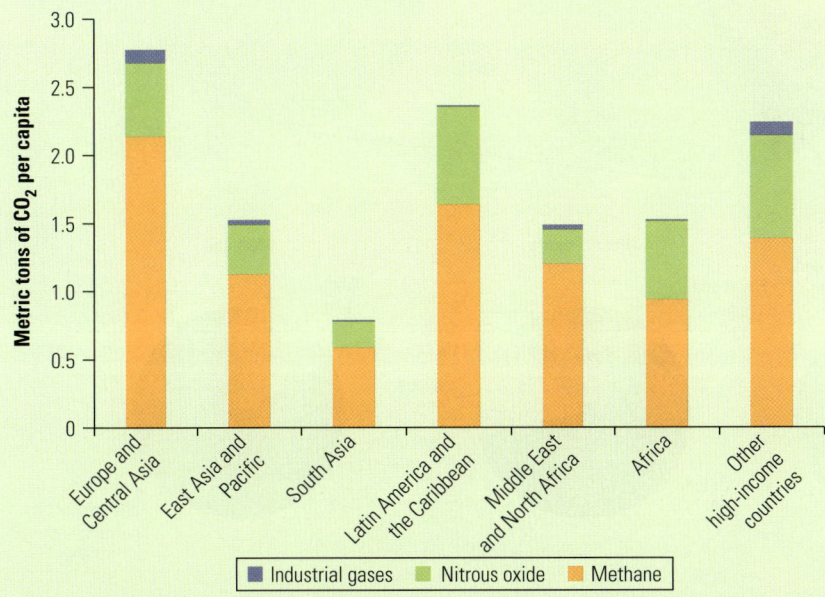

Source: Calculations based on PBL 2010 and EDGAR (Emission Database for Global Atmospheric Research) 4.1 data sets.
Note: CO$_2$ = carbon dioxide; t = tons.

forestry, and industrial processes; and 3 percent from landfills and waste treatment, as shown in figure S2.9.

ECA's per capita emissions of non-CO_2 GHGs is the highest in the world, mostly from methane released by gas flaring and venting, as figure S2.8 indicates. Non-CO_2 pollutants usually have life spans shorter than CO_2 but are much more potent warmers. For example, emitting one ton of methane or nitrous oxide would have the same immediate effect on warming as emitting 25 or 289 tons of CO_2 within a 100-year horizon. Efforts to limit these pollutants would therefore have a strong and quick effect in mitigating climate change.

Electricity and heat production accounts for the largest share of GHG emissions in ECA. The sector accounts for 31 percent of emissions, including 21 percent from power generation. When allocating electricity and heat to consuming sectors, industrial production contributes the largest share, at about 28 percent; followed by the built environment (urban) sector, at 21 percent; agriculture and forestry, 13 percent; and transport, 10 percent (see figure S2.10). These patterns

FIGURE S2.9
Annual Greenhouse Gas Emissions, by Source, Europe and Central Asia, 2008

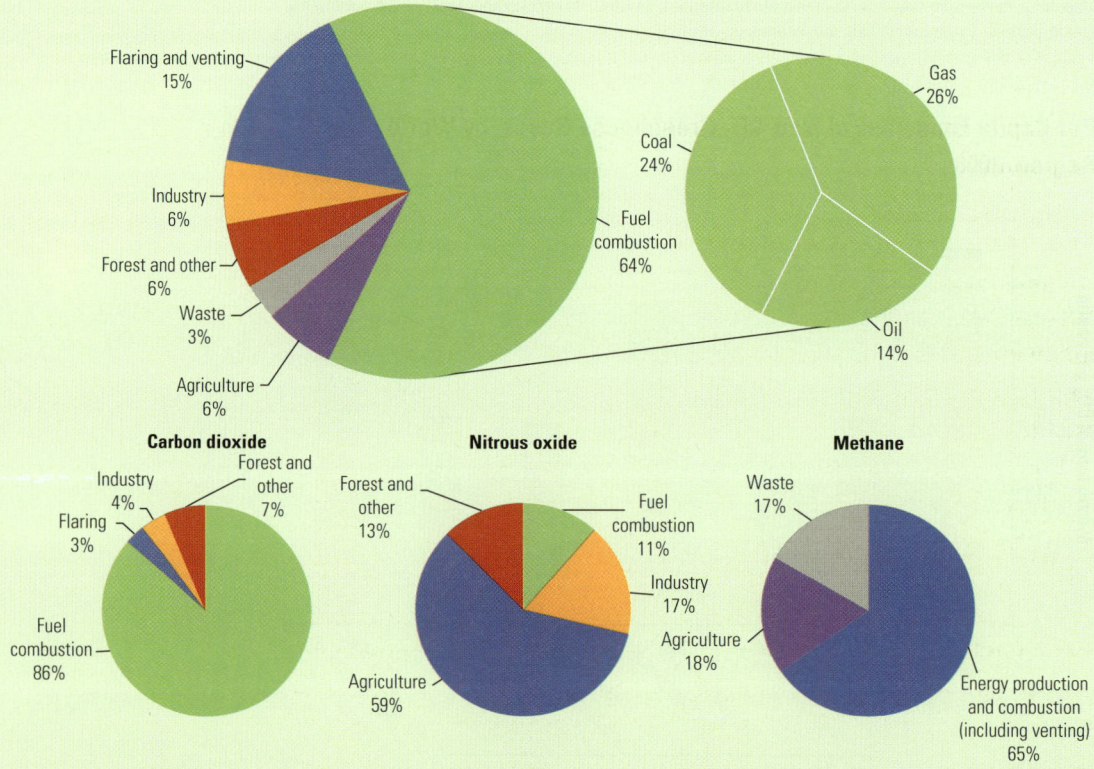

Source: Calculations based on IEA 2011, PBL 2010, and EDGAR (Emission Database for Global Atmospheric Research) 4.1 data sets.
Note: "Industry" refers to the noncombustion emissions from manufacturing of cement, lime, aluminum, and other materials. "Agriculture" comprises emissions from animals, animal waste, rice production, fertilizer use, and the like. "Forest and other" refers to direction emissions from forest fires plus emissions from decay of biomass.

FIGURE S2.10

Annual Greenhouse Gas Emissions, by Sector, Europe and Central Asia, 2008

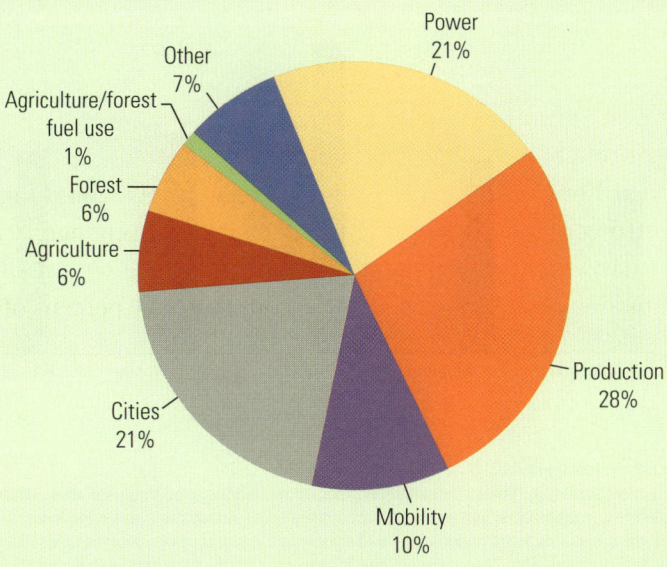

Source: Calculations based on IEA 2011.

Note: "Power" constitutes emissions from electricity production and other energy own use. "Production" refers to greenhouse gas emissions from manufacturing, construction, and other industrial activities. "Mobility" refers to emissions from transportation. "Built environment" comprises emissions from residential, commerce, public sector, and waste treatment. "Forest" emissions include those from forest or peat fires or deforestation, but not offsetting forest growth elsewhere.

vary somewhat by country. Electricity, heat, and other energy industries account for the largest share of emissions in most ECA countries. However, transport is the highest emitter in Albania, Armenia, Georgia, and the Kyrgyz Republic. Emissions from farming are significant in ECA but less so than in tropical regions. The region's emissions from agriculture were 331 million tons of CO_2 equivalent in 2008—approximately 6 percent of global emissions in this sector. Russia, Turkey, and Ukraine are the largest emitters.

Emissions from all sectors have increased in the past decade except for those from agriculture and forestry. The transportation sector has seen the largest increase in emissions, by 36 percent since 2000. In power generation and industrial production, emissions have increased by 15 percent and 9 percent, respectively. Emissions from urban sectors have increased by 11 percent since 1990 but have almost stabilized in the past decade. On the other hand, emissions from agriculture and forestry have steadily declined by 29 percent over the past decade, as shown in figure S2.11.

FIGURE S2.11

Trends in Greenhouse Gas Emissions, by Sector, Europe and Central Asia, 1990 vs. 2000–08

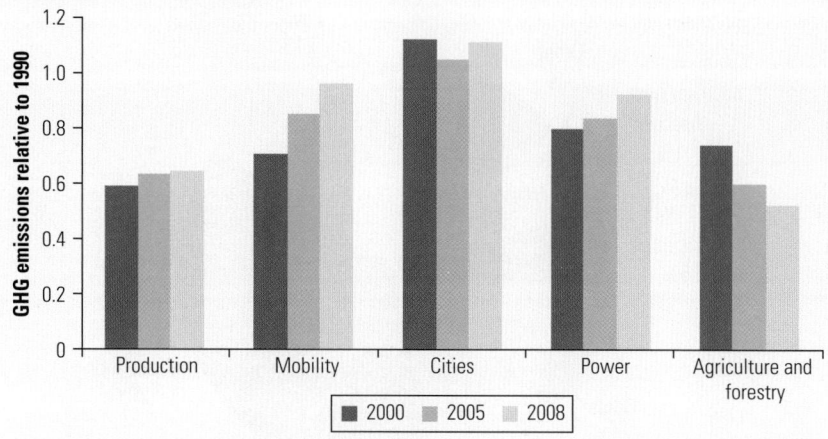

Source: Calculations based on IEA 2011.
Note: GHG = greenhouse gas. "Power" constitutes emissions from electricity production and other energy own use. "Production" refers to greenhouse gas emissions from manufacturing, construction, and other industrial activities. "Mobility" refers to emissions from transportation. "Built environment" comprises emissions from residential, commerce, public sector, and waste treatment. "Agriculture and forestry" emissions include those from forest or peat fires or deforestation, but not offsetting forest growth elsewhere.

Note

1. The IEA Caspian subregion includes Armenia, Azerbaijan, Georgia, and the five Central Asian countries: Kazakhstan, the Kyrgyz Republic, Tajikistan, Turkmenistan, and Uzbekistan.

References

EDGAR (Emission Database for Global Atmospheric Research). 4.1 data sets. http://edgar.jrc.ec.europa.eu/index.php

Fay, Marianne, Rachel I. Block, and Jane O. Ebinger. 2010. *Adapting to Climate Change in Eastern Europe and Central Asia.* Eastern Europe and Central Asia Report. Washington, DC: World Bank.

IEA (International Energy Agency). 2011. *World Energy Outlook 2010.* Paris: IEA.

UNDP (united Nations Population Division). World Population Prospects (database). UNDP, New York. http://www.un.org/esa/population. World Bank. 2010. *Lights Out? The Outlook for Energy in Eastern Europe and the Former Soviet Union.* Washington, DC: World Bank.

PART II

Promoting Growth and Ensuring Social Inclusion

Fears of unintended side effects have always accompanied major transformations. In 1941, the U.S. Temporary National Economic Committee wrote the following: "There is unmistakable evidence of a change in kind as well as severity of unemployment in the last depression. This change is characterized by the widespread use of electrical power and mass production methods which have shown a capacity to increase industrial activity on the upturn of the business cycle without a corresponding ability to absorb unemployed labor" (Cyert and Mowery 1987).

In 1964, a national commission warned about a "glut of productivity" caused by technological change that would reduce the demand for labor. In 1985, a prominent report predicted job losses in clerical and office occupations of 40 percent by 2000 (Cyert and Mowery 1987).

Clearly, electricity and automation did not lead to widespread unemployment in the postwar period. Rather than reducing labor demand, productivity growth from the emerging information technology (IT) sector created new jobs and growth. Job-loss scenarios for office professions assumed that the duties of these positions would remain unchanged and failed to realize that rising productivity would increase demand for such services. There were losers as well as winners along the way, and one should not underestimate the disruptions that major transformations can bring. But neither should one overstate them.

The low-carbon transition that will be required to avoid dangerous global warming will cause significant changes in national economies—though of a much smaller magnitude than the industrial or IT revolutions. Energy and other resources will need to be used far more efficiently. The energy system needs to gradually move toward low- and zero-carbon sources. Land use management will need to become far more sustainable. Shifts in relative prices will make some activities unprofitable and encourage others. They will also affect household budgets.

It is therefore important to ensure that there is broad public acceptance of climate action. One way to achieve this is through greater public discussion about the risks and rewards of action and inaction. Initial costs to avoid climate change impacts early will be lower than paying for damages or for the far more rapid emission reductions that would be required later.[1] Better-informed citizens will be more likely to support ambitious climate and energy policies. They will also be more likely to change their behavior to avoid damages, thus reducing the need for public policies (Stern 2010).

In addition, climate action will succeed only if smart policies maximize its economic and social benefits and minimize its costs. In other words, climate action needs to be closely aligned with the other pillars of the World Bank's regional strategy for Europe and Central Asia (ECA): growth and social inclusion. The following chapters discuss how this can be done. The main message is that to make climate action economically beneficial and socially benign does not require a lot of new or special policies. It requires the same conditions that are necessary to ensure "golden growth" (Gill and Raiser 2012): a good business climate, education and research systems that encourage innovation, flexible and mobile labor markets, and efficient governments that provide safety nets and social security.

CHAPTER 4

Economic Growth

Main Messages

- Europe and Central Asia (ECA) countries can attract a share of growing green foreign direct investment (FDI) and trade, especially from Western Europe. This requires general investment climate reforms. In addition, credible domestic green and climate policies make a country a more attractive investment destination in this rapidly growing sector: globally, in this decade, US$2.3 trillion is expected to flow into renewable energy alone.

- Globally, ECA accounts for only a small share of green innovation. This reflects generally weak national innovation systems that could better enable catch-up innovation and increase absorptive capacity, with opportunities for frontier innovation in some countries.

- Climate policies have created millions of jobs worldwide, in part reflecting the high labor intensity of energy efficiency and renewable energy technologies. But short-term employment creation should not dominate climate policy design. Low-carbon investments trigger medium- to long-term changes in efficiency and productivity that will be a more important determinant of economywide gains.

There is little agreement on the question of whether climate action will help or hurt economic growth. The main consensus appears to be that many of the actions necessary to avoid dangerous climate change will have a net cost but that this cost is manageable; its impact on firms and households can be moderated by well-designed policies; and dynamic economies will adjust to new cost structures relatively quickly. Most of the evidence comes from models rather than empirical research, with most estimates of the cost of effective climate action ranging between 0.5 percent and 1 percent of gross domestic product (GDP) per year during a transition time of two decades or so. There are more pessimistic views that peg the price tag higher. However, there are also those who believe that good environmental policy instruments—such as a carbon tax reinvested strategically to reduce other tax distortions—will stimulate innovation, economic activity, and productivity gains that will offset most or all of the cost of regulation.[2]

There is clearly no simple answer to this question. One reason is that country conditions vary greatly in terms of carbon intensity, economic capacity, and other relevant characteristics. Another reason is that different climate action policies will have different costs and benefits and these will change over time—sometimes quite rapidly. Energy efficiency measures are generally cost-effective within any reasonable time frame and will therefore be productivity and growth enhancing. Some elements of a cleaner energy strategy, such as reductions of gas flaring or pipeline leaks, share characteristics of energy efficiency investments and can pay off quickly. On the other hand, solutions such as carbon capture and storage, most renewable energy in low-energy-price countries, or electric vehicles will continue to require some form of subsidy for some time. Welfare gains such as lower health impacts, hard-to-quantify benefits from energy diversification, and the option value of postponing large capital investments can narrow the cost-benefit gap but will not necessarily close it. Large-scale deployment can then have a high opportunity cost in terms of growth. A sensible strategy for poorer countries is to concentrate on the more-affordable or already profitable actions and wait for large-scale deployment of costlier solutions until research and development (R&D) and widespread adoption in wealthier countries have brought down prices.

Yet, there is a case to be made that climate action will generate many economic opportunities. The renewable energy sector alone invested about US$263 billion in 2011, an increase of 6.5 percent over the previous year and 600 percent higher than five years earlier (Pew Charitable Trusts 2012). For governments, the major question is how actively they should promote promising green industries. The case for so-called vertical industrial policies that target specific firms (picking winners or rescuing losers) remains weak. A qualified case can be made for supporting broader sectors. For instance, a carbon tax by itself would, at least in the short term, promote the cheapest renewable energy technology at the expense of all others. However, a diverse mix of renewable options will be required to address the problem of intermittent availability. Support for solar or ocean power could thus be justified even if they remain relatively more expensive.

The strongest case can be made for economywide horizontal industrial policies that create the structural conditions for the private sector to take advantage of new opportunities and for households to strengthen their resilience through adjustments that are inevitable in a low-carbon transition. The industrial policies must be based on clear and stable domestic climate policies that signal to domestic producers that there will be a long-term market for their products and therefore a return on innovation and investment. Without appropriate incentives provided by climate policies, other growth-promoting instruments will be far less effective. Beyond this, the following sections briefly discuss three areas that will influence the growth benefits from climate action:

- Promoting foreign direct investment (FDI) and trade in sectors related to climate action to facilitate access to new technology and create economic opportunities in promising new industries

- Creating domestic innovation systems—from technical education to incentives for R&D and financing—that enable firms to adopt and diffuse advanced efficiency and clean-energy technologies and to develop locally adapted technologies and services

- Ensuring that job markets function well and support the shift from shrinking to growing industries

These growth-promoting priority areas are, of course, familiar. Creating a policy environment and business climate that is conducive to growth more generally (as described in the recent *Golden Growth* report [Gill and Raiser 2012]) will also make economic benefits from the low-carbon transition more likely.

Foreign Investment and Trade

Not all climate change mitigation-related products and services are technology intensive, but many are. Given Europe and Central Asia (ECA) countries' economic structure and domestic R&D capacity, the greatest opportunities in the production of green goods will likely be in medium to low technology. As with the car industry in Central Europe, initial production of components for foreign manufacturers can become increasingly sophisticated as countries develop expertise, deepen labor skills, and set up complementary supply systems. Investment from foreign firms in cleaner production generally and in the production of green goods specifically is one way to start this process in the fast-growing area of clean technology.

The main objective of attracting FDI is to stimulate economic activity and job growth in the country. In the environmental context, the benefits of green FDI include access for domestic companies to technology that is cleaner than locally available solutions. It allows recipient companies to leapfrog to state-of-the-art, "cleanest" technology. It will also generate spillovers from recipient firms to domestic competitors and suppliers.

Data and information about green FDI are scarce (Golub, Kauffmann, and Yeres 2011). Globally, total FDI flows to developing countries were about US$514 billion in 2010, of which US$87 billion went to ECA's low- and middle-income countries (World Bank, *World Development Indicators*). This compares with about US$130 billion in official global development assistance (about US$7 billion to ECA), of which about US$14 billion supported climate change mitigation-related activities.[3] There are no detailed, systematic data on the share of FDI in environmental goods and services sectors or on FDI that supports cleaner and more efficient production in other sectors. In Europe as a whole, this "green" FDI still represents a fairly small share of overall FDI—about 5 percent of all FDI projects but growing almost fivefold between 2005 and 2009.[4]

How can ECA countries increase their share of incoming FDI in climate change mitigation-related activities such as clean-energy technologies or energy-efficient goods? In general, to attract green investments, it helps to have a green reputation. Firms will look at factor costs and the ease of doing business. More specifically, they will also be more willing to invest in places that have clear and consistent environmental policies. Such policies ensure a high level of support for their activities as well as a domestic market for their products. Further, they signal to investors that the country is a good place to invest in clean production and in the production of clean products.

Costa Rica has shown that such branding can be successful. The Central American country started to market itself as an environmental and biodiversity hot spot in the early 1990s. It pursued comprehensive environmental regulations, setting aside large protected areas and promoting sustainable development of tourism infrastructure. It even implemented innovative new policy instruments such as payments for environmental services. The reward has been recognition as one of the world's primary ecotourism destinations. By the mid-1990s, tourism became the country's largest foreign exchange earner, and by 2009, it accounted for 17 percent of FDI inflows.

More specifically, countries can consider incentives for green FDI and remove barriers that prevent it. Incentives for green FDI do not differ significantly from those for other types of investment. However, they can be more targeted to capture a share of a fast-growing market. China and other countries have used instruments including reduced corporate taxes, tax holidays to encourage clean-energy technology transfer, investment allowances, accelerated depreciation for environmentally friendly goods, or exemptions from import tariffs on inputs (Bakker 2009). Similarly, policy makers need to ensure that there are few restrictions on environmental FDI. Research by the Organisation for Economic Co-operation and Development (OECD) suggests that such barriers tend to be low, including in several ECA countries except for the Russian Federation, as shown in figure 4.1. Assessments for other countries in the region would be useful.

The effects of trade on domestic greenhouse gas (GHG) emissions are ambiguous, and there is relatively little evidence on this for ECA (Onder 2011). By increasing overall economic activity, trade expansion without technical change would increase emissions (a scale effect). By changing relative prices in favor of exported goods, a country will produce more of those goods (a composition effect). The effect on emissions depends on what is exported. For the Balkan countries, the share of imported and exported manufacturing products is roughly balanced at 95 percent versus 91 percent.[5] In the Commonwealth of Independent States (CIS) economies, about half of exports are minerals, while 95 percent of imports are manufactured goods.[6] The composition effect thus depends on the relative

FIGURE 4.1

Restrictions on Two Dimensions of Green FDI and Economywide Restrictions, Selected Countries, 2010

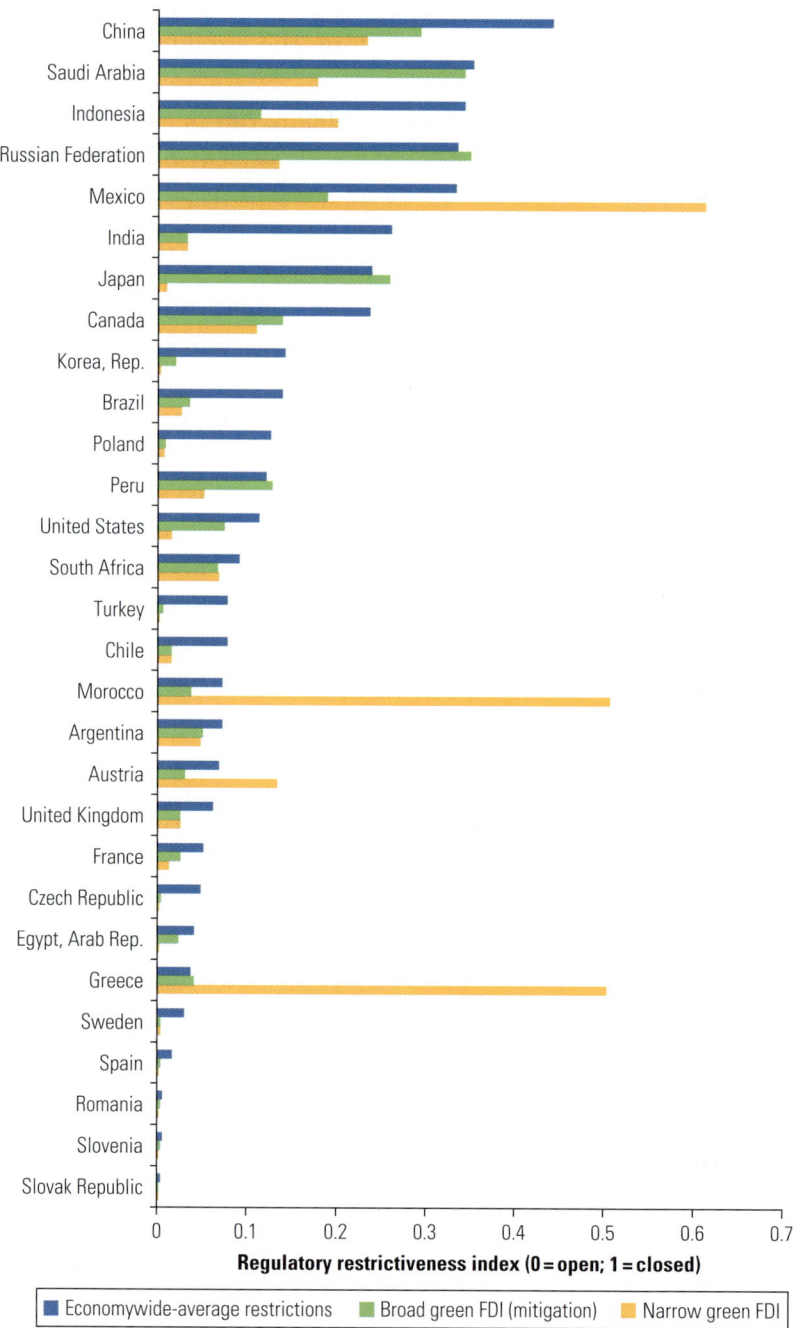

Regulatory restrictiveness index (0 = open; 1 = closed)

■ Economywide-average restrictions   ■ Broad green FDI (mitigation)   ■ Narrow green FDI

Source: Golub, Kauffmann, and Yeres 2011, based on the OECD Regulatory Restrictiveness Index (http://www.oecd.org /investment/fdiindex.htm) and the World Bank's Investing Across Borders (http://iab.worldbank.org/) indicators.
Note: "Narrow" indicates FDI to produce environmental goods and services. "Broad" indicates FDI for cleaner production. "Economywide" refers to all FDI for comparison. FDI = foreign direct investment; OECD = Organisation for Economic Co-operation and Development.

energy intensity of manufacturing versus mineral extraction. Trade will also lead to a change in the types of inputs used in production (a technique effect). Trade, like FDI, often provides access to cleaner and more efficient technology and other inputs. This has no doubt contributed to a reduction of emission intensities in ECA firms.

Another trade-related issue is how domestic emission policies will affect firms exposed to trade. For domestic consumption, both locally produced and imported goods are subject to the same policies. Therefore, producers of appliances would be subject to the same efficiency standards, and car producers to the same emission restrictions. Regarding production-related emission policies, there is concern that more-stringent domestic emission policies will hurt the international competitiveness of pollution-intensive industries. The empirical evidence is mostly from industrialized countries and suggests that this effect is usually small and short term (Copeland 2012). Other cost factors or agglomeration economies have a far greater impact on firm location decisions. Regarding emissions from energy use, the increased short-run compliance costs are likely offset by longer-term efficiency gains.

The third trade-related question, finally, is how local firms are affected by emission policies in export markets. European Union (EU) climate policies impose a price on carbon through the cap-and-trade system and a host of national policies that price emissions or make energy more expensive. These policies impose a cost on EU firms. Policy makers worry that this cost will shift energy and emission-intensive production to countries without, or with more lenient, emission regulations. Indeed, many EU countries have a large "CO_2 trade deficit," which means that the EU imports more carbon-intensive products than it exports (Peters et al. 2011). Not surprisingly, China has piled up the largest "CO_2 trade surplus" as its economy's global share of manufacturing has increased. ECA countries for which estimates are available divide about equally into net importers of carbon dioxide (CO_2) embedded in traded goods, and net exporters, with Russia having by far the largest carbon exports, followed by Ukraine, Kazakhstan, and Poland, as figure 4.2 illustrates. About half of the exports of oil, gas, and products from six energy-intensive industries in Azerbaijan, Kazakhstan, and Russia go to the EU. Most of the net carbon importers are EU-10 or candidate countries.[7]

It is unlikely that climate change policies have been the major cause of the relocation of manufacturing industries and the resulting imbalances in the trade of virtual carbon. Empirical evidence is scarce, but model estimates suggest a so-called carbon leakage of no more than 5–30 percent. Cost advantages related to labor and other

FIGURE 4.2

Average Annual Net Exports of Embedded CO$_2$ Emissions, Selected Europe and Central Asia Countries, 1990–2008

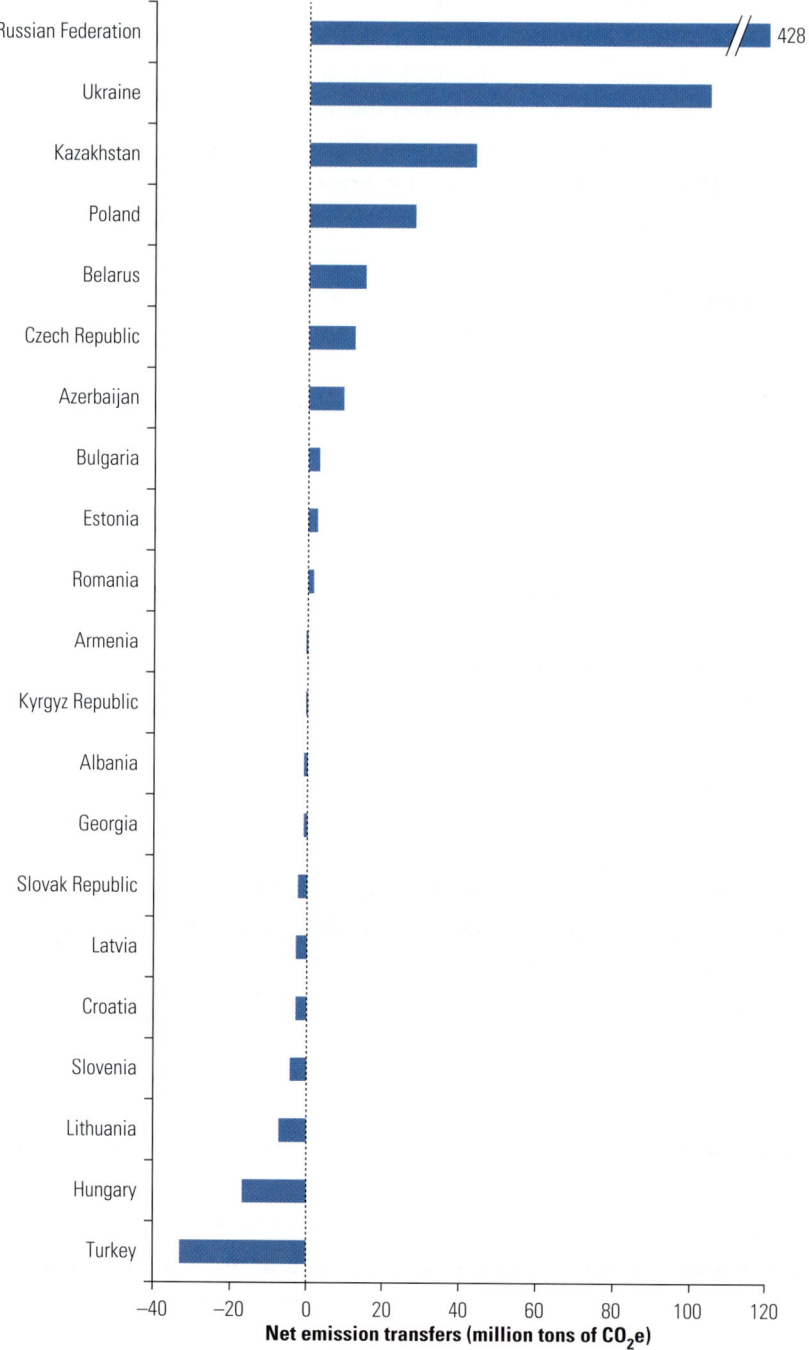

Source: Peters et al. 2011.
Note: Bar for Russia is truncated (428 MtCO$_2$e). Negative values = country is a net importer of CO$_2$ emissions; CO$_2$ = carbon dioxide; MtCO$_2$e = million tons of CO$_2$ equivalent.

production factors, and the emergence of tightly integrated production networks in East Asia, were far more important. Nevertheless, in 2006, France proposed to introduce so-called border tax adjustments into international climate negotiations (Umweltbundesamt 2009). These would impose a charge on goods imported into the EU that is equivalent to the cost of carbon regulation faced by domestic producers.[8]

As long as foreign firms face emission-related charges that are no higher than those paid by domestic firms, the consensus is that border tax adjustments would be legal under World Trade Organization rules. So far, none has been introduced, although in 2012 the EU introduced requirements for including foreign air carriers in its emission trading system. One reason why the EU has not yet introduced border tax adjustments is that their implementation would be complex. Any so-called prior burdens or environmental charges imposed by the exporting countries would need to be subtracted because they represent equivalent measures. This would lead to an extremely complex adjustment system whose cost may well exceed the benefits for countries imposing such taxes. Simplified approaches—based on average energy intensity and fuel mix, for instance—would be poorly targeted.

Unilateral introduction of emission-based border taxes by industrialized countries could have significant impacts on developing and emerging countries' manufacturing sectors, including those in Turkey or the CIS countries, especially if the taxes are based on the export countries' emission intensities rather than the generally lower ones in importing countries (Mattoo et al. 2009). (See box 4.1.) Such charges would shift some of the costs of emission reductions from high-income countries to low- and middle-income countries through terms-of-trade effects (Copeland 2012).

Alternative policy responses are preferable. One is to encourage adoption of cleaner production technology in exporting countries—for instance, through technical assistance or export guarantees for green capital goods. Another is to support the introduction of emission regulations in exporting countries, in the form of either a domestic carbon tax or cap-and-trade system, or an export tax. Choosing either option would remove the argument for border tax adjustments in importing countries. It would also keep the revenue from carbon pollution charges in the exporting country, where it could be reinvested to reduce emissions or to lower other production costs. To facilitate the design and introduction of market-based instruments for carbon mitigation, the World Bank–coordinated Partnership for Market Readiness provides support for low- and middle-income countries. In ECA, Turkey and Ukraine participate in this program, which includes setting up monitoring, reporting, and verification systems; piloting market exchanges; and planning for emission trading systems.[9]

BOX 4.1

Differentiated Climate Action, Carbon Leakage, and Anti-Leakage Measures

Unilateral climate policies can have spillovers to other countries. An example is the EU's emission reduction commitment of 14 percent by 2020 relative to 2005—currently the only binding regional climate agreement. Economic modeling captures policy-induced interactions between countries and economic sectors in the global economy and illustrates how climate action in the EU affects emissions and welfare elsewhere (see Kasek et al. 2012).

The analysis divides the world into three groups: (a) the 27 EU countries; (b) Annex I countries (A1) that have emission reduction obligations under the Kyoto Protocol, including Russia and Ukraine; and (c) non-Annex I countries (non-A1) that have no formal obligations, such as the South Caucasus and Central Asian countries.[a] In the reference scenario (figure B4.1.1), the EU

FIGURE B4.1.1

Projected CO$_2$ Emission Reductions in Annex 1 Countries Relative to Other Country Groups by 2020

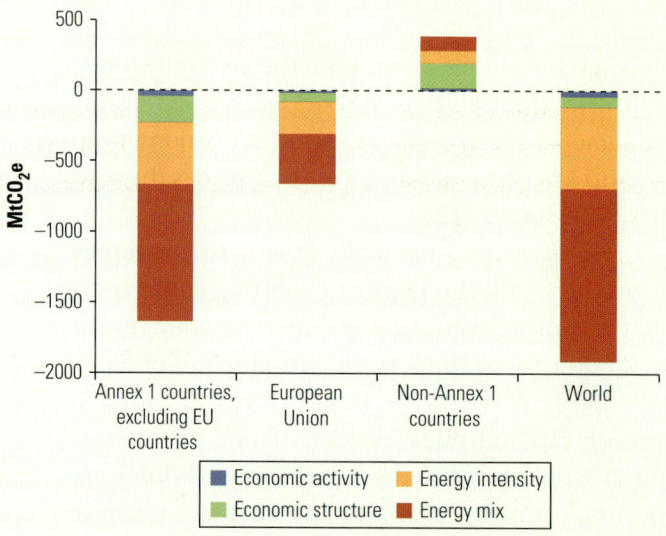

Source: Kasek et al. 2012.
Note: The figure decomposes emission changes by 2020 in a reference (Kyoto Protocol) scenario as opposed to a business-as-usual (no climate action) scenario. Annex 1 countries are those with emission reduction obligations under the Kyoto Protocol. EU = European Union; MtCO$_2$e = million tons of CO$_2$ equivalent.

continued

BOX 4.1 *continued*

fulfills its commitment by 2020, the other A1 countries reduce their emissions by 4 percent, and the remaining (non-A1) countries do not introduce carbon reduction targets. This scenario reduces global CO_2 emissions by 6 percent relative to business-as-usual (with no climate action). This scenario causes a modest economic loss in the EU and A1 countries, with real GDP less than 1 percent lower than under business-as-usual (without considering co-benefits such as health improvements).

Under this scenario, global emissions fall by almost 2 billion tons of carbon dioxide equivalent (CO_2e) annually by 2020, mostly in A1 countries outside of the EU that accounted for about a third of global emissions from fossil-fuel burning in 2009. The EU, which emitted 12 percent of global CO_2 in 2009, would lower emissions by more than 600 million tons. About a quarter of the emission reductions comes from decreasing energy intensity as countries invest in energy-saving technology. Another 5 percent comes from an economic shift toward less-energy-intensive sectors. This effect is small in the EU, where these shifts have mostly already taken place.

Two-thirds of emission reductions comes from a change in the energy mix, through a gradual replacement of coal with cleaner fuels and higher imports of electricity from non-A1 countries. Increased demand for electricity from countries not required to cut emissions causes their emissions to rise. This is the essence of the carbon leakage problem: strict policies in one place encourage emission-intensive industries to relocate to places with fewer or no restrictions.

The loss of competitiveness from unilateral climate policies can be addressed by imposing duties equivalent to the domestic carbon charge for emissions embedded in imported goods (border tax adjustment on imports, or BTA). Model simulations show that such duties reduce carbon leakage significantly, but they are controversial (as further discussed in the main text). Less effective is output-based allocation of free emission permits to energy-intensive and trade-exposed sectors. Such permits reduce the leakage rate slightly but contradict the initial objective of making domestic production cleaner. A Clean Development Mechanism (CDM) scenario can either avoid or increase carbon leakage, depending on implementation.[b] All of these policies affect opportunities in countries not adopting strong climate policies. The welfare effect for non-A1 countries is most detrimental under the BTA regime and modest under the CDM scenarios.

What becomes clear is that unilateral carbon abatement policies are less effective than coordinated global climate action because some part of the emissions reduction in the EU or A1 countries will be offset by an increase in emissions elsewhere. The stricter the unilateral policy, the higher the leakage rate. The more aggressive the countermeasures, the greater the welfare and output loss elsewhere. Therefore, unilateral action will always be a second-best option.

Source: Contributed by Leszek Kasek.
a. The South Caucasus countries include Armenia, Azerbaijan, and Georgia. The Central Asian countries include Kazakhstan, the Kyrgyz Republic, Tajikistan, Turkmenistan, and Uzbekistan.
b. For more about the Clean Development Mechanism, see http://unfccc.int/kyoto_protocol/mechanisms/clean_development_mechanism/items/2718.php.

Innovation[10]

There has been a large increase in green innovations worldwide as countries have put a greater emphasis on environmental protection. The vast majority of green innovations have been in industrialized countries, with Germany, Japan, and the United States accounting for 60 percent of the global patents related to GHG mitigation. Developing and emerging economies account for a very small share of those innovations. The number of green patents has increased in Latin America and East Asia but has been stagnant in the ECA region, as shown in figure 4.3.[11] Within ECA, Russia accounts for the largest share over the past two decades (with 19 of the region's 45 green patents), followed by the Czech Republic and Hungary, as figure 4.4 indicates.

Patent data reflect frontier innovation, creating new products or techniques. More important for many countries is catch-up innovation, which involves adoption and adaptation of existing technology. One way to assess the economic opportunities from catch-up innovation is to look at the trade structure. The share of imports of green products is about the same in ECA, Latin America, and East Asia and slightly higher than in high-income countries, as shown in figure 4.5.

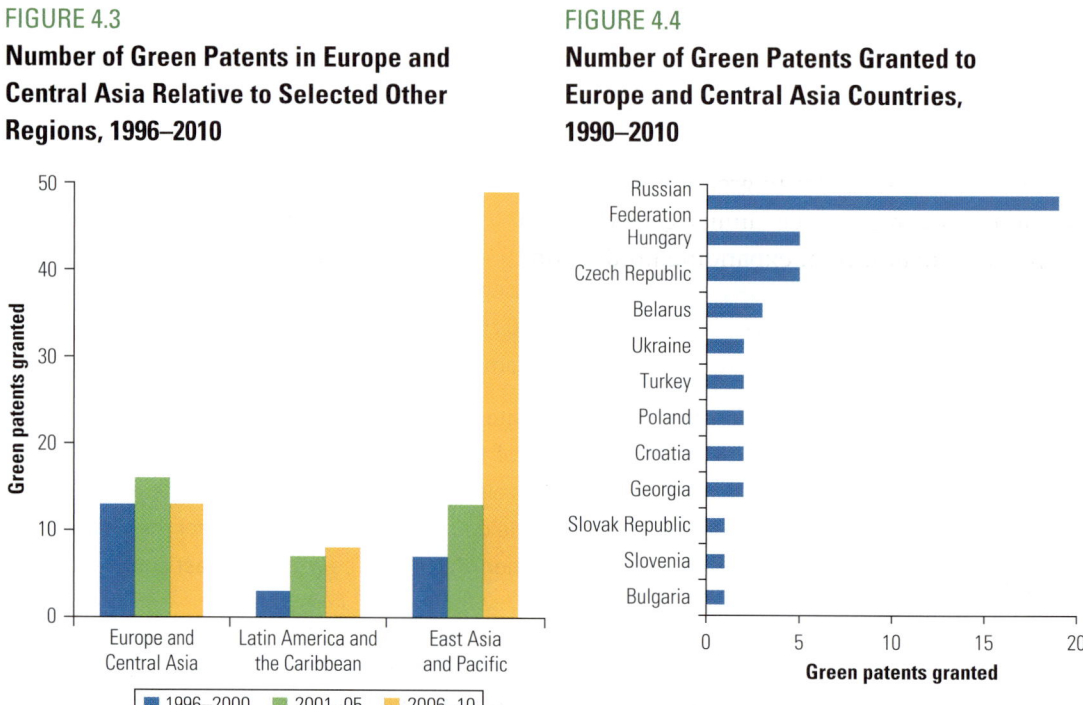

FIGURE 4.3

Number of Green Patents in Europe and Central Asia Relative to Selected Other Regions, 1996–2010

FIGURE 4.4

Number of Green Patents Granted to Europe and Central Asia Countries, 1990–2010

Source: PATSTAT database of U.S. Patent and Trademark Office (USPTO); see Dutz and Sharma 2012.
Note: Numbers are for total USPTO-granted patents in OECD Green Technology Areas. OECD = Organisation for Economic Co-operation and Development.

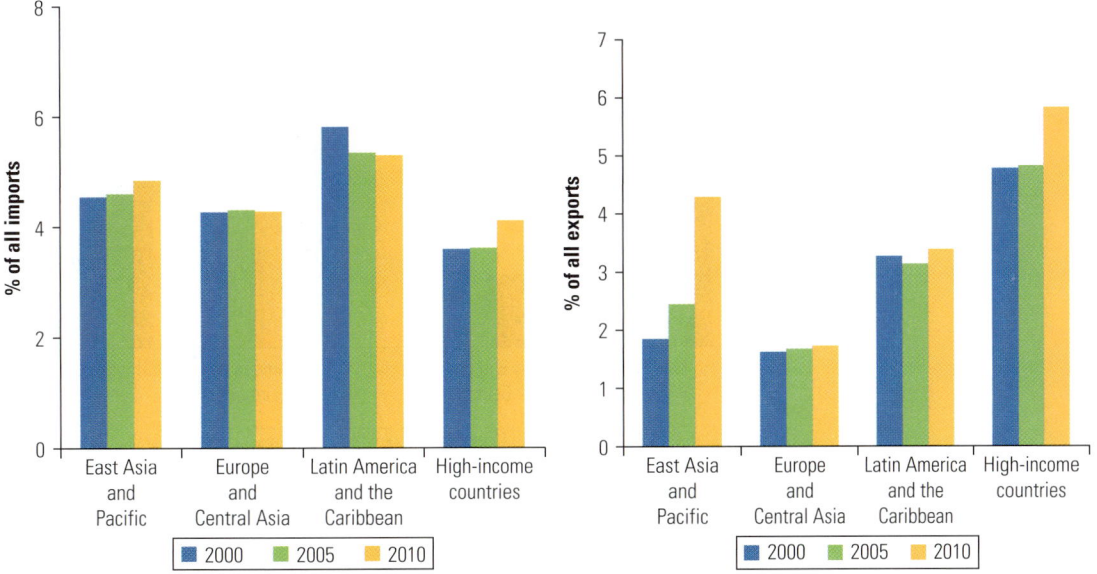

Figure 4.5

Imports of Green Goods and Services in Europe and Central Asia Relative to Selected Other Regions, 2000–10

Figure 4.6

Exports of Green Goods and Services in Europe and Central Asia Relative to Selected Other Regions, 2000–10

Source: United Nations Comtrade database and proximity matrix based on 6-digit Comtrade; see Dutz and Sharma 2012.

However, their export share is far lower in ECA and, in contrast especially to East Asia, is not expanding (see figure 4.6).

To gauge the potential for expanding these exports, one can look at exports that are similar to green products. The idea is that a country that exports many such similar goods and services should have a comparable advantage in expanding production into green products. Figure 4.7 shows that—relative to Latin America and East Asia, which have a far higher share of green and close-to-green exports— ECA's potential for green export expansion is only about twice as large as the actual green export share.

Firms increase investments in green innovation in response to market demand. Environmental and climate policies will generate such market demand for green goods by requiring pollution control technology or higher-efficiency consumer products, for instance. Evidence from Western Europe suggests that "demand-pull" policies, such as feed-in tariffs for renewable energy and other environmental regulations, are indeed a major driver for environmental innovations. ECA's relatively low share of green products may well reflect weaker environmental standards. Firm surveys suggest that existing or future environmental regulations, followed by market demand

FIGURE 4.7

Exports of Green and Close-to-Green Goods and Services in Europe and Central Asia Relative to Selected Other Regions, 2000–10

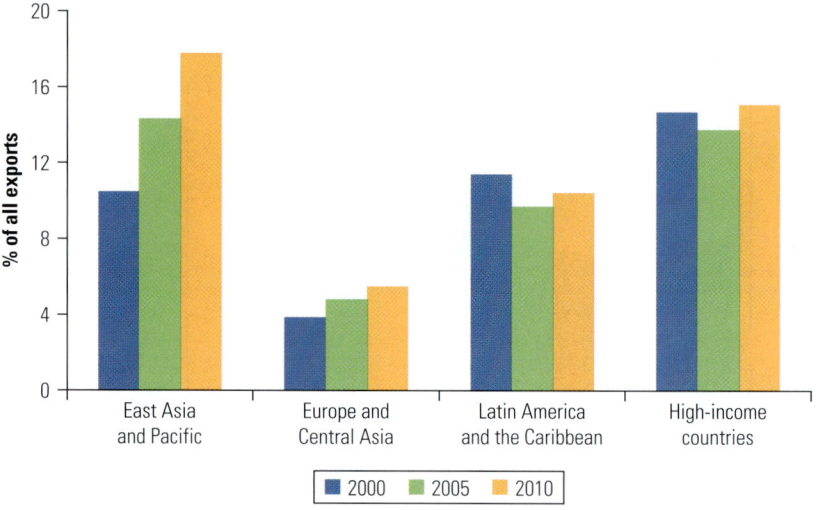

Source: United Nations Comtrade database and proximity matrix based on 6-digit Comtrade; see Dutz and Sharma 2012.

from customers, are the main drivers of introducing environmental innovations. Incentives that "push the supply" of innovations will still be necessary because firms tend to underinvest in knowledge creation and R&D generally (because they can usually capture only a small share of the value of the innovation) and in environmental technology in particular (because environmental damages are rarely fully priced).

Green innovation policies need to be tailored to country needs and characteristics. They will require a mix of all three types of innovation support (as summarized in table 4.1), implemented with the help of suitable policy instruments:

- *Frontier innovation* needs to focus on countries with advanced technological capabilities. Apart from strengthening of intellectual property rights and early-stage technology finance, public R&D funding is most important. However, support for environmental R&D is low in the ECA countries for which data are available—a fraction of 1 percent in Russia, for instance.

- *Catch-up innovation* involves stimulating the adoption of technologies that are new to firms. An important instrument is support for early adopters and demonstration programs. For example, several Eastern European countries, including the Czech Republic,

TABLE 4.1

Public Policies that Promote Innovation, by Type

Innovation policy type	Intended beneficiaries	Policy instruments
1. Promoting frontier innovation (innovation finance and other policies for development and commercialization of new-to-the-world knowledge)	Firms with sufficient technological capabilities; financiers	Government-funded R&D (public labs, matching grants, soft loans, and tax credits for private firms) Patents and other support for intellectual property rights (IPRs) Support for early-stage technology development (ESTD) finance, including support for private capital (angels, early-stage VC) Prizes and advanced market commitments (AMCs)
2. Promoting catch-up innovation (policies to facilitate access to new-to-the-firm knowledge and to stimulate technology absorption)	All firms; public labs and universities; all citizens	Open trade, FDI, IPRs, diaspora, and ICT policies Patent buyouts and compulsory licenses Patent pools and open-source mechanisms Public procurement, standards, and regulations Support for finance to early adopters and demonstrations
3. Developing absorptive capacity (policies to strengthen skills and more broadly to spur the accumulation of new knowledge by entrepreneurs and firms)	All firms; workers and managers; researchers; trainers	Education and life-long learning policies Enterprise-based worker, management, and entrepreneurship training as well as other technical and vocational education and training (TVET) Facilitation of connectivity through global alliances and supplier development links to global value chains Rule of law, contract enforcement, competition, bankruptcy and re-entry facilitation; urban policies ("sticky" cities to attract and retain talent)

Source: Dutz and Sharma 2012, which discusses these policy instruments in detail in the context of innovation for climate action.

Note: Although some policy instruments such as public procurement, standards, and regulation are relevant for all three policy areas, they are listed in the areas deemed most important to stimulate green innovation in most developing countries. R&D = research and development. VC = venture capital. FDI = foreign direct development. ICT = information and communication technology.

benefited from a pilot training course by the Certified European Passive House Designer project that trains architects, engineers, and building designers in the construction of buildings that require 90 percent less energy than conventional ones.[12]

- *Absorptive capacity* is promoted by more general educational and training programs and by making it easier to access global information, value chains, and entrepreneurial practices. An example is Mexico's Green Supply Chains Program, which brought together 14 multinational and 146 small and medium-size enterprises to exchange knowledge and build buyer-supplier links for environmental products.

ECA countries have so far not played a large role in innovation for environmental and climate action technologies and services. Innovation research suggests some lessons for more effective innovation policies (Popp 2011; Dutz and Sharma 2012):

- A major reason for the region's underperformance is the relatively low level of public support. Both demand-pull signals (in the form of effective climate change policies) and supply-push incentives (such as support for innovation systems) are inadequate. Increased policy support will be critical for ECA to benefit from green growth opportunities.

- Current technology leaders will continue to dominate cutting-edge R&D activities, but ECA can benefit greatly from catch-up innovation and improved absorptive capacity that stimulates investment and growth in locally relevant areas of environmental management. Adaptive R&D must be a core component of the region's overall innovation system, as box 4.2 discusses.

- Firms in ECA take insufficient advantage of their proximity to green-technology leaders in Western Europe. China and India have shown how to mobilize knowledge generated abroad to build successful renewable energy companies—for instance, through licensing agreements with industry leaders. ECA can do much better in exploiting such beneficial spillovers.

- Development or introduction of new technologies is only the first step. These technologies also need to spread widely within a country. Too often they run into additional barriers such as limited

---

BOX 4.2

Broader Innovation System Reforms Promote ECA's Competitiveness

Innovation in the areas of climate change mitigation specifically and environmental management more generally constitute just one subsector of a country's innovation system. A recent World Bank report on innovation in ECA argues that the region needs to move much closer to the scientific and technological frontier if its economies are to become more diversified and competitive (Goldberg et al. 2011).

The report identifies four specific priorities that apply equally to innovation for climate action:

- Enhancing integration in the global R&D community by supporting collaboration between local and foreign researchers and by attracting FDI in R&D in ECA countries

- Restructuring research and development institutions to make them more effective in supporting commercialization of R&D

- Promoting risk taking by reforming and strengthening financial support instruments such as investment grants or tax exemptions

- Improving the business climate to facilitate trade, FDI, and entrepreneurial start-ups

Source: Goldberg et al. 2011.

financing or limited ease of use. Measures to promote innovation diffusion should complement ECA's policies for innovation creation and initial adoption.

Jobs[13]

Innovation makes existing firms more productive and generates new business opportunities. These results, in turn, create employment. Employment generation is considered one of the main benefits of a policy-induced transformation of energy systems—creating jobs in renewable energy, energy efficiency, and the natural resources sector. Many such jobs have been created in countries with ambitious climate policies, but both the net employment effects and the cost-effectiveness of such policy-induced job creation are somewhat ambiguous. What emerges from empirical evidence is that employment gains by themselves are a poor criterion for evaluating climate action projects or policies. Such projects and policies should be implemented if they are profitable based on a cost-benefit analysis that considers wages as well as indirect benefits from environmental improvements, learning, and R&D. Employment effects should be seen in the context of the broader transition to a low-carbon economy where the most significant payoffs may be indirect and longer term. Labor market policies should aim at preparing workers to take advantage of new opportunities wherever they emerge and cushioning adjustment shocks in industries that are adversely affected (see chapter 5).

There is no simple definition of a green job or one that is induced by climate action. Green employment overall, which includes traditional environmental sectors such as conservation or pollution control, currently constitutes a fairly small share of total employment—about 1.7 percent in Europe, for instance, and around 1 percent in most low- and middle-income countries. These jobs range from highly skilled R&D to unskilled construction and agricultural jobs, though there is some evidence from the United States that green sector wages are 13 percent higher than the average (Muro, Rothwell, and Saha 2011). However, most low- to medium-skilled green jobs do not require unique or highly specialized skills, so they are unlikely to command a wage premium.

Statistics on job creation in the energy efficiency sector are scarce, in part because energy-efficient construction and energy-related services are typically provided among many other services by established companies. For example, a construction firm will offer to add home insulation in addition to conventional building-related services. Better evidence exists for renewable energy manufacturing, installation, and operations and maintenance. In the EU, this sector employs more than 1.1 million people, about one tenth of those (111,000) in the EU-10. The three largest subsectors account for 70 percent of these jobs: solid biomass with 273,000, solar photovoltaics with 268,000, and wind with 253,000. Across countries, the sector's importance varies, with the biggest absolute job numbers in large Western European countries, as shown in figure 4.8. As a share of total employment, the Nordic and Baltic countries stand out.

Many of these jobs are indirect, in installation or maintenance, rather than in the production of clean-energy capital goods. In Germany, only about 18,000 of 130,000 solar energy jobs have been in photovoltaic module manufacturing.[14] This also implies that even countries that are not at the technology frontier can realize large employment gains in the clean-energy and energy efficiency sectors as domestic policies and, increasingly, market forces promote such investments.

The large number of jobs created in the clean-energy field also points to a weakness. At least some renewable energy technologies

FIGURE 4.8

Employment and Shares of Employment in the Renewable Energy Sector in the EU, 2010

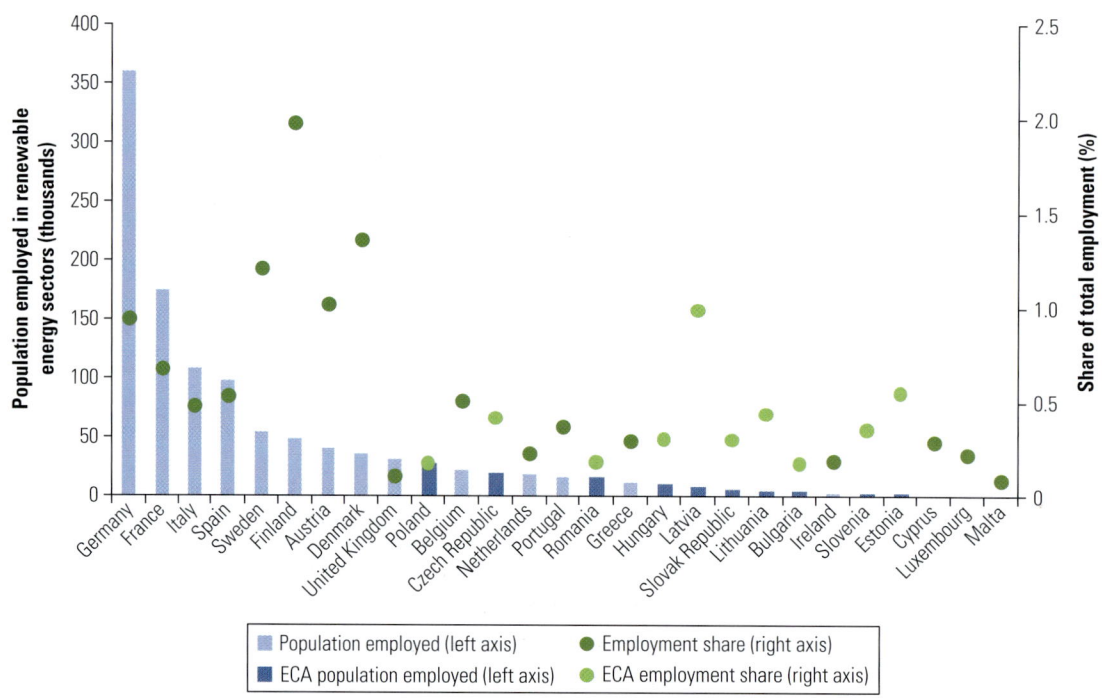

Sources: European Commission n.d.; *EurObserv'ER Report's* 11th Annual Overview Barometer (http://www.eurobserv-er.org/observer.asp).
Note: EU = European Union.

are currently far more labor intensive than traditional ones, especially when viewed in relation to energy production. Solar photovoltaics, for instance, generate between 7 and 10 times the employment per megawatt of coal, gas, or even advanced wind energy (Kammen, Kapadia, and Fripp 2006). This larger labor input requirement is probably a larger barrier to market competitiveness than equipment prices, which have been falling rapidly. Employment creation alone is therefore an insufficient argument for renewable energy, also because there may be other employment-intensive activities that would have higher payoffs. The main emphasis should be on raising efficiency and productivity to generate growth, which in turn creates jobs—not only in new energy-related sectors but also in the rest of the economy. For example, better management and more efficient use of forest resources in ECA would increase carbon sequestration and thus contribute to climate change mitigation. It could also generate 3 million additional jobs (as further discussed in chapter 10).

As in any major economic transformation, gains from job creation in new sectors could be partly offset by losses in traditional industries. These are discussed in chapter 5 along with policies that create the labor market conditions that facilitate such a transition.

Notes

1. Nordhaus (2008) is one of many papers to point out that the earlier that one tackles climate change, the lower the costs will be.
2. This argument goes back to Porter and van der Linde (1995).
3. See "OECD DAC Statistics on Climate-Related Aid": http://www.oecd.org/dac/aidstatistics/FactsheetRio.pdf.
4. Ernst & Young (2010) defines green FDI as investments in industries or services with a focus on renewable and clean technology.
5. The Balkan countries include Albania, Bosnia and Herzegovina, Croatia, Kosovo, the former Yugoslav Republic of Macedonia, Montenegro, Serbia, and Slovenia.
6. The Commonwealth of Independent States (CIS), created in December 1991, currently includes Armenia, Azerbaijan, Belarus, Georgia, Kazakhstan, Moldova, Russia, Tajikistan, Turkmenistan, Ukraine, and Uzbekistan. For more, see http://www.cisstat.com/eng/cis.htm.
7. The EU-10 countries are those countries that joined the EU in 2004: Bulgaria, the Czech Republic, Estonia, Hungary, Latvia, Lithuania, Poland, Romania, the Slovak Republic, and Slovenia.
8. This is not a new idea; see Markusen (1975).
9. For more about the World Bank's Carbon Finance Unit, see http://wbcarbonfinance.org.
10. This section is based on Dutz and Sharma (2012).
11. This finding is based on patent applications in 13 GHG mitigation technology fields filed in 76 countries, with "high quality" restricted to

patents filed in more than one country. See Dutz and Sharma (2012) for details.

12. For more about the Certified European Passive House Designer project, see http://eu.passivehousedesigner.de/.

13. This section draws on Bowen (2012).

14. Estimate is for 2011, according to the German Solar Business Association.

References

Bakker, Anuschka, ed. 2009. *Tax and the Environment: A World of Possibilities.* Amsterdam: IBFD.

Bowen, Alex. 2012. "'Green' Growth, 'Green' Jobs, and Labor Markets." Policy Research Working Paper 5990, World Bank, Washington, DC.

Copeland, Brian R. 2012. "International Trade and Green Growth." Policy Research Working Paper 6235, World Bank, Washington, DC.

Cyert, Richard M., and David C. Mowery, eds. 1987. *Technology and Employment: Innovation and Growth in the U.S. Economy.* Washington, DC: National Academies Press.

Dutz, Mark, and Siddharth Sharma. 2012. "Climate Change Mitigation Technologies and Innovation in Europe and Central America." Background paper, Poverty Reduction and Economic Management, World Bank, Washington, DC.

Ernst & Young. 2010. "Waking Up to the New Economy: 2010 European Attractiveness Survey." Survey report, Ernst & Young Global Limited, New York.

European Commission. n.d. Eurostat (online database). http://epp.eurostat.ec.europa.eu.

Gill, Indermit, and Martin Raiser. 2012. *Golden Growth: Restoring the Lustre of the European Economic Model.* Washington, DC: World Bank.

Goldberg, Itzhak, John Gabriel Goddard, Smita Kuriakose, and Jean-Louis Racine. 2011. *Igniting Innovation: Rethinking the Role of Government in Emerging Europe and Central Asia.* Washington, DC: World Bank.

Golub, Stephen S., Celine Kauffmann, and Philip Yeres. 2011. "Defining and Measuring Green FDI: An Exploratory Review of Existing Work and Evidence." Working Paper on International Investment 2011/2, Organisation for Economic Co-operation and Development, Paris.

Kammen, D. M., K. Kapadia, and M. Fripp. 2006. "Putting Renewables to Work: How Many Jobs Can the Clean Energy Industry Generate?" Report of the Renewable and Appropriate Energy Laboratory, University of California, Berkeley.

Kasek, Leszek, Olga Kiuila, Krzysztof Wojtowicz, and Tomasz Zylicz. 2012. "Regional Economic Effects of Differentiated Climate Action, Carbon Leakage, and Anti-Leakage Measures." Background paper, Europe and Central Asia Region, World Bank, Washington, DC.

Markusen, James R. 1975. "International Externalities and Optimal Tax Structures." *Journal of International Economics* 5 (1): 15–29.

Mattoo, Aaditya, Arvind Subramanian, Dominique van der Mensbrugghe, and Jianwu He. 2009. "Can Global Decarbonization Inhibit Developing Country Industrialization?" Policy Research Working Paper 5121, World Bank, Washington, DC.

Muro, Mark, Jonathan Rothwell, and Devashree Saha. 2011. "Seizing the Clean Economy: A National and Regional Green Jobs Assessment." Metropolitan Policy Program study, Brookings Institution, Washington, DC.

Nordhaus, William D. 2008. *A Question of Balance: Weighing the Options on Global Warming Policies*. New Haven, CT: Yale University Press.

Onder, Harun. 2011. "Climate Change and Trade Implications for ECA: Some Preliminary Ideas." Unpublished manuscript, Europe and Central Asia Region, World Bank, Washington, DC.

Peters, G. P., J. C. Minx, C. L. Weber, and O. Edenhofer. 2011. "Growth in Emission Transfers via International Trade from 1990 to 2008." *Proceedings of the National Academy of Science* 108 (21): 8903–08. http://www.pnas.org/cgi/doi/10.1073/pnas.1006388108. Accessed December 20.

Pew Charitable Trusts. 2012. "Who's Winning the Clean Energy Race? Growth, Competition and Opportunity in the World's Largest Economies." G-20 Clean Energy Factbook, The Pew Charitable Trusts, Washington, DC.

Popp, David. 2011. "The Role of Technological Change in Green Growth." Report prepared for the Green Growth Knowledge Platform, Maxwell School of Citizenship and Public Affairs, Syracuse University, Syracuse, NY.

Porter, Michael, and Claas van der Linde. 1995. "Toward a New Conception of the Environment-Competitiveness Relationship." *Journal of Economic Perspectives* 9 (4): 97–118.

Stern, Nicholas H. 2010. "Imperfections in the Economics of Public Policy, Imperfections in Markets, and Climate Change." *Journal of the European Economic Association* 8 (2–3): 253–88.

Umweltbundesamt (German Federal Environment Agency). 2009. "Border Tax Adjustments for Additional Costs Engendered by Internal and EU Environmental Protection Measures: Implementation Options and WTO Admissibility." Umweltbundesamt, Dessau, Germany.

World Bank. Various years. *World Development Indicators*. Washington, DC: World Bank.

CHAPTER 5

Social Inclusion

Main Messages

- Climate policies will raise the price of energy and the cost of polluting. Energy- and emission-intensive firms will become less competitive and may reduce employment. As in other structural transformations, good labor market policies and greater labor mobility let workers move from sunset to sunrise industries. Social safety nets help those who are unable to make the transition.

- Energy price reform, motivated by energy security as well as fiscal and climate objectives, could create economic distress, particularly for low-income households. More-targeted social safety nets will avoid hardships but should be complemented by support for household-level energy efficiency improvements that are a more sustainable solution in the long term.

- Lower public health burdens are a direct social benefit from climate action that reduces emissions from fossil-fuel plants. These plants cause an estimated US$19 billion in annual health damages in Europe and Central Asia (ECA). Economywide climate action can often be more cost effective in reducing local pollution than increasingly expensive plant-level reductions in emissions.

It is tempting to design policies intended to achieve two desirable outcomes at once. On rare occasions this can work. Fuel taxes, for instance, raise funds for financing transport infrastructure (or for subsidizing public transit) and also encourage fuel efficiency in cars, reducing local and climate pollution.[1] However, more often it is better to break down a problem into its components and address each with a separate policy instrument. This is the case with green jobs.

As chapter 4 argued, rather than subsidizing green job creation directly, it is preferable to have relatively neutral climate and energy policies aimed at emission reductions and a separate set of labor market policies that encourage skill development and flexible hiring. Robert Stavins, an environmental economist, describes a related example of this general principle in the context of the world's first large-scale pollution cap-and-trade system (Stavins 2009; see also Bennear and Stavins 2007):

> In 1990, when Congress sought to cut sulfur dioxide (SO_2) emissions from coal-fired power plants by 50% to reduce acid rain, Senator Robert Byrd (West Virginia) argued against the proposal for a national cap-and-trade system, because it would displace Appalachian coal mining jobs through reduced demand for high-sulfur coal. He recommended instead a national requirement for all plants to install scrubbers, which would have increased costs nationally by $1 billion per year in perpetuity.

> Fortunately, Senator Ted Kennedy (Massachusetts) recognized that these two problems (acid rain and displaced miners) called for two separate policy instruments. Simultaneous with the passage of the Clean Air Act amendments of 1990, which established the path-breaking SO_2 allowance trading program, Congress passed a job training and compensation initiative for Appalachian coal miners, at a one-time cost of $250 million. Acid rain was cut by 50%, $1 billion per year was saved for the economy, and sensible and meaningful aid was provided to the displaced miners. Two different policies were used to address two different purposes. Sometimes that is the wisest course.

This chapter discusses the potential social impacts of climate policies and the complementary policies that help soften those impacts. The first section complements the discussion of employment growth by looking at potential job losses caused by policies that make energy more expensive. The Europe and Central Asia (ECA) region has already gone through a period of significant industrial restructuring and energy price reforms. Still more can be done to promote economic efficiency and to contribute to climate goals. Countries vary greatly in vulnerability to energy price increases as well as their ability to respond and adapt to the labor market shocks that the price increases could trigger. In most cases, these shocks will be relatively modest and will play out over a long transition period. They will unlikely be more severe than the shocks from market opening or major economic crises, for instance. There is broad experience in managing such impacts by making labor markets efficient and by using more active labor policies and social safety nets where necessary.

Energy price increases caused by subsidy reform or environmental policies affect not only firms but also households more directly, as the second section on distributional impacts discusses. As a result, the energy budget share can exceed sustainable levels among low-income households that may not be able to reduce already minimal consumption. As with labor market impacts in ECA, existing social policy instruments and safety nets can help buffer these impacts. But equally important is to help households reduce their energy consumption through investments in energy efficiency, especially in building insulation and appliances.

Climate action can also have direct welfare benefits for vulnerable populations, like increasing the comfort in buildings that are more efficiently heated or improving public transport and reducing its cost for those who can't afford a car. Another important benefit, discussed in the final section of this chapter, is the reduction in health impacts from fossil-fuel emissions. Reducing energy use and replacing fossil fuels with renewable energy would significantly lower mortality and health expenditures in many ECA countries.

Labor Markets[2]

Job creation in green sectors shows only one side of the employment implications of low-carbon growth policies. A price on carbon emissions changes the relative prices across energy technologies and can make energy generally more expensive, so gains in clean energy may coincide with job losses in coal mining or fossil-fuel power plants. Energy-intensive firms in other industries such as chemicals or paper production will face a higher energy price that reflects the real cost to the environment. If they can't pass on these costs or if they face foreign competition not subject to carbon pricing, jobs could be lost. Furthermore, targeted subsidies for low-carbon technologies that are still uncompetitive at market prices divert funds that could be invested elsewhere, possibly generating more employment. This is an important consideration if employment creation is the main objective—for instance, in green stimulus programs (Strand and Toman 2010).

The employment implications of climate action are part of a necessary economic transition, similar to other transformations that are usually market driven rather than policy induced. These transformations cause both losses and gains in different industries. Such a transition would see three types of effects (Fankhauser, Sehlleier, and Stern 2008):

- *Short-term* job creation and job losses would occur in sectors directly affected by climate policies.

- *Medium-term* economywide impacts would occur as other sectors are affected by energy price increases, either directly or indirectly through supply chains, for instance. Price changes will trigger changes in the mix of inputs in production (for example, the use of more capital to reduce energy use).

- *Long-term* effects would occur as the economy responds and adjusts to the new prices of various products. This adjustment results in changes in the composition of durable and investment goods such as household appliances or machine tools and encourages more innovation and research and development (R&D).

Of most concern to policy makers is usually the immediate effect. Most ECA economies have comparatively higher shares of employment and value added in energy-intensive industries than their Western European counterparts, as shown in figure 5.1. Other industries that use less energy depend on the products from these energy-intensive companies, accounting for about 16 percent of cleaner sectors' inputs in the Czech Republic, for instance. Countries where many people rely on energy-intensive industries for jobs will be more vulnerable to energy price increases than countries where these industries play a smaller role. The illustrative analysis in this section considers the impact of both a rise of energy prices to market costs of US$0.125 per kilowatt-hour (kWh) and a carbon tax of US$15 per ton of carbon dioxide equivalent (CO_2e) (see figure 5.6 in the "Distributional Impacts" section below).

FIGURE 5.1

Employment and Value Added in Energy-Intensive Sectors of Europe and Central Asia Countries, 2007

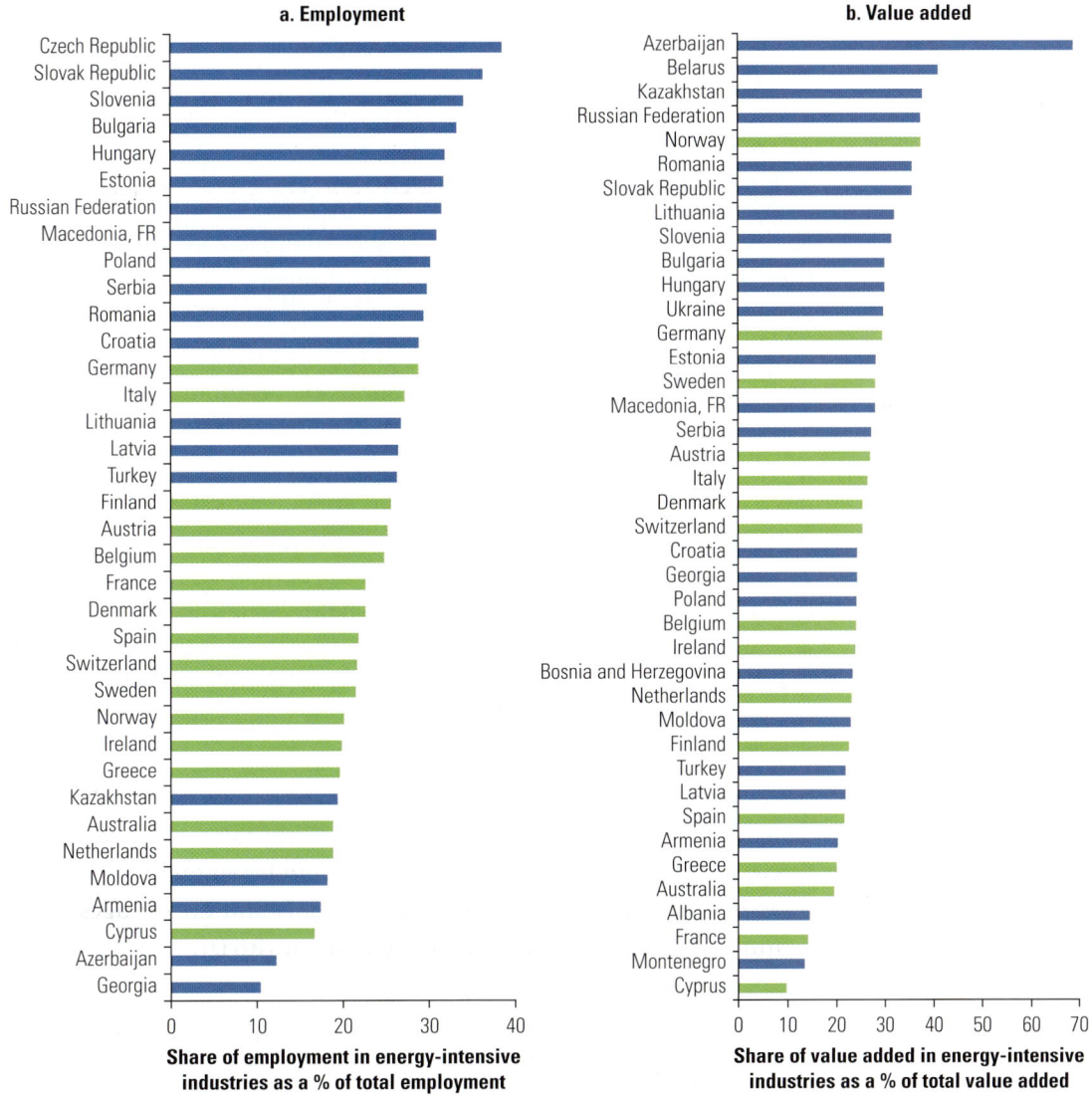

Source: United Nations Statistics Division (http://unstats.un.org/unsd/).
Note: Energy-intensive industries comprise the following sectors: mining, quarrying, other utilities, construction, and transportation. More disaggregated data for energy-intensive manufacturing sectors are not consistently available.

Countries differ not only by vulnerability but also in their ability to respond to labor market impacts. Social and labor market policies cushion negative effects by creating incentives for mobility to growing (and greener) sectors. Labor market adaptability to energy price shocks is determined by four main factors:

• *Labor market costs and flexibility* such as employment protection legislation or minimum wage levels

- *Skills development and innovation* that equip workers to transition to new jobs

- *Active labor market policies* that enhance labor supply (for example, training); increase labor demand (such as through public works or hiring and wage subsidies); and improve the functioning of the labor market (for example, employment services)

- *Social protection* that provides safety nets

Assessing both vulnerability to energy price increases and adaptability to labor market impacts suggests two broad categories of countries in ECA: (a) Bulgaria, Croatia, Hungary, Latvia, Lithuania, Poland, and the Slovak Republic will likely be less affected by climate policies that raise energy prices. Their energy prices are already quite high, and most derive a smaller share of economic output from energy-intensive industries. (b) Azerbaijan, Georgia, Kazakhstan, and Ukraine, in contrast, are more vulnerable to carbon pricing but also have labor market characteristics that support adaptation to energy price changes. In contrast to those two categories, the former Yugoslav Republic of Macedonia, the Russian Federation, and Serbia combine high vulnerability with low adaptability, as figure 5.2 illustrates.

Countries that are most at risk of adverse employment impacts from climate action that raises energy prices can respond by reducing their vulnerability or by increasing adaptability. They can reduce vulnerability in one of two ways: by increasing energy efficiency in the industrial sector (as discussed in chapter 7), or by accelerating the shift out of energy-intensive industrial sectors—an economic shift that has been ongoing in many countries independently of climate policies. The latter is obviously not a sensible strategy where other factors make those sectors highly competitive or essential to the national economy.

Increasing adaptability involves strengthening the overall business environment and increasing labor market flexibility so that workers in sunset industries can find jobs in growing sectors. Many ECA countries rank low in the World Bank's Doing Business Indicators[3] and have overly restrictive employment protection legislation, so firms are reluctant to hire. Experience in Western European countries such as Denmark shows that greater labor market flexibility combined with a strong social protection system tends to raise employment levels.

Support for skills training is a second way to increase adaptability. Relatively few green jobs require unique skills, but acquiring new skills will be important for workers losing jobs in shrinking industries. Training is also important when structural transformation occurs within rather than between companies—for instance, when a

FIGURE 5.2

Employment Vulnerability and Adaptability of Europe and Central Asia Countries, 2008

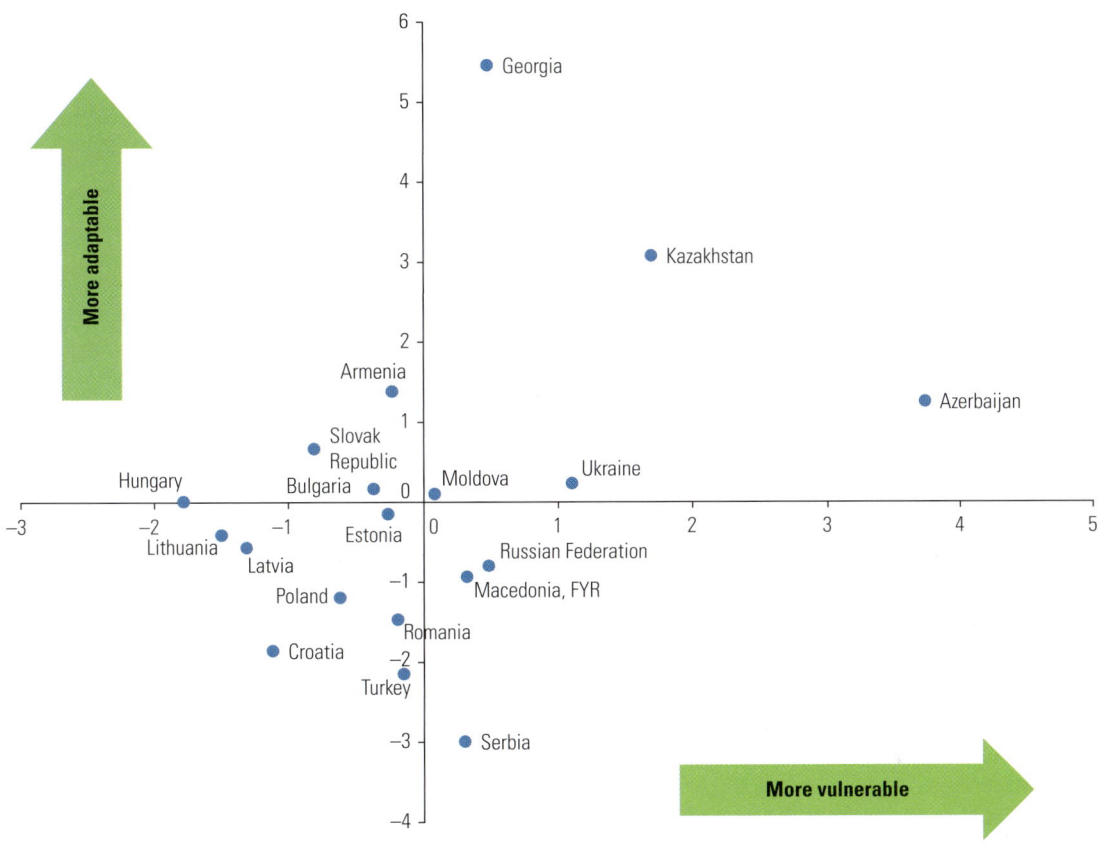

Source: Calculations based on UN data. See Oral, Santos, and Zhang 2012 for details.

Note: Countries with high vulnerability (higher values) are those with higher shares of employment and value added originating from energy-intensive industries, as well as those that would experience a higher energy price increase. Countries with high adaptability (higher values) are the ones with high labor market flexibility, high skills, high social protection readiness, and higher spending on active labor market policies. The value 0 represents the average position in terms of vulnerability and adaptability across all countries in our sample. This analysis takes into account the overall manufacturing sector. A more detailed breakdown of the manufacturing sector that results in a similar analysis, but covering fewer countries, is in Oral, Santos, and Zhang (2012, annex 1).

firm switches from making internal combustion motors to high-efficiency electric ones.

Although firms and workers have the greatest incentive for training and retraining, public assistance can also facilitate green-technology skills development through national education systems. Many ECA countries face significant challenges in this area. A large share of businesses rate insufficient or inappropriate worker skills as a "major" or "very severe" constraint. Skills constraints appear most severe in some of the Commonwealth of Independent States (CIS) countries such as Belarus, Kazakhstan, Moldova, Russia, and Ukraine. Beyond flexibility and skills, labor market policies such as hiring incentives or matching services can ease some of the challenges of an economic transition. However, experience with these kinds of policies has been mixed.

Policy makers are understandably split over the job market implications of climate policies. Some expect large job growth in modern,
competitive industries. Others fear the adjustment costs of increasing
energy prices. But, with some exceptions among technology leaders
and heavily fossil-fuel-dependent countries, the impacts will likely be
relatively modest in both directions and will be further softened by
their occurrence over a number of decades (UNEP 2011; ILO 2009;
Dupressoir et al. 2007; also see box 5.1).

BOX 5.1

Impact of Climate Policies on Employment Patterns in Poland Relative to Overall Structural Transformations

The evidence is mixed on whether environmental policies affect the poor more than the rich.
Studies in the United States and Australia show that a carbon tax would hit lower-income households more severely, while studies in Italy and Spain showed the impact to be neutral. A carbon
tax in China would be progressive, reflecting differences in goods consumed at the household
level across income groups. In general, impacts on household income would largely stem from
people who lost jobs in energy-intensive companies subject to a carbon tax, for instance.

Poland's energy mix has a high share of coal, so a significant carbon tax will have an impact on
the cost structure of energy-intensive firms and consequently also on employment. A study of the
impact of carbon policies, building on the World Bank's recent Poland Low Carbon Growth Study,
compares a business-as-usual (BAU) scenario with a green scenario. The BAU scenario is not
static because Poland continues to undergo significant economic transformations unrelated to climate policies. The results suggest that the BAU scenario will decrease employment of lower-
skilled workers as fewer low-paid jobs in agriculture and industry get created. Green policies have
only a small additional influence. In the longer term, the impact would be relatively neutral. Also, in
the longer term, the effects of labor market turnover would be lower on the poor than on the rich.

However, the analysis also shows that local impacts can be severe. Under the green scenario, the Slaskie (Upper Silesia) and Mazowieckie regions see the highest job impacts. But
although gains roughly balance losses in Mazowieckie, the Slaskie region—home to much of
Poland's heavy industry and mining—experiences net job losses. With the implementation of
green policies, Slaskie would account for 36 percent of all the jobs lost but for only 20 percent of
the jobs gained in Poland by 2030 in comparison to the BAU scenario, as shown in figure B5.1.1.

Poland could do more to ensure that its labor markets support reallocation of jobs across sectors as the structure of the economy changes. The country still has relatively high nonwage labor
costs and about average employment protection laws. Poland does spend significant resources
on active labor market policies compared with other countries. However, its social protection

continued

BOX 5.1 *continued*

FIGURE B5.1.1

Projected Job Losses and Gains in Poland, by Region, under a Green Scenario Relative to a BAU Scenario, 2020 and 2030

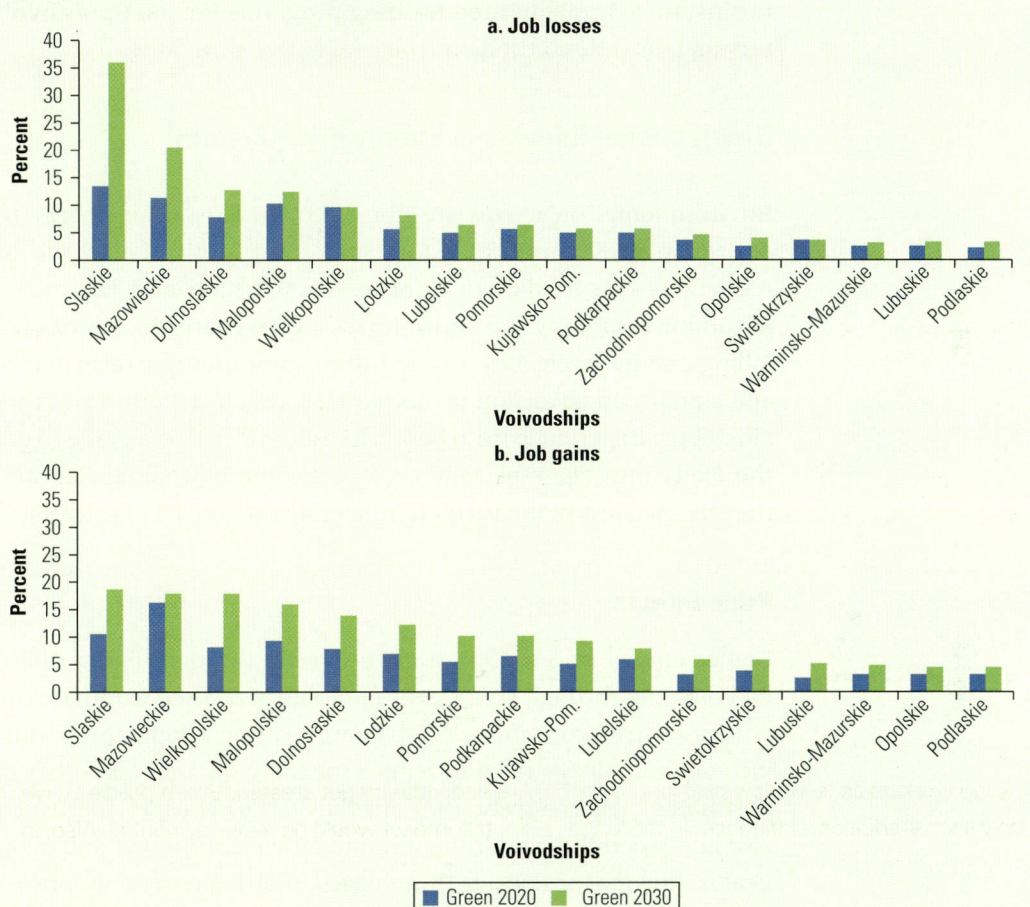

Source: Calculations based on the 2009 household budget survey for Poland.
Note: BAU = business as usual, that is, a scenario without new green policies.

spending relative to gross domestic product (GDP) is the lowest in the region at 1.6 percent, and a quarter leaks to the richest 20 percent of the population. Unemployment benefits are not generous compared with other countries, although they are paid over a long duration of 18 months. Equally important, Poland has made strides in improving its education system as reflected in improvements in its Organisation for Economic Co-operation and Development (OECD) educational attainment rankings. In addition, the country could improve its innovation systems to support uptake of green technologies in its firms. These structural improvements are at least as important as more direct employment policies in aiding labor market adjustments.

Sources: Jorgensen, Kasek, and Shkaratan 2012; Oral, Santos, and Zhang 2012.

Rather than designing climate policies specifically to create green jobs, policy makers should generally ensure a good business environment, a flexible labor market, and a functioning social protection system. Where more active programs are used, it is important to coordinate climate policies with labor market policies aimed at supporting green job creation. Some states in the United States, for instance, implemented training programs for solar photovoltaic technicians only to then cut solar incentive programs.

Distributional Impacts of Energy Price Reform[4]

Environmental objectives are not the main reason for energy price reform in ECA countries. More important motivations have been reductions of subsidies to improve fiscal health and tax increases to fund investments. Past price increases have already contributed to falling energy intensities. In the future, countries may also decide to add a pollution or carbon tax to reduce local air pollution and carbon emissions. Experience from past price reforms will be a guide to gauge the likely impacts, especially on low-income households, as well as the effectiveness of measures to moderate the impact of price shocks.

Price Effects

Higher energy prices affect welfare directly through higher utility or fuel bills and through rising costs of products and services that require energy in their production. When energy eats up a higher proportion of income, households need to reduce spending on food and other basic necessities.[5] Alternatively, they may switch to cheaper fuels such as coal and wood that are dirtier, harming the environment and people's health. Previous studies in ECA suggest that higher energy prices are likely to hurt the poor more than the rich. Poor households generally spend a larger share of income on energy. They also have fewer alternatives such as access to natural gas and district heating. Households whose electricity payments make up a larger share of their budgets are also more likely to be behind in payments (Lampietti, Sudeshna, and Amelia 2007; Zhang 2011). Cost recovery in the form of strengthened collection will therefore affect the poor more than others.

Electricity prices in the region's countries have steadily increased over the past decade as a result of power market reforms. Depending on the country, these reforms raised tariffs to cost-recovery levels, to market levels, or to rates that reflect some of the social costs of energy use including those related to climate change. On average, residential electricity tariffs in real terms jumped by 40 percent over the past decade, rising from US$0.047 per kWh in 2000 to US$0.066 per kWh

FIGURE 5.3

Residential Electricity Prices in Europe and Central Asia, by Subregion, 2000–10

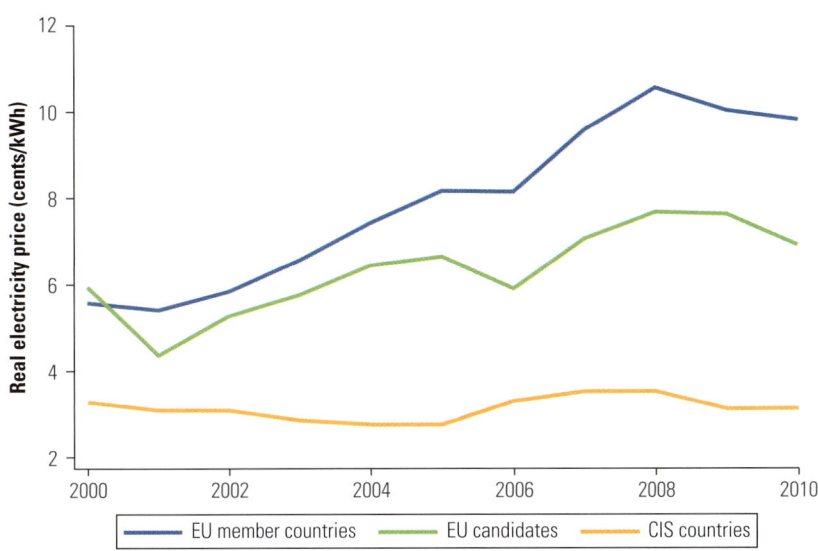

Source: Calculations based on ERRA n.d.

Note: EU member countries include Bulgaria, the Czech Republic, Estonia, Hungary, Latvia, Lithuania, Poland, Romania, and the Slovak Republic. EU candidate countries in this analysis include Albania, Bosnia and Herzegovina, Croatia, FYR Macedonia, Montenegro, Serbia, and Turkey. CIS (Commonwealth of Independent States) countries include Armenia, Azerbaijan, Georgia, Kazakhstan, the Kyrgyz Republic, the Russian Federation, and Ukraine. Other ECA countries are not included in the analysis because of limited data availability. kWh = kilowatt hour.

in 2010. Across subregions, the picture varied, as figure 5.3 shows. There has been much progress in tariff reform, with average real residential prices increasing by 75 percent in the European Union (EU)-10 and 17 percent in the EU candidate countries during 2000–10.[6] Tariffs in CIS countries[7] are still substantially lower and have increased only slightly over the period.[8]

Despite recent progress, residential electricity is still underpriced in many ECA countries. Substantial cross-subsidization exists between residential and nonresidential customers, especially in Albania, FYR Macedonia, Russia, and Ukraine, as shown in figure 5.4. Because the costs of serving industrial customers are typically lower, these customers should have lower tariffs. However, in almost half the region's countries, industrial tariffs are currently higher than residential tariffs. In Ukraine, for instance, industrial customers pay three times more for electricity than residential customers, much higher than the EU-27 average of 70 percent.[9] A common benchmark for the long-run marginal cost of power generation is determined by the average cost of building and operating a gas-fired combined cycle power plant. The benchmark suggests that the average cost recovery price of electricity supply in ECA is around US$0.125 per kWh.[10] More than half of the region's countries are still below that level.

FIGURE 5.4

Residential Electricity Prices and Cross-Subsidization between Residential and Nonresidential Consumers in Selected Europe and Central Asia Countries, 2011

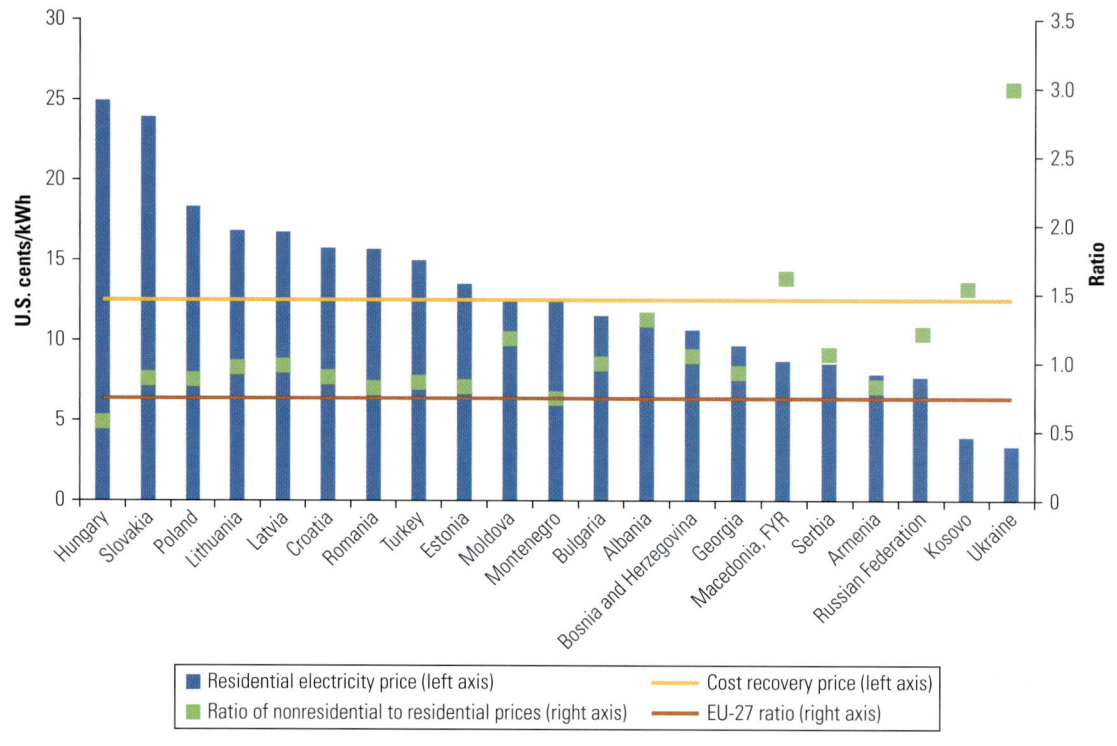

Source: Calculations based on the ERRA n.d. and European Commission n.d.

Note: The blue columns show average residential electricity prices in 2011. The green dots show the ratio between nonresidential and residential electricity prices. The red line indicates that the average price ratio between nonresidential and residential consumers is 0.73 in EU-27 countries. The orange line is the considered cost-recovery price of electricity supply at US$0.125 per kWh in ECA. Many countries had block tariff structures; prices in the chart show weighted averages of electricity prices; kWh = kilowatt hour.

Electricity prices will likely rise further in all ECA countries for at least three main reasons:

- *Fiscal pressure.* Energy subsidies are expensive. According to the International Energy Agency (IEA), electricity subsidies in Russia reached US$22.26 billion in 2010, the highest in the world (IEA 2011c). When viewed as a percentage of GDP, Uzbekistan had the biggest subsidies for electricity, at almost 6 percent of GDP. Electricity subsidies in Kazakhstan, Turkmenistan, and Ukraine collectively reached US$4.76 billion in 2010 and ranged from 1 percent to 3 percent of GDP.[11] Most countries in the region face significant fiscal deficits and increasing government debt. Electricity subsidies further strain public finance. The need to restore fiscal stability makes continued price liberalization an urgent issue for governments.

- *Investment needs.* Demand for electricity was expected to increase at an annual rate of 3.1 percent between 2005 and 2010 in ECA in the absence of greater gains from energy efficiency measures (World Bank 2010). Because much of the supply capacity is old and needs to be retired or restored, enormous capital investments are needed to maintain supply. The estimated investment amounts to US$3.3 trillion, including US$1.3 trillion for primary energy development and US$2 trillion for the power and heat infrastructure. To fully meet capital needs, tariffs have to rise to cover the incremental costs.

- *New norms.* EU and EU accession countries will have to conform to EU climate and energy policies, including a 20 percent reduction of greenhouse gas emissions (compared with 1990 levels), a 20 percent increase in energy efficiency, and a 20 percent share of renewable energy, all by 2020. Many CIS countries also plan to invest in higher energy efficiency and renewable energy generation. These results require up-front investments, often funded through electricity prices. In fact, all members of the Energy Community of South East Europe,[12] as well as Armenia, Belarus, and Kazakhstan, have adopted or proposed incentive programs (such as feed-in tariffs) to guarantee higher prices for renewable electricity (Fischer and Preonas 2012).

Raising electricity prices to cost recovery levels of US$0.125 per kWh combined with a carbon charge of US$15 per ton of CO_2 equivalent would raise electricity prices considerably in a number of ECA countries, as figure 5.5 indicates. Some CIS countries (such as Kazakhstan, the Kyrgyz Republic, Tajikistan, and Ukraine) need to make large adjustments to reach cost recovery. Some EU and EU candidate countries that rely more heavily on fossil fuel-fired generation (such as Bosnia and Herzegovina, Estonia, and FYR Macedonia) would be significantly affected by a carbon price. At the other end of the scale, Hungary, Latvia, and Lithuania already have prices higher than the "social cost of electricity" assumed in this scenario. In countries such as Albania, Azerbaijan, Kosovo, the Kyrgyz Republic, FYR Macedonia, and Uzbekistan, where arrears and nonpayment remain a persistent challenge, cost recovery also requires the removal of implicit subsidies by strengthening collections (World Bank 2010).

With these potential price hikes, households would have to allocate a higher share of their expenditures to electricity, as shown in figure 5.6. These figures assume that households do not respond to price increases by consuming less energy, so they represent an upper-bound estimate of actual impacts.[13] In the EU-10 countries, electricity bills for average-income households would remain the same or increase by no more than 0.42 percent of their budget. Households in Hungary, Latvia, and Lithuania (where no price shock is expected) would not have to pay more, while Bulgarians and Estonians, for instance, would pay an additional US$27 per month. In EU accession countries,

Potential Electricity Price Hikes with Subsidy Removal and a Carbon Tax, Europe and Central Asia Countries, 2009

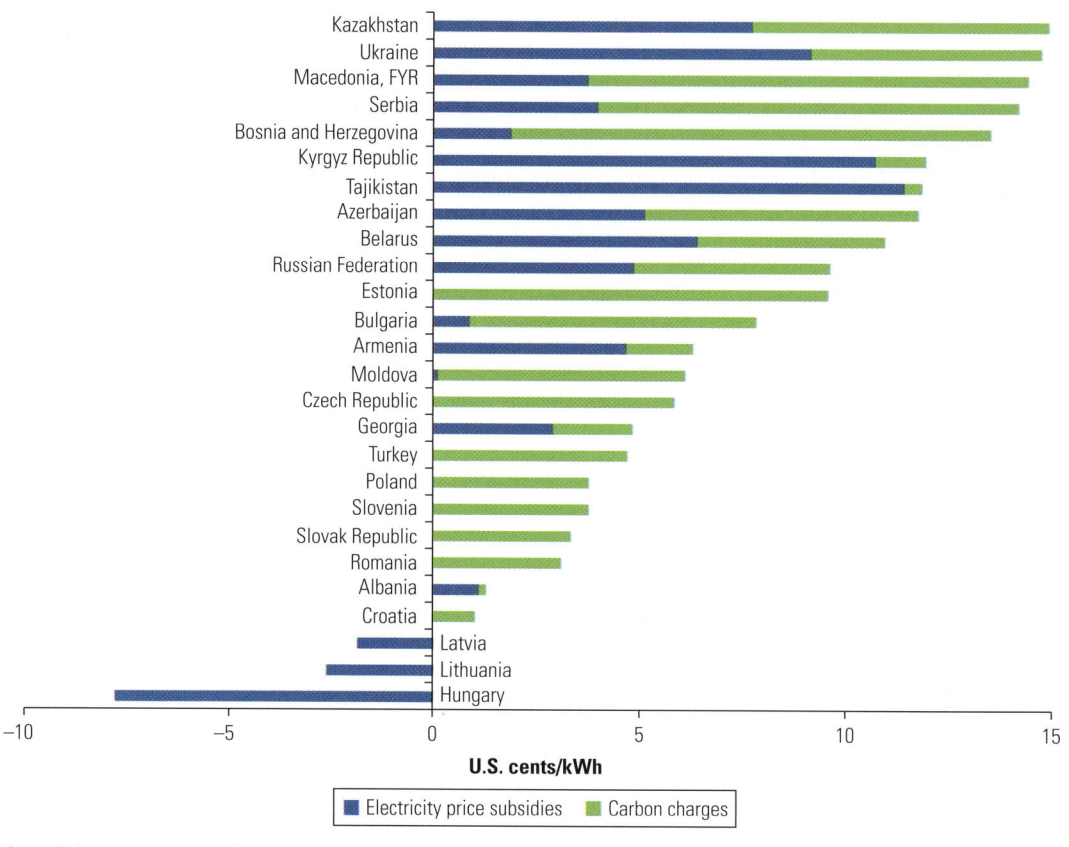

Source: Calculations based on IEA CO_2 emissions database (www.iea.org) and ERRA n.d.
Note: Electricity price charges are determined by the gap between prevailing electricity price and the average long-term cost recovery price at US$0.125 per kWh. Carbon subsidies are determined by the CO_2 intensity of power generation and a carbon price of US$15 per ton of CO_2 equivalent. No data were available for ECA countries excluded from the figure. kWh = kilowatt hour.

the increase is between 0.02 and 1.4 percent of total household expenditures, or from US$3 per month in Albania to US$88 per month in FYR Macedonia and Serbia. In the CIS countries (excluding Tajikistan), budget shares would increase 0.3–7 percent, or between US$6 in Belarus to US$44 per month in Russia. In Tajikistan, the welfare consequence would be by far the largest at US$51 per month or 17 percent of household budget. Higher-income households typically spend a smaller share of their total expenditures on electricity (Albania and FYR Macedonia are exceptions), so poorer households will experience a sometimes significantly higher increase in electricity budget shares.

These increases in expenditures are significant, but are they also unaffordable? A commonly used benchmark is that households should not have to spend more than 10 percent of their income on an adequate supply of electricity (Bagdadioglu et al. 2009; World Bank 2010; Ruggeri Laderchi, Olivier, and Trimble 2012). Current

FIGURE 5.6

Estimated Change in Total Household Expenditure on Electricity after Price Increase, by Expenditure Quintile, Selected Europe and Central Asia Countries

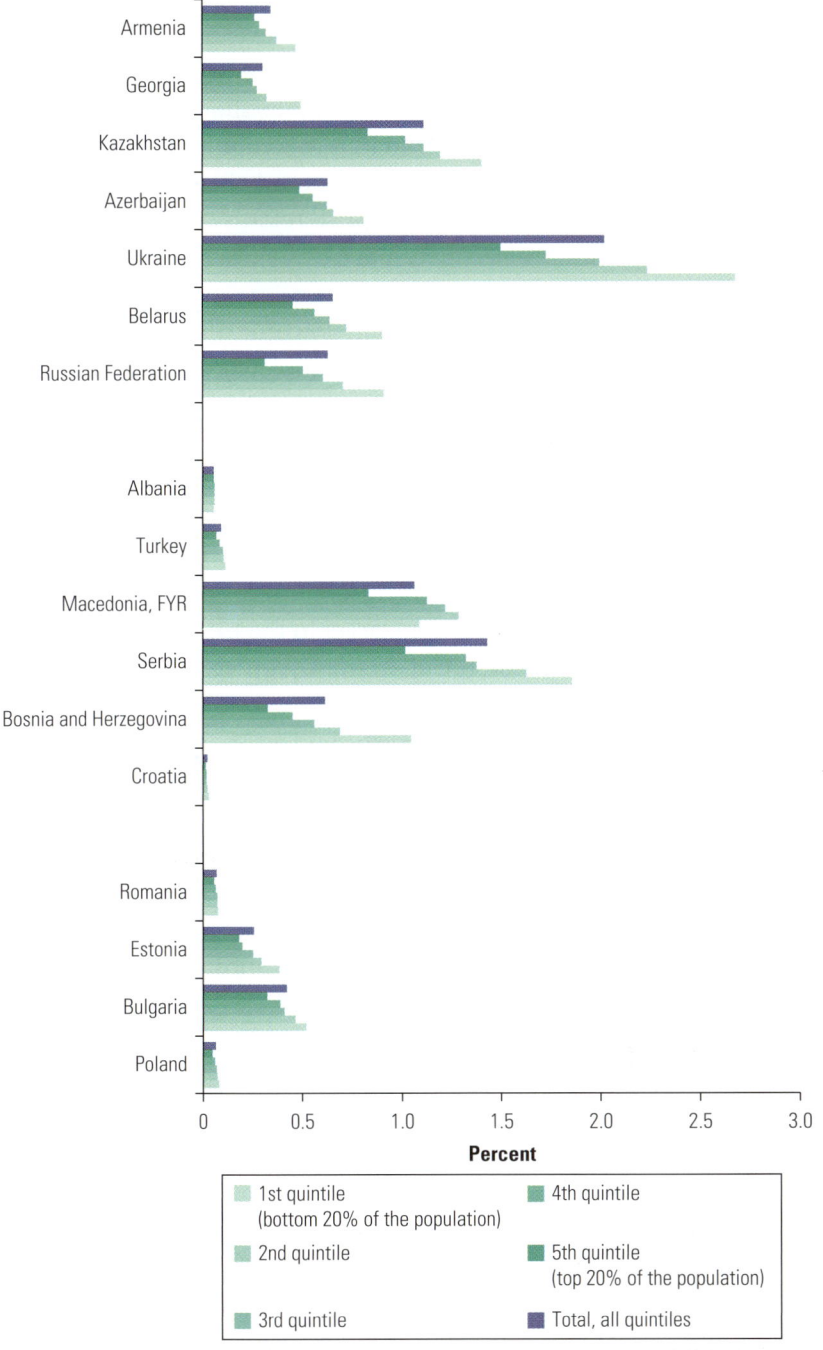

Source: Calculations based on World Bank's ECAPOV database.

Note: The magnitude of price increase is based on the estimation previously illustrated in figure 5.5. Assuming zero price elasticity, the figure illustrates an upper-bound estimate for the impact of price increases on household welfare.

actual household expenditure shares for average-income households in ECA range from below 2 percent in Kazakhstan, Tajikistan, and Ukraine to above 7 percent in Bulgaria, Montenegro, and Serbia, as shown in figure 5.7. On average, households in EU-10 and EU accession countries spend 5 percent and 6 percent of total expenditure on electricity, respectively. CIS countries have a lower share, at 3 percent, largely due to high subsidies. Figure 5.7 also shows that poorer households typically allocate a higher share of expenditure for electricity (except in Albania and FYR Macedonia), although in all countries, their expenditure shares are currently below 10 percent.

The impact on affordability in this price reform scenario is proportional to the increase in electricity prices, as shown in figure 5.8. In EU-10 countries, the budget share of electricity would increase by 1.8 percentage points on average. Some lower-income households in Bulgaria and Estonia would exceed the affordability benchmark. The impact is substantially higher in EU accession and CIS countries, where household expenditure on electricity would, on average, increase by 4.5 and 7.1 percentage points, respectively. Several countries would face drastic adjustments, including Bosnia and Herzegovina, FYR Macedonia, and Serbia in the Western Balkans, which rely on electricity for heating and are heavily dependent on coal for power generation. The impact of price increases is even larger in the CIS countries, where electricity subsidies are widespread. In the Kyrgyz Republic, Tajikistan, and Ukraine, complete removal of electricity subsidies combined with a carbon charge would raise electricity budget shares by 10, 8, and 4 times the current level, respectively. In the remaining countries, electricity expenditure shares remain below 10 percent on average but are close to or above 10 percent for the poorest households.

Overall, higher-income households would lose more in absolute terms, but as a percentage of total expenditure, lower-income groups bear a disproportionately higher share of the burden (see table 5.1). The relative price increases for the poor are 25–50 percent higher than those for the average households. The additional monthly cost due to the electricity price increases (previously indicated in figure 5.5) is approximately US$10 for an average-income household in the EU-10 states, US$41 in EU accession countries, and US$24 in the CIS countries. This is equivalent to 0.12 percent of monthly household expenditure in the EU-10, 0.54 percent in the EU accession countries, and 3.4 percent in the CIS.

In absolute terms, Western Balkan countries would see the largest price impacts.[14] Their households use electricity as the main source for heating because of limited access to natural gas or district heating. Household electricity consumption in that subregion is almost twice

FIGURE 5.7

Current Share of Electricity Expenses in Total Household Expenditure, by Expenditure Quintile, in Europe and Central Asia Countries

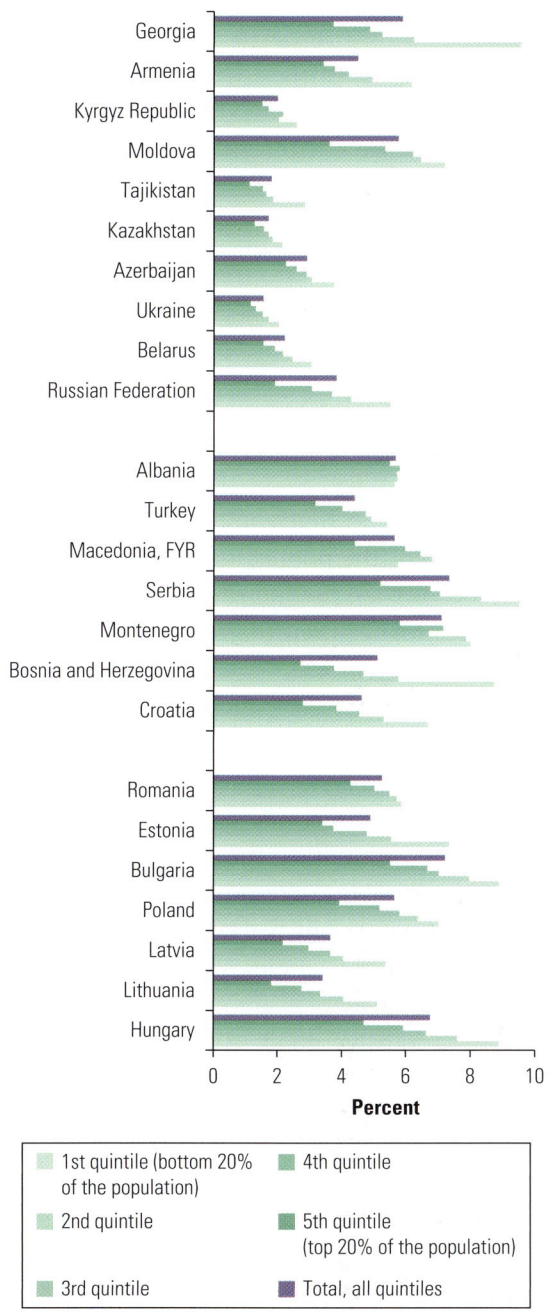

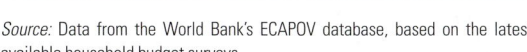

Source: Data from the World Bank's ECAPOV database, based on the latest available household budget surveys.

FIGURE 5.8

Estimated Share of Electricity Expenses in Total Household Expenditure after Potential Price Increase, by Quintile, in Europe and Central Asia Countries

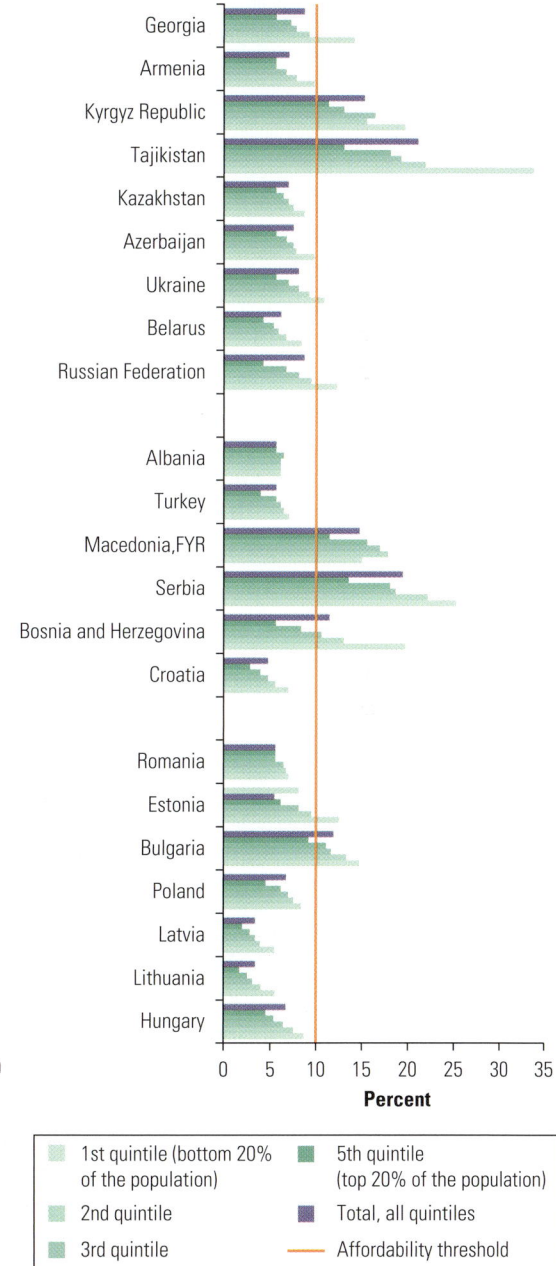

Source: Calculations using data from the World Bank's ECAPOV database.
Note: The magnitude of price increase is based on estimation illustrated in figure 5.3. Assuming zero price elasticity, the figure illustrates an upper-bound estimate for the impact of price increases on household electricity budget share.

TABLE 5.1

Estimated Impact of Potential Electricity Tariff Increases, by Subregion, on Europe and Central Asia Households

Subregion	Average			Bottom 20%	
	Electricity expenses as a share of total expenditure (%)	Consumer surplus change (US$/month)	Consumer surplus change as a share of total expenditure (%)	Electricity expenses as a share of total expenditure (%)	Consumer surplus change as a share of total expenditure (%)
EU	6.8	10.1	0.12	8.9	0.15
	(1.1)	(4.5)	(0.06)	(1.4)	(0.08)
EU Accession	10.5	41.2	0.54	13.4	0.69
	(2.4)	(16.4)	(0.24)	(3.3)	(0.30)
CIS	10.1	24.0	3.42	14.3	5.10
	(1.7)	(5.0)	(1.90)	(2.7)	(3.0)

Source: Calculations based on the World Bank's ECAPOV database.
Note: Standard deviations are reported in parentheses. EU = European Union. CIS = Commonwealth of Independent States.

as high as in EU or CIS countries. In relative terms, the largest welfare impact would occur in the CIS countries, where prices are expected to rise by the highest percentage.

In all subregions, the lower-income households would suffer a greater impact on their welfare. For the poorest one-fifth of households in EU and EU accession countries, the increase as a percentage of total expenditure is 1.8 times higher than that of the wealthiest, and 2.2 times higher in the CIS.[15]

Estimates of *indirect effects* of electricity price increases, channeled through higher prices of other energy-intensive commodities, are not available for ECA countries. However, studies on other countries and regions suggest that such indirect effects may also place a higher burden on low-income households. In the United States, for example, the poorest fifth of households are responsible for an average of about 2 metric tons of CO_2 emissions per US$1,000 worth of income, whereas the richest fifth of households are responsible for about 0.7 metric tons. Wealthier households are nearly three times more efficient because energy-intensive goods, such as food, take up a large percentage of a low-income person's budget (Grainger and Kolstad 2010). Studies in the EU and New Zealand also find that the indirect impact of a carbon tax is likely to be regressive (Brännlund and Nordström 2004; Kerkhof et al. 2008; Wier et al. 2005; Callan et al. 2009).

These estimates assume that households can respond to electricity price increases only by paying more. In reality, they will have two other options: they can consume less electricity, and they can switch to other fuels, as further discussed in box 5.2. Reducing consumption is easier for wealthier households. A recent study on Turkey reveals

BOX 5.2

Fuel Substitution as a Response to Energy Price Increases

The Turkish example illustrates the degree of fuel substitution for heating or cooking, typically to coal or wood (Zhang 2013). Burned indoors in inefficient stoves, these fuels can cause respiratory health problems, especially for women and children. Although sustainably harvested wood is a renewable resource, household-level coal burning adds to CO_2 emissions even when it replaces electricity generated from coal. Before 2008, both retail gas and electricity tariffs were regulated in Turkey with cross-subsidies from industrial consumers to households. In 2008, as part of market reforms designed to improve efficiency and attract outside investors, the government pegged retail gas and electricity rates to fuel costs, inflation, and the exchange rate. Retail gas prices rose by a cumulative 75 percent in one year because of higher import prices that are fully passed on to consumers. On the other hand, Turkey produces all the lignite coal it uses, and domestic lignite prices remained stable, as figure B5.2.1 shows.

Heating represents more than 80 percent of Turkish households' total energy consumption. Historically, solid fuels such as coal and fuelwood were the dominant heating fuel. In the late 1980s, natural gas was promoted as a cleaner fuel for residential heating to reduce major air pollution problems in Istanbul and Ankara. Since then, the expansion of natural gas distribution networks for residential areas has promoted nationwide gasification. As a result, natural gas

FIGURE B5.2.1

Change in Relative Residential Fuel Prices from 2007 in Turkey

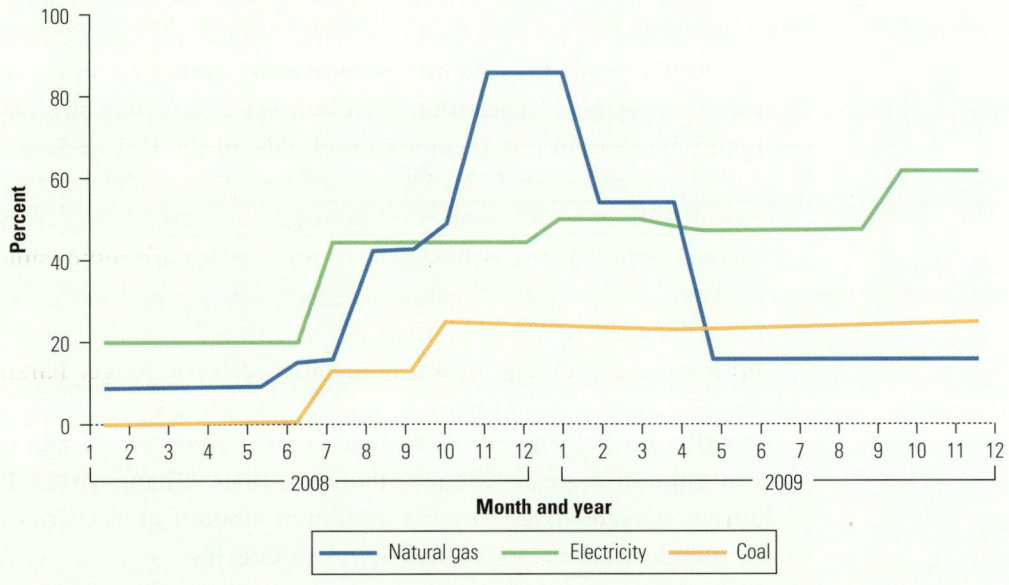

Sources: Turkish Ministry of Energy and Natural Resources and Turkish Electricity Distribution Company.

continued

BOX 5.2 *continued*

consumption has increased rapidly, rising from zero in 1990 to one-third of demand in the residential sector, mostly for heating. However, the change in relative prices between coal and gas since energy tariff reform in 2008 has prompted concerns that the country may again face air pollution problems as consumers return to coal for heating. Indeed, while residential gas consumption fell by 33 percent after the price hikes in 2008, coal consumption increased by 20 percent, as figure B5.2.2 illustrates. On average, a 1 percent increase in gas prices increases the probability of switching to coal or wood by 0.7 percent; for households that already partially rely on coal or wood, a 1 percent increase in gas prices will on average increase solid fuel consumption by 3 percent, all else being equal.

FIGURE B5.2.2

Residential Energy Consumption in Turkey, by Source, 2000–09

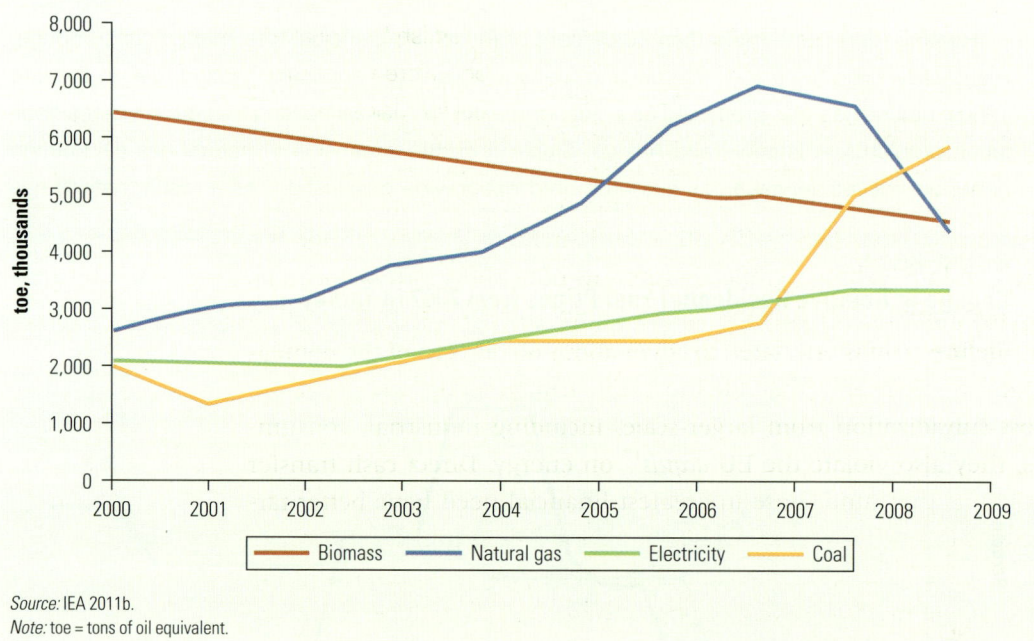

Source: IEA 2011b.
Note: toe = tons of oil equivalent.

that rich households are three times more responsive in adjusting consumption to price changes than the poor (Zhang 2011). Poor households already consume a minimum amount of electricity and are therefore unable to respond by consuming less.

Policy Responses

To buffer the social impacts of energy price reform, many ECA countries have implemented complementary social protection programs. In the past, income transfers and lifeline tariffs were the dominant policy instruments (Lampietti, Sudeshna, and Amelia 2007; Ruggeri Laderchi, Olivier, and Trimble 2012). The former requires efficient targeting, while the latter causes benefit leakage to those who may not need it. However, in each case, the primary objective is to reduce expenses.

A third, more sustainable approach is to help poor households reduce their energy consumption—for instance, by supporting improved home insulation or energy-efficient appliances. This requires higher administrative efforts but avoids long-term recurrent assistance and reduces fuel consumption overall. Although it takes time to improve energy efficiency, social assistance programs can provide immediate relief and ensure a smooth transition in the short term. In most cases, both strategies (short-term relief through social protection and a long-term transition through energy efficiency) are needed to help households cope with higher energy prices.

Lifeline Tariff and Direct Cash Transfers

Most ECA countries have put in place some form of social assistance program to ensure energy affordability. These programs include lifeline tariffs (low rates for a small but essential amount of energy) and direct cash transfers, which are sometimes earmarked for energy consumption. Lifeline tariffs offer the benefits of high coverage but often end up leaking significant amounts of benefits to those who do not need them.[16] In Albania, for example, the subsidy embedded in the lifeline tariff is estimated to cover about 80 percent of the population. To the extent that lifeline tariff arrangements are based on cross-subsidization from larger-scale, including industrial, consumers, they also violate the EU *acquis*[17] on energy. Direct cash transfer programs that fund those in greatest financial need have better targeting but lower coverage, while categorical cash transfer programs, in which eligibility is determined by belonging to a certain group such as veterans or public sector employees, have higher coverage but often poor targeting because eligibility status is often not connected to income levels.

Targeting of energy social assistance programs varies in the ECA region. The poorest households in EU countries receive 30–60 percent of the benefits, the richest households less than 10 percent. Social programs in the CIS countries are, in contrast, often poorly targeted. For example, the housing and utility allowances in Ukraine compensate households for utility expenditures that are above 20 percent of income. However, fewer of the poor qualify because they spend mostly on food and are less likely to be connected to utilities. In Kazakhstan, the richest households receive the same share of social benefits as the poor (Ruggeri Laderchi, Olivier, and Trimble 2012).

In the wake of the global financial crisis, many ECA countries have started reforming existing social assistance schemes to address the increasing needs for social benefit assistance and growing pressure on government fiscal balance. Ongoing reforms aim to improve program targeting and to increase the transparency and efficiency of service delivery. The measures taken include removing benefits that are not directly linked to income, creating a unified registry of beneficiaries, and consolidating small social assistance programs.[18] Some countries, such as FYR Macedonia and Moldova, have adopted easy-to-implement and often flat payment schemes to ensure quick disbursement in times of stress. A flat payment also delinks benefits from energy consumption to preserve incentives for energy saving.[19]

Finally, the business community and civil society have also been playing a role in helping poor communities to deal with increased energy costs. In Turkey, some consumers obtain assistance for the payment of their electricity bills from social solidarity foundations as well as private foundations and charity groups. In Bulgaria, an Austrian company called EVN worked closely with nongovernmental organizations (NGOs) and local communities to mitigate the impact of rising electricity prices while also improving bill collection (Ruggeri Laderchi, Olivier, and Trimble 2012).

Energy Efficiency Programs

Although energy price reform is usually motivated primarily by fiscal concerns, it is also an instrument to raise energy efficiency. Efficiency improvements are, in turn, an effective response to improve energy affordability. Assistance in the form of preferential loans and grants to low-income households make energy savings investments, such as home insulation and energy efficiency appliances, more affordable. Educational or information programs provide practical information about energy-saving measures tailored to low-income households.

Energy efficiency programs for low-income households in ECA should be a priority for at least two reasons:

- *Direct cash transfer is expensive and insufficient.* Analysis shows that many of the existing welfare transfer programs are too small to serve their targeted groups effectively once subsidies are completely phased out, while expanding these programs could be prohibitively expensive. Energy efficiency investment, on the other hand, avoids recurrent expenses and provides a cost-effective long-term solution. Results of efficiency programs in other regions demonstrate such an impact (see box 5.3).

- *There are large opportunities to reduce energy consumption in households.* For example, most buildings in the region still use two or three

BOX 5.3

Energy Efficiency Programs Targeting Low-Income Households

Brazil End-Use Energy Efficiency Program

In 2005, the Brazilian government established an energy efficiency program that requires utilities to invest 0.5 percent of their annual revenues in improving customers' energy efficiency, of which 50 percent should go toward low-income households. Eligibility for the program is determined by consumption levels, connection type, and enrollment in other social assistance schemes. The utilities are in charge of designing and implementing the efficiency projects but are also allowed to subcontract energy service companies.

So far, most of the efforts have targeted low-income households and include these solutions:

- Replacing old, inefficient refrigerators

- Installing compact fluorescent lightbulbs

- Replacing inefficient electrically heated showers with more efficient ones or solar water heaters

- Informing households about efficient use of electricity

Investments are sometimes covered by the utilities and sometimes shared with households. In the latter case, utilities offer financing schemes including rebates and monthly payment (integrated with electricity bills).

Between 2005 and 2007, over 5 million compact fluorescent lightbulbs and 60,000 efficient refrigerators were installed under the program. Because refrigerators and lighting account for the bulk of the electricity consumption of low-income households in Brazil (90 percent), the program achieved significant reductions in energy consumption. According to field assessments, electricity consumption of refrigerators and lighting on average has been reduced by around 70 percent and 22.7 percent, respectively. As a result, peak demand for power has decreased by 15–20 percent.

U.S. Weatherization Assistance Program

In the United States, energy efficiency programs for low-income households have always existed in parallel with direct financial aid. The Weatherization Assistance Program, established in 1976, is one of the most successful and largest efficiency programs in the United States. The program provides low-income households with weatherization services, initially targeting heating (insulation and heating systems) and broadening over time to include cooling, appliances, and lighting. Eligibility for the program is based mainly on income levels, using thresholds defined according to the national poverty guidelines. The weatherization services are managed by local agencies and include a visit by an energy auditor, then the installation of the chosen energy-saving measures, and finally the verification of the works done by an inspector.

continued

BOX 5.3 *continued*

Based on regular monitoring and evaluation, it is estimated that the program has achieved average annual energy savings of about 30,000 British thermal units (Btu) per year per household—a 30 percent reduction that, in total, saves the equivalent of the Republic of Georgia's entire primary energy supply. Thus, the program also reduced CO_2 emissions by around 2.6 metric tons per household per year. At the same time, beneficiary households have saved between US$300 and US$400 per year. A recent cost-benefit analysis suggests that for every US$1 invested, US$1.80 is returned in reduced energy bills, and US$0.71 is returned to ratepayers, households, and communities through increased local employment, reduced uncollectible utility bills, improved housing quality, and better health and safety.

Source: WEC 2010.

times as much heat as buildings in comparable climates in Western Europe (see chapter 9). Investments in efficiency and insulation can therefore substantially reduce consumption.

The payoffs of efficiency improvements in ECA could be large even with modest investments such as basic insulation or caulking windows. A US$50 investment could lead to as much as a 10 percent reduction in energy demand and a 2 percent reduction in the energy budget share in the region. The benefits would be highest for the lowest-income households and most significant in CIS countries.

As with direct cash transfers, the difficulty in energy efficiency programs for low-income households is defining eligibility criteria to minimize administrative costs and to ensure that subsidized resources end up with those who need them the most. For example, programs that provide tax incentives to support retrofit measures in France and Italy have boosted retrofit investment in the residential sector, but they also have resulted in high free-rider rates, particularly for replacement of windows (Neuhoff et al. 2011). Eligibility and targeting design depends on the national (or even local) context. One recommendation is to develop local partnerships facilitated through utilities like the Brazilian efficiency program. Local stakeholders (local authorities, NGOs, and so forth) have better knowledge of their territories and where to focus the efforts (WEC 2010).

Although a number of easy-to-achieve opportunities might be available, addressing energy efficiency comprehensively requires longer-term investments, and it takes time for the benefits to reach the households. An effective social assistance package should therefore consist of both welfare transfers that offer immediate relief and efficiency programs that provide sustainable long-term solutions.

FIGURE 5.9

Estimated Gains from Removing Energy Subsidies while Compensating Poor Households and Improving Energy Efficiency in Europe and Central Asia Countries

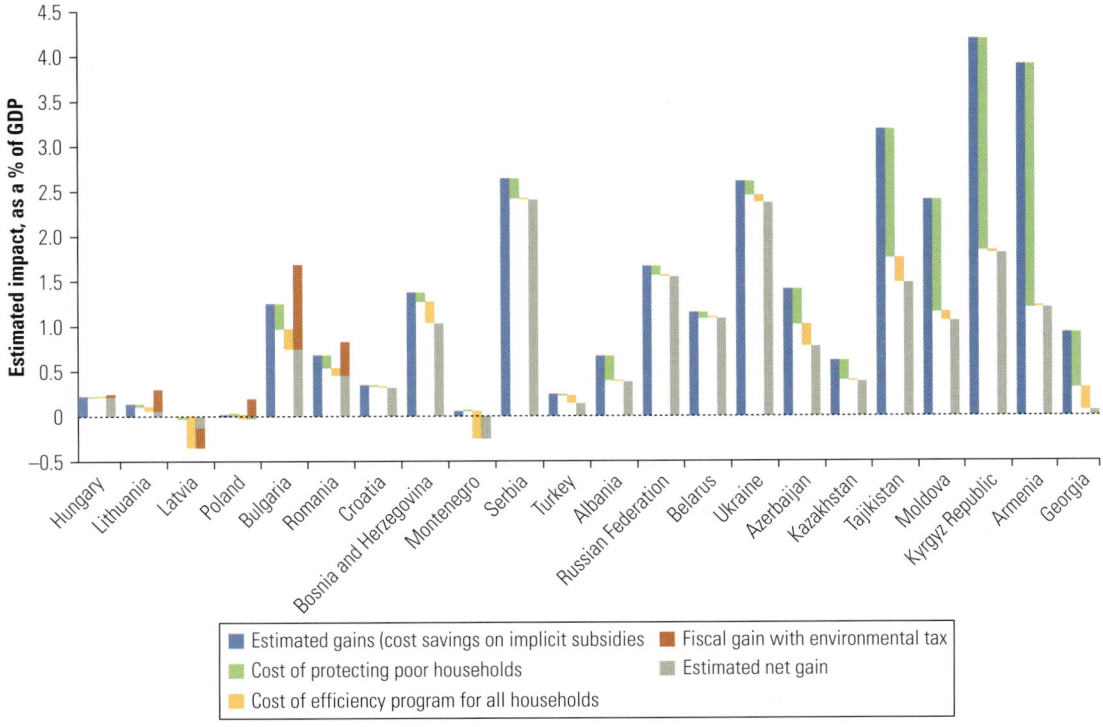

Source: Ruggeri Laderchi, Olivier, and Trimble 2012.

When such a strategy is implemented together with the removal of energy subsidies, analysis shows that all ECA countries (except Latvia and Montenegro) could expect to have a net fiscal gain of around 0.5–1.0 percent of GDP annually, as shown in figure 5.9.

Health Impacts

A shift to more efficient energy use and cleaner energy sources also has direct social benefits. Fossil-fuel combustion in power plants, industrial facilities, and motor vehicles is the main source of CO_2 emissions that cause global warming. However, it also causes emissions of particulate matter ($PM_{2.5}$ and PM_{10}), nitrous oxides (NO_x), sulfur dioxide (SO_2), organic aerosols, and black carbon, among others. Less fossil-fuel use reduces the probability of climate change damages in the future. It also reduces damages to health and quality of life from air pollution today.

These damages, expressed as economic costs (see box 5.4), are large, and 94 percent of them come from premature deaths caused by respiratory disease, congestive heart failure, and cancer that can

BOX 5.4

How Air Pollution Damages Are Estimated

Estimates of health damages from air pollution are based on empirical models that relate emissions from power plants to health impacts on people living in the vicinity. A spatial emission model predicts how pollutants disperse around the plant, generating estimates of ambient concentrations for each place. Maps of population distribution then yield the number of people exposed to different pollution loads. A dose-response function derived from detailed epidemiological studies then estimates the impacts.

These impacts are often expressed as disability adjusted life years (DALYs)—essentially the number of years lost due to poor health, disability, or early death. This number is then translated into monetary terms using a value of a statistical life (VSL), usually derived from an estimate of how much people would be willing to spend to reduce their mortality risk. This is the most critical step in translating health impacts into economic terms (for example, in Ashenfelter 2006).

The estimates presented here use a value of €60,000 (US$78,000) in 2010 prices—compared with €40,000 in 2000 prices in the original pan-European study. This is far lower than the US$6 million in 2000 prices used in a recent U.S. study.

Source: Markandya, Bigano, and Porchia 2010; NRC 2010.

be attributed to emissions from fossil-fuel burning in power plants or cars (NRC 2010). According to estimates from the *Global Burden of Disease and Risk Factors* study, particulate pollution caused 2.4 million premature deaths worldwide in 2000, although that number includes mortality attributed to indoor air pollution from burning biofuels in poor countries (Lopez et al. 2006). Wealthy countries are not spared from severe impacts on the health of their citizens along with other costs. In 2005, coal-fired power plants in the United States generated approximately US$62 billion in aggregate damages, with individual plants causing between US$0.002 and US$0.12 of damages per kWh depending on their pollution intensity (NRC 2010). These damages are highly concentrated: the dirtiest 10 percent of plants produced 25 percent of electricity but accounted for 43 percent of all damages.

According to another study, despite emission regulations that are already relatively strict, air pollution damages from the utility sector may equal more than a third of its total value added (Muller, Mendelsohn, and Nordhaus 2011). All of these estimates do not even include other damages such as acid rain, ozone pollution, eutrophication of water bodies, haze and reduced visibility, or the health impacts and mortality from occupational risks and accidents in the fossil-fuel mining and power sectors.

Air pollution fell significantly in ECA countries after transition when many polluting facilities were decommissioned for economic reasons. Over the past decade, the economy rebounded but with a somewhat cleaner stock of factories and vehicles. The new EU members, for example, had to introduce more-stringent emission standards. Fine particulate matter, $PM_{2.5}$, which is associated with cardiovascular and lung disease, dropped in most of the region's countries, as did sulfur oxide emissions, as shown in figure 5.10.

FIGURE 5.10

Change in $PM_{2.5}$ and SO_x Emissions in Europe and Central Asia Countries, 1999–2009

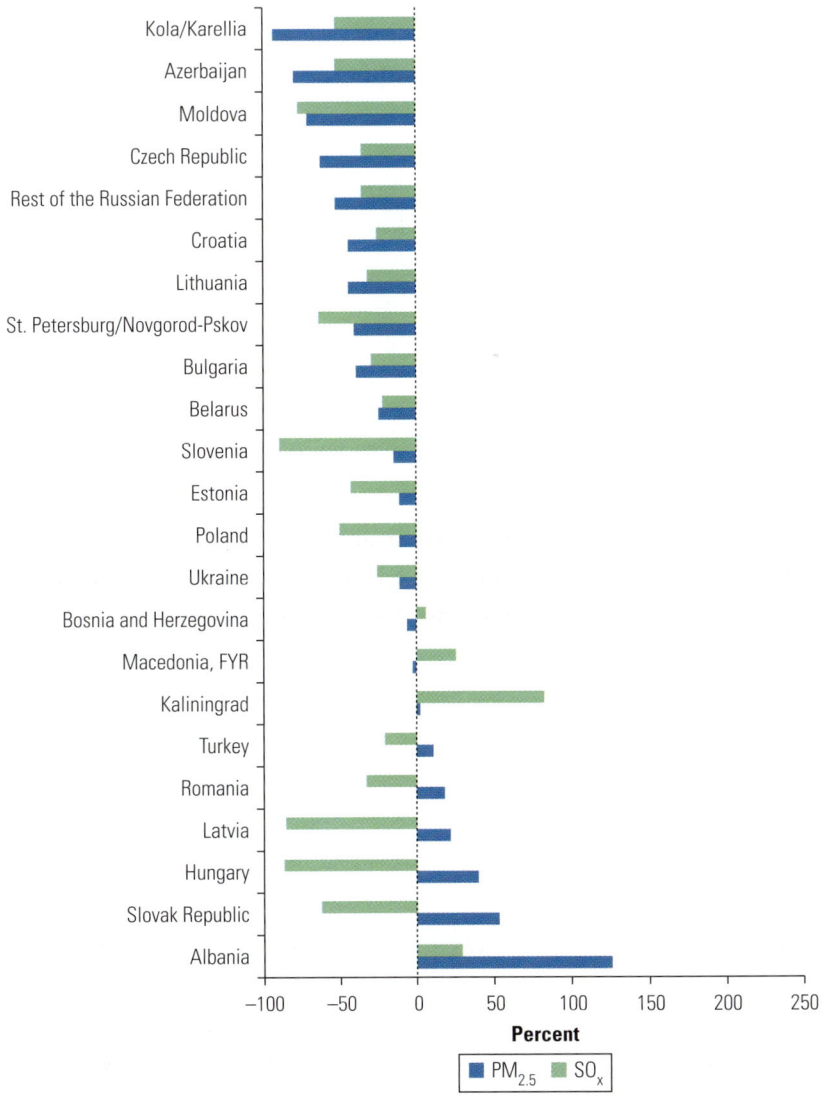

Source: EMEP 2011.
Note: Trends of emissions as used in European Monitoring and Evaluation Programme (EMEP) models. Russia is split into several zones. There were insufficient data for excluded ECA countries. $PM_{2.5}$ = particulate matter; SO_x = sulfur oxide.

Estimates of damages from air pollution caused by fossil-fuel burning in ECA countries are generally lower than in the United States, largely because of differences in the assumed value of a statistical life (see box 5.4). Damages vary greatly by country depending on the technology and fuel used to generate power. Power plants using heavy oil emit the largest amount of pollutants, followed by coal and natural gas. New estimates produced for this report range from as low as US$0.002 per kWh for electricity from natural gas in several countries to as high as US$0.052 per kWh for oil-fired power in Hungary, as table 5.2 indicates.[20] These estimates are fairly uncertain because of the wide variety of power plants in operation as well as data limitations. Most estimates are based on detailed pollution-dispersion modeling. For Serbia, the South Caucasus, and Central Asian countries, additional assumptions needed to be made.[21] With these caveats in mind, the new estimates suggest that total health damages from fossil-fuel burning in the ECA region amount to US$19 billion per year, as figure 5.11 illustrates. About a third of these damages, amounting to almost US$6 billion, occur in Russia, followed by Poland with US$3.3 billion and Turkey with US$1.8 billion.

Climate action that reduces energy consumption or promotes a shift to cleaner energy sources therefore also brings large ancillary benefits from reduced air pollution. The magnitude of these co-benefits depends on the scale and characteristics of climate change mitigation strategies. In Western Europe, a reduction of CO_2 emissions by 4–7 percent under various scenarios of Kyoto Protocol implementation would reduce SO_2 emissions, for instance, by 5–14 percent (Van Vuuren et al. 2006). The benefits from reduced local and regional air pollution would offset about half the cost of the climate policies (€2.5–€7 billion of the €4–€12 billion price tag). Other studies found somewhat lower benefits of about 20–30 percent of climate change mitigation costs. These estimates suggest that climate action could sometimes be a more attractive way to reduce immediate, local, and severe environmental health impacts than installing increasingly expensive technologies at the plant level that reduce harmful pollutants.

Better information collection and more consistent analytical methods would assist efforts to assess the health benefits of climate change mitigation in ECA and to determine the costs and benefits of national climate action versus local emission-reduction efforts. More detailed estimates of damages from air pollution should be generated in country-specific analyses using more detailed data, spatial modeling, and locally agreed-upon estimates of the value of a statistical life.

More comprehensive and transparent monitoring of air pollutants creates incentives for air pollution control and, indirectly, also for climate change mitigation. Moscow has developed a comprehensive

TABLE 5.2

Estimated Health Damages from Power Generation, by Fuel Source, in Europe and Central Asia Countries, 2009

U.S. cents per kilowatt-hour

Country	Coal	Oil	Gas
Albania	..	2.8	..
Armenia[a]	..	..	0.9
Azerbaijan[a]	..	1.3	0.3
Belarus	2.1	3.8	1.2
Bosnia and Herzegovina	1.5	2.8	0.9
Bulgaria	1.8	3.4	1.5
Croatia	2.3	4.1	1.8
Czech Republic	2.8	5.0	2.1
Estonia	1.1	2.2	0.5
Georgia[a]	1.0	1.6	0.4
Hungary	..	5.2	2.5
Kazakhstan[a]	2.7	4.1	1.5
Kyrgyz Republic[a]	0.5	..	0.2
Latvia	1.5	2.8	0.8
Lithuania	1.9	3.4	1.2
Macedonia, FYR	1.3	2.5	0.9
Moldova	2.9	4.8	1.8
Poland	2.3	4.2	1.5
Romania	2.5	4.3	2.1
Russian Federation	1.7	2.5	0.6
Serbia[a]	1.5	2.8	1.1
Slovak Republic	2.7	4.8	2.2
Slovenia	2.7	4.8	2.1
Tajikistan[a]	..	..	0.2
Turkey	1.6	3.0	0.8
Turkmenistan[a]	..	..	0.7
Ukraine	2.3	3.9	1.2
Uzbekistan[a]	..	1.0	0.3

Source: Markandya and Golub 2012.
Note: These are mean estimates; .. = negligible.
a. Countries for which damages were imputed because earlier direct damage estimates were unavailable.

and transparent air pollution monitoring system.[22] China's air pollution transparency index, a comparative ranking of Chinese and other cities, is an example of public information disclosure that encourages improvements (RUC Law and IPE 2010). The Intergovernmental Panel on Climate Change (IPCC) of the United Nations Framework Convention on Climate Change (UNFCCC) and the World Health

FIGURE 5.11

Total Estimated Health Damages from Fossil-Fuel Power Generation in Europe and Central Asia Countries, 2009

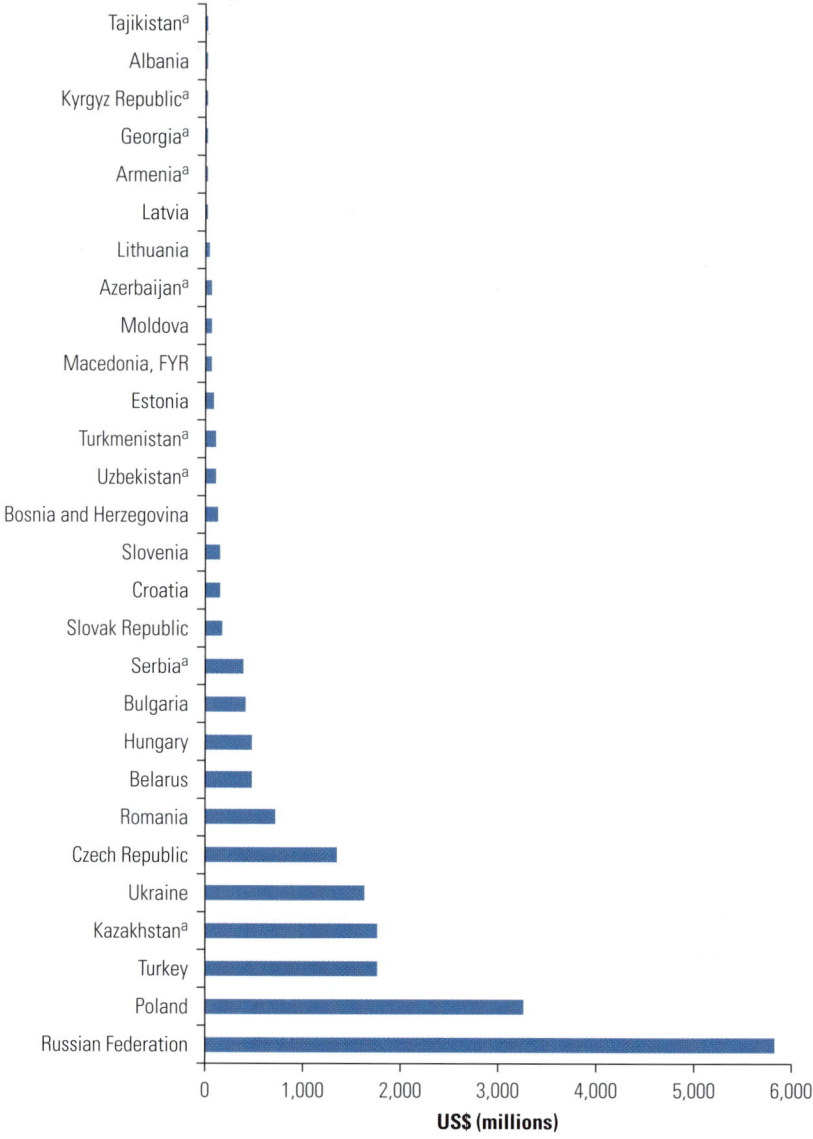

Source: Markandya and Golub 2012.
a. Countries for which damages were imputed because earlier direct damage estimates were unavailable.

Organization (WHO) have developed standard methods for estimating health and climate co-benefits (Smith and Haigler 2008). More generally, a sustainable energy policy must take account of all costs of energy production and consumption, including nonmarket costs that are too often ignored.

Notes

1. Frankel (2009) proposes that a fuel tax actually addresses seven public issues: traffic congestion, traffic accidents, local air pollution (health benefits), global climate change, national security (cutting energy dependence), trade deficits (by reducing oil imports), and a budget deficit.
2. This section is based on Oral, Santos, and Zhang (2012).
3. For more about the World Bank's Doing Business Indicators (database), see http://www.doingbusiness.org/.
4. This section builds on Ruggeri Laderchi, Olivier, and Trimble (2012) and Zhang (2013).
5. In Armenia, households' spending on food dropped by 8 percent, and spending on health care dropped by 50 percent following a decade-long substantial increase in gas prices (Ruggeri Laderci, Olivier, and Trimble 2012).
6. The EU-10 countries include Cyprus, the Czech Republic, Estonia, Hungary, Latvia, Lithuania, Malta, Poland, the Slovak Republic, and Slovenia. EU candidate countries include Albania, Bosnia and Herzegovina, Croatia, Kosovo, FYR Macedonia, Montenegro, Serbia, and Turkey.
7. CIS countries include Armenia, Azerbaijan, Belarus, Georgia, Kazakhstan, the Kyrgyz Republic, Moldova, Russia, Tajikistan, Turkmenistan, Ukraine, and Uzbekistan.
8. To some extent, differences in electricity prices between subregions also reflect the difference in fuel mix and the state of energy supply infrastructure.
9. The EU-27 countries include all current EU members.
10. The long-run marginal cost of generation is estimated to be US$0.125 per kWh. This estimate is based on construction of a gas-fired combined cycle power plant and includes costs associated with transmission and distribution. Refer to World Bank (2009) and Ruggeri Laderchi, Olivier, and Trimble (2012) for details.
11. IEA Energy Subsidies Database: http://www.worldenergyoutlook.org /subsidies.asp.
12. Albania, Bosnia, Bulgaria, Croatia, Georgia, Kosovo, FYR Macedonia, Moldova, Montenegro, Romania, Serbia, and Ukraine are party to (or observers of) the Energy Community of South East Europe.
13. The financial burden of higher electricity prices will be lower when consumers respond to price increases by reducing electricity consumption.
14. The Western Balkan countries include Albania, Bosnia and Herzegovina, Croatia, Kosovo, FYR Macedonia, Montenegro, Serbia, and Slovenia.
15. These estimates should be viewed as an upper bound on the welfare impact but a lower bound on the regressivity of the price reform. A higher electricity price will likely induce a household to reduce electricity consumption and thus its financial burden. But because poor households will be less able to respond to price changes (they consume the bare minimum), assuming zero price elasticity for all will underestimate the differential impact across income groups.

16. Ruggeri Laderchi, Olivier, and Trimble (2012) provide a detailed review of the mixed results of implementing lifeline tariffs in ECA.

17. The EU *acquis* refers to the body of European Union law, of which Energy is one of the 31 chapters.

18. For example, in Moldova, the government consolidated social assistance programs by linking a new heating allowance program with an existing means-tested program.

19. For more details of energy social assistance systems in ECA, see Ruggeri Laderchi, Olivier, and Trimble (2012).

20. See Markandya and Golub (2012). These updated estimates are based on prior work described in Markandya, Bigano, and Porchia (2010).

21. The South Caucasus countries include Armenia, Azerbaijan, and Georgia. The Central Asian countries include Kazakhstan, the Kyrgyz Republic, Tajikistan, Turkmenistan, and Uzbekistan.

22. For more about the Moscow City Government's air-quality monitoring effort, see http://www.mos.ru/en/authority/activity/ecology/index .php?id_14=22254.

References

Ashenfelter, O. 2006. "Measuring the Value of a Statistical Life: Problems and Prospects." *Economic Journal* 116 (510): C10–C23.

Bagdadlioglu, N., A. Basaran, S. Kalaycioglu, and A. Pinar. 2009. "Integrating Poverty in Utilities Governance." United Nations Development Programme Report, Ankara, Turkey.

Bennear, Lori Snyder, and Robert N. Stavins. 2007. "Second-Best Theory and the Use of Multiple Policy Instruments." *Environmental and Resource Economics* 37 (1): 111–29.

Brännlund, R., and J. Nordström. 2004. "Carbon Tax Simulations Using a Household Demand Model." *European Economic Review* 48 (1): 211–33.

Callan, T., Sean Lyons, Susan Scott, Richard S. J. Tol, and Stefano Verde. 2009. "The Distributional Implications of a Carbon Tax in Ireland." *Energy Policy* 37 (2): 407–12.

Dupressoir, S. et al. (There are 23 authors) 2007. "Climate Change and Employment: Impact on Employment in the European Union-25 of Climate Change and CO_2 Emission Reduction Measures by 2030." Study report, European Trade Union Confederation, Brussels.

EMEP (European Monitoring and Evaluation Programme). 2011. "Trends of Emissions as Used in EMEP Models." Convention on Long-Range Transboundary Air Pollution, Umweltbundesamt, Vienna.

ERRA (Energy Regulators Regional Association). n.d. Tariff Database. ERRA, Budapest. http://www.erranet.org/Products/TariffDatabase.

European Commission. n.d. Eurostat online database. http://epp.eurostat. ec.europa.eu.

Fankhauser, S., F. Sehlleier, and N. Stern. 2008. "Climate Change, Innovation, and Jobs." *Climate Policy* 8 (4): 421–29.

Fischer, C., and L. Preonas. 2012. "Feed-In Tariffs for Renewable Energy: Effectiveness and Social Impacts." Background paper, World Bank, Washington, DC.

Frankel, Jeffrey. 2009. "Energy and the Environment: Policy Advice for the New Administration." Joint Session of the American Economics Association and the Association of Environmental & Resource Economists, San Francisco, January 3.

IEA (International Energy Agency). 2011a. "Development of Energy Efficiency Indicators in Russia." IEA Energy Paper, IEA, Paris.

———. 2011b. *Energy Balances of OECD Countries*. Paris: IEA and Organisation for Economic Co-operation and Development.

———. 2011c. *World Energy Outlook 2010*. Paris: Organisation for Economic Co-operation and Development and IEA.

ILO (International Labour Organization). 2009. *World of Work Report 2009: The Global Jobs Crisis and Beyond*. Geneva: ILO.

Jorgensen, Erika, Leszek Kasek, and Maria Shkaratan. 2012. "Distributional Effects of the Transition to a Low-Emissions Economy in Poland: Employment Impact by Income and by Region." Background paper, Europe and Central Asia Region, World Bank, Washington, DC.

Kerkhof, A. C., Henri C. Moll, Eric Drissen, and Harry C. Wilting. 2008. "Taxation of Multiple Greenhouse Gases and the Effects on Income Distribution: A Case Study of the Netherlands." *Ecological Economics* 67 (2): 318–26.

Lampietti, J. A., G. B. Sudeshna, and B. Amelia. 2007. *People and Power: Electricity Sector Reforms and the Poor in Europe and Central Asia*. Directions in Development Series. Washington, DC: World Bank.

Lopez, A. D., C. D. Mathers, M. Ezzati, D. T. Jamison, and C. J. L. Murray, eds. 2006. *Global Burden of Disease and Risk Factors*. Washington, DC: Oxford University Press and World Bank.

Markandya, A., A. Bigano, and R. Porchia. 2010. *The Social Costs of Electricity: Scenarios and Policy Implications*. Cheltenham, UK: Edward Elgar Publishing.

Markandya, A., and A. Golub. 2012. "Health Impacts of Fossil Fuel Use in ECA Countries." Background paper, Metroeconomica Economic and Environmental Consultants, Bath, UK.

Muller, Nicholas Z., Robert Mendelsohn, and William Nordhaus. 2011. "Environmental Accounting for Pollution in the United States Economy." *American Economic Review* 101 (5): 1649–75.

Neuhoff, K., H. Amecke, A. Novikova, and K. Stelmakh. 2011. "Using Tax Incentives to Support Thermal Retrofits in Germany." Research report, Climate Policy Initiative, Berlin.

NRC (National Research Council). 2010. *Hidden Costs of Energy: Unpriced Consequences of Energy Production and Use*. Washington, DC: National Academies Press.

Oral, Isil, Indhira Santos, and Fan Zhang. 2012. "Climate Change Policies and Employment in Eastern Europe and Central Asia." Background paper, Europe and Central Asia Region, World Bank, Washington, DC.

RUC Law (Renmin University School of Law) and IPE (Institute of Public and Environmental Affairs). 2010. "Air Quality Information Transparency Index." RUC Law and IPE, Beijing. http://www.ipe.org.cn/upload /report-aqti-en.pdf.

Ruggeri Laderchi, Caterina, Anne Olivier, and Chris Trimble. 2012. "Balancing Act: Cutting Subsidies, Protecting Affordability and Investing in the Energy Sector in Eastern Europe and Central Asia." Energy study, Europe and Central Asia Region, World Bank, Washington, DC.

Smith, Kirk R., and Evan Haigler. 2008. "Co-Benefits of Climate Mitigation and Health Protection in Energy Systems: Scoping Methods." *Annual Review of Public Health* 29: 11–25.

Stavins, Robert N. 2009. "Green Jobs." Blog article for "An Economic View of the Environment," Harvard Kennedy School and Belfer Center for Science and International Affairs. http://www.robertstavinsblog.org/2009/03/07 /green-jobs/.

Strand, Jon, and Michael Toman. 2010. "'Green Stimulus,' Economic Recovery, and Long-Term Sustainable Development." Policy Research Working Paper 5163, World Bank, Washington, DC.

UNEP (United Nations Development Programme). 2011. "Towards a Green Economy: Pathways to Sustainable Development and Poverty Reduction." UNEP, Paris.

Van Vuuren, D. P., J. Cofala, H. E. Eerens, R. Oostenrijk, C. Heyes, Z. Klimont, M. G. J. den Elzen, and M. Amann. 2006. "Exploring the Ancillary Benefits of the Kyoto Protocol for Air Pollution in Europe." *Energy Policy* 34 (4): 444–60.

WEC (World Energy Council). 2010. "Measures Focused on Low Income Households: WEC-ADEME Case Studies on Energy Efficiency Measures and Policies." WEC, London.

Wier, Mette, Katja Birr-Pedersen, Henrik Klinge Jacobsen, and Jacob Klok. 2005. "Are CO_2 Taxes Regressive? Evidence from the Danish Experience." *Ecological Economics* 52 (2): 239–51.

World Bank. 2009. "Project Implementation Completion Report of the Renewable Energy Project in Turkey." World Bank, Washington, DC.

————. 2010. *Lights Out? The Outlook for Energy in Eastern Europe and the Former Soviet Union*. Washington, DC: World Bank.

Zhang, Fan. 2011. "Distributional Impact Analysis of Energy Price Reform in Turkey." Development Policy Research Working Paper, World Bank, Washington, DC.

————. 2013. "Energy Price Reform, Secondary Demand Effects and Household Fuel Substitution." Development Policy Research Working Paper, World Bank, Washington, DC.

Why Climate Action Is a Harder Sell in ECA

Relative to other regions, Europe and Central Asia (ECA) will be more affected by emission pricing (because of high energy intensities and high dependency on fossil fuels) and less affected by climate change impacts. Public support for climate action is therefore generally lower than in Western Europe. But locally, future impacts will be severe, as recent heat waves, droughts, and floods already indicate. Support for climate action is indeed higher in those ECA countries that are more affected by climate change. Better information dissemination and education could help to further shift public opinion in support of climate action.

Despite significant progress in some countries, climate action in the ECA region has lagged behind developments in many other regions.[1] Map S3.1 gives some indication of why decision makers and the public in ECA might be less supportive of climate action. It groups countries on two dimensions:

- *Source vulnerability* is determined by fossil-fuel resources; employment in the coal, oil, and gas sectors; renewable energy potential; and scope for carbon sequestration in soils and forests. The economies of countries highlighted at the bottom of the map matrix would be less affected by the imposition of a global price on carbon; the economies of countries in the top row would be most affected.

Contributed by Mame Fatou Diagne and Lourdes Rodriguez-Chamussy.

Vulnerability to Higher Carbon Prices and Effects of Global Warming

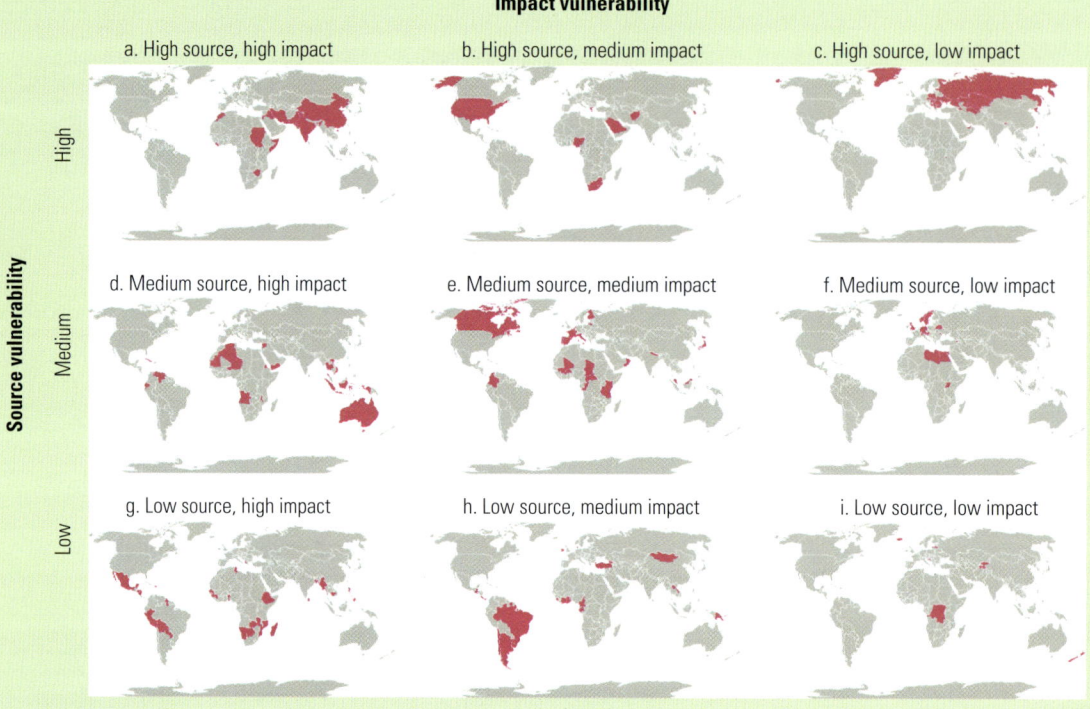

Source: Buys et al. 2009.
Note: "Source vulnerability" concerns levels of fossil-fuel resources, related employment, and renewable energy potential. High source vulnerability indicates greater likelihood to be affected by a global price on carbon. "Impact vulnerability" concerns potential exposure to effects of climate change on agriculture, extreme weather events, and sea-level rise. High impact vulnerability indicates greater likelihood of being threatened by climate change effects.

- *Impact vulnerability* indicates countries' likely exposure to the effects of climate change on agriculture, extreme natural hazard events, and sea-level rise. Countries on the left set of maps are more threatened, countries on the right, less.

Almost all of the ECA region's countries are clustered in the top right map. They are highly dependent on fossil fuels and, on average, have lower renewable energy potential than many other countries. Although predicted impacts are severe in some parts of the region, ECA countries are also generally less threatened by climate change impacts than are countries such as Bangladesh or Mexico. The region's reluctance to embrace ambitious climate action—as reflected in recent survey data reviewed below—is therefore not entirely irrational.

Perceptions and Attitudes toward Climate Change in ECA: From Concern to Action?

The climate change module of the 2010 Life in Transition Survey explores the perceptions and attitudes of Europe's[2] and Central

Asia's people with respect to climate change (EBRD and World Bank 2011). For comparison, it also contains information on five Western European countries (France, Germany, Great Britain, Italy, and Sweden). Climate change is a source of concern for most people in most countries, although there is variability: Higher-concern countries include Azerbaijan, Germany, Moldova, Slovenia, and Sweden. Lower-concern countries include Great Britain, Poland, and Russia. Although over 80 percent of Moldovans and 77 percent of Swedes agree that climate change is a serious problem currently facing the world, this opinion is shared by only 37 percent in Great Britain and 40 percent in Poland. When asked to rank the most important problems in the world today, respondents in ECA rank climate change in fifth position on average, after poverty, terrorism, infectious disease, and economic downturns, and before armed conflicts, nuclear weapons, or the increasing world population. This is in contrast with Germany or Sweden, where climate change comes second only to poverty as the most important problem facing the world.

People in ECA consider themselves to be less informed about the causes and consequences of climate change than their neighbors in Western Europe. But concern about climate change is not unfounded: it is related to genuine risks and can translate into policy under certain conditions. There is a positive relationship between the average degree of concern in ECA countries and their vulnerability to climate change, as measured by the vulnerability index created for the region's countries that captures each country's exposure, sensitivity, and adaptive capacity to climate change (figure S3.1). Concern about climate change is greater in countries with higher exposure and sensitivity to climate change.

Although people in ECA are no less concerned about climate change than their Western European neighbors, they are much less likely to have taken personal action: only 23 percent of respondents in the region say they have "personally taken actions aimed at helping to fight climate change," versus 60 percent in Western European countries. In fact, there is a positive correlation between a country's per capita income and the extent of reported climate change mitigating actions taken by the general population: EU-10 countries and Croatia show levels of action that are comparable to Western European countries, whereas much lower levels of action are reported in Central Asia and the Caucasus. In Armenia, Azerbaijan, and Georgia, only 3 percent of the population report taking personal action. The correlation is nevertheless imperfect, with Russia notably standing among the countries with the lowest proportion of people reporting taking action personally.

FIGURE S3.1

Concern about Climate Change among Populations of Europe and Central Asia Countries, 2010

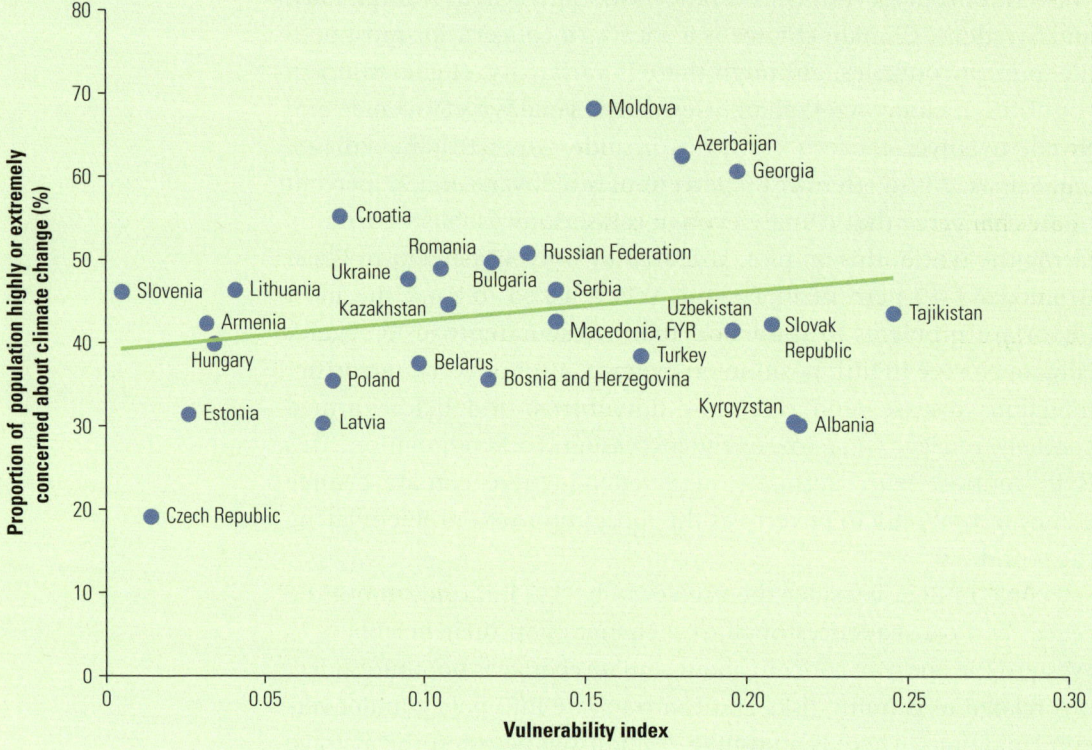

Source: Diagne and Rodriguez-Chamussy 2012 using the vulnerability index from Fay and Patel 2008.

At the individual level, taking personal actions to help fight climate change also correlates with measures of economic welfare such as assets or consumption per capita, and has no significant relationship with education, age, or gender. The most commonly taken specific actions in ECA are also cost saving (11 percent reduced water consumption, and 10 percent reduced energy consumption at home). The greater number of specific actions taken by Western Europeans indicate greater public coordination (for example, 45 percent started separating waste for recycling, versus only 8 percent in ECA); greater choice by richer consumers (for example, 13 percent in Western Europe purchased a more environmentally friendly car versus 1.9 percent in ECA); or better public services (with much larger numbers choosing public transportation or reducing the use of a car).

Another way in which people can take action against climate change is by demanding climate change mitigation through the political process. The proportion of people who support the environment as a first priority for government spending is relatively low in most

ECA countries, and most are not willing to pay extra taxes to fight climate change. There is, however, a positive relationship between individual information, citizen engagement, concern about climate change, and willingness to pay: controlling for economic characteristics and education, the probability that individuals may be willing to pay higher taxes or give part of their income to fight climate change is higher when they live in a country more vulnerable to climate change, feel well informed on the causes and consequences of climate change, and have high levels of concern.

The evidence thus suggests that translating concern about climate change or climate change vulnerability into mitigation action (through individual behaviors or collective action) hinges on agency and public coordination, the expansion of economic opportunities, and the availability of information at the individual level.

Notes

1. As reflected, for example, in the national communications to the UNFCCC and Copenhagen Summit commitments.
2. "Europe," in this instance, includes Eastern Europe and five Western European countries.

References

Buys, Piet, Uwe Deichmann, Craig Meisner, Thao Ton That, and David Wheeler. 2009. "Country Stakes in Climate Change Negotiations: Two Dimensions of Vulnerability." *Climate Policy* 9 (3): 288–305.

Diagne, Mame Fatou, and Lourdes Rodriguez-Chamussy. 2012. "Attitudes toward Climate Change in Europe and Central Asia." Background paper, World Bank, Washington, DC.

EBRD (European Bank for Reconstruction and Development) and World Bank. 2011. "Life in Transition: After the Crisis." Life in Transition Survey II report, EBRD, London; World Bank, Washington, DC. http://www.ebrd.com/pages/research/publications/special/transitionII.shtml.

Fay, Marianne, and Hrishi Patel. 2008. "A Simple Index of Vulnerability to Climate Change." Background paper, World Bank, Washington, DC.

PART III

Sectoral Priorities

There are two main ways to think about climate action. The first, favored by most economists, is to see climate change as a single, large externality problem. Markets for energy and natural resources determine price based on production and transport costs and supply and demand. But these prices do not currently reflect the impacts of fossil-fuel burning and natural resource degradation on the atmosphere and the resulting future damages from climate change. There is little that economists agree on, but almost all would say that the best solution to this problem is to put a price on greenhouse gas emissions that reflects these damages.[1] With a price on carbon emissions, firms and households will find the most efficient ways to reduce energy and resource use without the need for additional policies aimed at changing technology use or behavior.

In practice, with few exceptions, countries have not been willing to introduce such a carbon tax. One problem is the difficulty of determining the carbon price given uncertainties about the magnitude of damages: representative estimates range from US$33 to US$88 per ton of carbon dioxide (CO_2) and rising over time to reach a 450 parts per million (ppm) stabilization target (Aldy et al. 2010). Another problem is a lack of popular support for a carbon tax—perhaps because of a general aversion to taxes, opposition to the idea that firms and households should be allowed to just pay to pollute, and skepticism that such a scheme would actually reduce pollution. Instead, what has been implemented to a limited extent is cap-and-trade: a system with an overall emission limit and tradable pollution permits whose price is determined in a carbon market. The largest existing cap-and-trade system is the European Union's (EU) Emissions Trading System (ETS) for energy-intensive economic sectors. Whether such a system, which under certain conditions is equivalent to a carbon tax, will be effective remains to be seen. The EU ETS has helped reduce emissions in Europe but has recently seen carbon prices plummet because of an overallocation of permits. Carbon markets will play an important role in climate change mitigation in the future, but they will likely be confined to some countries and selected sectors.

A second approach to climate action, therefore, adopts a less efficient but more pragmatic perspective. It accepts that addressing climate change will require a large number of smaller steps—taxes, quotas, standards, incentives, and so on—that together will add up to significant reductions in emissions. There are two common ways to illustrate this approach:

- In so-called stabilization wedges, the gap between emissions in a business-as-usual scenario and one consistent with stabilization is broken down into smaller, manageable steps such as changing the power sector fuel mix or raising energy efficiency in buildings (Pacala and Socolow 2004). Each wedge would then require a separate set of policy instruments.

- A marginal abatement cost (MAC) curve (further discussed in chapter 2) evaluates numerous technical mitigation options and sorts them by cost of avoided emissions and cumulative emission reduction. Some actions make immediate economic sense. Some will be attractive only if there is a price on carbon emissions or an equivalent subsidy. And some are uneconomical even at any reasonable carbon price.

Stabilization wedges and MAC curves reflect the fact that climate action will involve numerous individual policies, some of which may not even be primarily motivated by climate objectives.

The five following chapters discuss priorities in the main sectors of the economy:

- A lower carbon *power* sector (chapter 6)

- More efficient industrial *production* (chapter 7)

- Lower emission *mobility* within cities and regionally (chapter 8)

- Less-energy-intensive buildings and public services in *cities* (chapter 9)

- Fewer emissions from land use on *farms* and in *forests* (chapter 10)

MAP 6.1

Power Plants in Europe and Central Asia, 2012

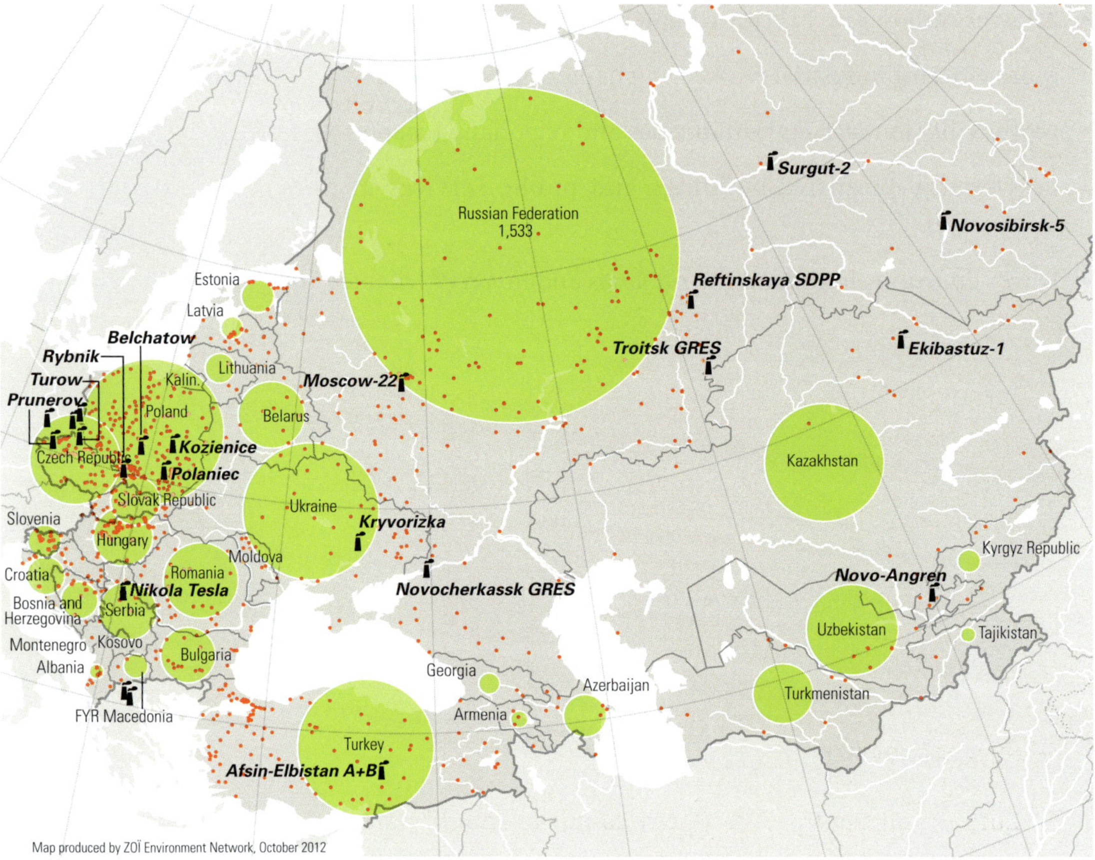

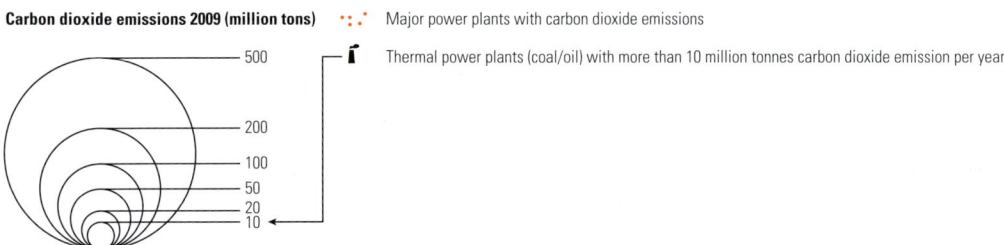

Russian Federation
1,533

Surgut-2

Novosibirsk-5

Reftinskaya SDPP

Troitsk GRES

Ekibastuz-1

Estonia

Latvia

Belchatow

Rybnik

Turow

Prunerov

Kalin.

Lithuania

Moscow-22

Poland

Belarus

Kazakhstan

Czech Republic

Kozienice

Polaniec

Slovak Republic

Ukraine

Kryvorizka

Kyrgyz Republic

Slovenia

Hungary

Moldova

Novocherkassk GRES

Novo-Angren

Croatia

Romania

Uzbekistan

Tajikistan

Nikola Tesla

Bosnia and
Herzegovina

Serbia

Montenegro

Kosovo

Albania

Bulgaria

Georgia

Azerbaijan

Turkmenistan

FYR Macedonia

Armenia

Turkey

Afsin-Elbistan A+B

Map produced by ZOÏ Environment Network, October 2012

Power

Carbon dioxide emissions 2009 (million tons)

500
200
100
50
20
10

Major power plants with carbon dioxide emissions

Thermal power plants (coal/oil) with more than 10 million tonnes carbon dioxide emission per year

Source: Carbon Monitoring for Action (CARMA), Center for Global Development, Washington D.C., United States (http://carma.org); IEA World Energy Outlook 2011 (http://www.worldenergyoutlook.org)

CHAPTER 6

Power

Main Messages

- Europe and Central Asia (ECA) countries should pursue all cleaner energy sources, but a successful low-carbon transition will likely rely mostly on natural gas in the short term and renewables in the medium to long term.

- Natural gas will be the main bridge fuel, but solely switching from coal to gas will still result in dangerous global warming later this century. In the short run, the impact of natural gas can be reduced by making pipeline transmission far more efficient. Emissions from some aging pipelines could add up to a fifth of total emissions from burning gas.

- Capturing carbon at coal, gas, or industrial plants and storing it underground could contribute to climate action if technical, economic, and social barriers can be overcome. But the uncertain prospect of carbon capture and storage (CCS) should not justify unchecked expansion of fossil-fuel use today. Further, although nuclear power has a significant share in the energy mix of several ECA countries, rising costs suggest that technological breakthroughs may be necessary to further increase nuclear power's share of electricity production.

- A long-term strategy for a low-carbon transition requires an increasing role for renewable energy sources. In high-income countries, investments will need to further reduce the costs of renewables so that they become competitive with fossil-fuel alternatives, and so that future renewable expansion will be market driven rather than policy induced. ECA countries should focus on locally competitive renewables and design sensible support strategies for a gradual and economically sustainable expansion as costs come down. All of the region's countries have some form of renewables support, but not all incentives are well designed.

The European Union (EU) target of a 20 percent renewable energy share by 2020 seems ambitious, but it is only a first step toward a low-carbon energy future. On the small Danish island of Bornholm in the Baltic Sea, a group of 16 firms and academic institutions, including the Tallinn University of Technology, is testing a prototype smart grid and electricity market that can handle a 50 percent share of renewables supplied by a large number of distributed producers.[2] To the south, in Germany, engineers are exploring how more than 90 percent of the country's electricity could be generated from renewables by 2050 (Umweltbundesamt 2010; Nitsch et al. 2012). Academic work suggests that, theoretically, all the world's energy needs could be supplied by wind, water, and solar power someday (Jacobson and Delucchi 2011). It is impossible to predict what the electricity system of the future will look like. However, while fossil fuels will contribute a significant share for several more decades, it is clear that the future system will look very different from the one that has sustained economic growth for the past 100 years.

Energy efficiency by itself will not lead to a sufficient fall in total energy consumption, and thus a sufficient fall in emissions, because reductions in energy intensity are offset by rising demand fueled by economic growth. Therefore, along with strong energy efficiency policies, it is also necessary to lower the emission intensity of energy consumption. This requires a switch to cleaner and eventually renewable energy sources. The Europe and Central Asia (ECA) region is currently not well positioned to pursue a clean energy path. It is highly dependent on fossil fuels, which provide 85 percent of the total energy supply. It has large fossil-fuel reserves that generate export earnings and have been intensively exploited: in 2010, the region accounted for 9.3 percent of proved oil reserves, 31.7 percent of proved natural gas reserves, and 34.8 percent of proved coal reserves (BP 2011).

At the same time, the electricity supply system in ECA needs massive investments to replace aging infrastructure and add new capacity. The scope of this task has been laid out in a recent report on the outlook for energy in the region (World Bank 2010b). An estimated 500–600 gigawatts (GW) of generating capacity will need to be built or replaced between 2010 and 2030 even if energy intensities continue to decrease. Including transmission and distribution improvements, this could require US$1.5 trillion in investments. Governments and investors will have to make decisions about long-lived investments in the face of great uncertainty, as described in chapter 3—regulatory uncertainty concerning future environmental and climate change regulation; technological uncertainty about the future price of different power generation technologies; and climate uncertainty as future temperature and rainfall patterns affect hydropower and thermal electricity generation.

The large-scale replacement and addition of power generation capacity in ECA provides an opportunity to reshape the region's electricity systems. Apart from standard investment criteria such as technical feasibility and economic viability, this process should be guided by the principles for dealing with the abovementioned uncertainties in power sector planning. Predictable climate policies will encourage cleaner, more flexible electricity generating systems that are more reliable in the face of environmental changes.

This chapter discusses the four main options for reducing emissions in ECA's electricity sector:

- The most immediate task is to diversify away from highly emission-intensive, coal-fired power to a greater share of *natural gas*, especially if inefficiencies in gas supply networks can be reduced.

- Where coal is expected to remain important, options for *carbon capture and storage* (CCS) should be explored.

- *Nuclear energy* provides a large share of zero-emission energy in several countries, some of which plan to expand it further.

- Finally, although the share of *renewable energy* in the region is still small, it will increase quite rapidly in the EU-member countries and eventually also in the rest of ECA.

As this chapter shows, none of these four major options provide an easy or quick solution.

Natural Gas[3]

Natural gas plays a large role in the energy sector of all ECA countries and will continue to do so in coming decades. The region contains two of the main gas production centers in the world—namely the Russian Federation (Siberia) and Central Asia (Kazakhstan, Turkmenistan, Uzbekistan)—and an extensive pipeline system connects these to all centers of consumption. Producing, processing, and transporting gas within this regional system generates substantial greenhouse gas (GHG) emissions. These emissions upstream from the point of use are of concern in that they are large in themselves and offset some of the advantage of gas over fuels like coal that emit more carbon dioxide (CO_2). However, there is believed to be considerable potential for reducing GHG emissions from the region's gas system, at relatively low cost.

History and geography have combined to make most of the gas-importing countries in the region extremely dependent upon the exports from within the region, particularly from Russia. The countries of Central and Eastern Europe obtained only 29 percent of their supplies from domestic production in 2010 and were dependent on imports from Russia and Central Asia for the rest of their needs. Turkey's history and geography have enabled it to have diverse sources of supply, complementing sources from ECA with large imports from North Africa and the Middle East. Gas is likely to maintain its key role in ECA's energy systems in coming decades for the following reasons:

- Rigidities of existing energy infrastructure

- Limited competition from coal and nuclear power

- Lack of alternative fuels in the domestic and residential sectors

- Slow penetration of renewables

- Environmental policies that favor gas use

In the past few years, natural gas from shale formations has been identified as a new and large potential source of supply, particularly in Central and Eastern Europe, and this could further entrench the use of gas in these areas while changing the patterns of supply.

Conventional Gas

The gas sector in ECA is a major contributor to the region's aggregate GHG emissions. Russia's GHG emissions from the gas sector in 2009 were 199 million tons of CO_2 equivalent (CO_2e) (excluding distribution), constituting about 11 percent of national emissions, according to the United Nations Framework Convention on Climate Change (UNFCCC). By comparison, the gas industry in North America accounts for around 3 percent of the continent's total GHG emissions. Two types of GHG emissions are of concern: methane[4] that is leaked or vented from gas systems and the CO_2 emitted by energy use for gas production, processing, and transport. Leakage and venting of methane is a concern because of its relatively high Global Warming Potential, which is 25 times greater than that of CO_2.[5] Use of gas to provide the energy for gas systems is mostly of concern where gas must be transported over long distances, either by pipeline or as liquefied natural gas (LNG).

The total GHG emissions from systems bringing natural gas from the field to the point of use are described using a life-cycle model. In this methodology, the corresponding GHG emissions are attributed to each element of the supply chain. The natural gas supply chain is shown in figure 6.1. Gas that is produced from gas wells is subjected to processing to bring it up to standard quality specifications for the main transmission lines. During processing, hydrocarbon fractions heavier than methane, and any nonmethane gases

FIGURE 6.1
Life-Cycle GHG Emissions from Natural Gas

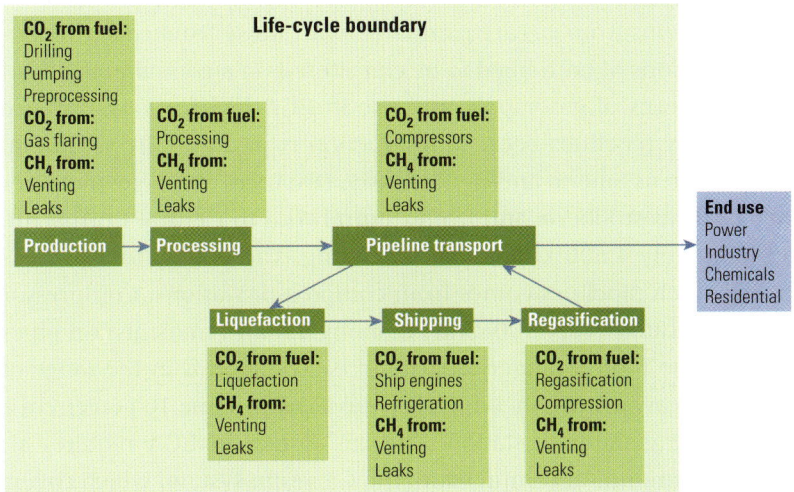

Source: Levitsky, Howorth, and Zhang 2012.
Note: CO_2 = carbon dioxide; CH_4 = methane.

(including CO_2) are separated. Gas entering the pipeline system will then be transported to the point of final use, either directly through pipelines or by a combination of pipeline and LNG, which is most often transported using special trucks or sea vessels. Large natural gas users, such as power stations, will mostly receive gas directly from a main high-pressure pipeline, while smaller users will obtain gas at lower pressure through a local gas distribution system.

Defining the life-cycle emissions of gas raises two methodological issues. First, the boundaries for the life cycle need to be fully defined. A life-cycle analysis could take the supply chain to include further activities upstream or downstream. For example, the natural gas life cycle shown in figure 6.1 could be extended back to include the GHG emissions arising from production of the steel used in the gas pipelines. Similarly, in some analyses, the life cycle is completed not at a point of delivery but when the gas is combusted to produce energy at the final point of use such as a power station or a domestic heater.

The second major methodological question is how to deal with joint products. During processing, natural gas liquids must be removed from the raw gas. The liquids are taken for separate use. The GHGs emitted in processing could be divided between the natural gas and the co-products, and this can be done according to either physical quantity (for example, weight or volume) or energy content. It could also be argued that because the gas liquids arise only because of the activity of gas production, all of the GHGs arising from them should be attributed to the pure natural gas leaving the processing plant.[6] An important aspect of using life-cycle emissions is that when comparing emissions from different products or processes, the scope of the life cycle must be consistent. For example, when the GHG emissions from gas are compared with those from coal, the life-cycle emissions of both need to be considered on a comparable basis.

The patterns of gas supply and use in ECA involve long-distance transport by pipelines because production is primarily in peripheral regions such as Siberia and Central Asia. Distances are large: pipelines from northwest Siberia and from Central Asia to Central Europe are approximately 4,600 kilometers (km) long. The energy to move gas through such pipelines comes from compressor stations. Compressor units are located every 100–200 km along a major transmission pipeline, and the typical power capacity of each station is in the range of 50–200 megawatts (MW), which is equivalent to some 10 percent of a large power generating station. The gas system in ECA requires an enormous amount of compression; the Gazprom system, which covers all of Russia, has 47 GW of installed compressor capacity, which is approximately equivalent to the power generation capacity of Ukraine. The compressors are powered by the natural gas coming through the pipeline and are thus major sources of CO_2. Compressors are also a source of leaks and vents because gas passing through them must go through many valves. Valves can leak gas because of imperfect seals ("fugitive emissions"). Maintenance of compressor stations may also require some gas to be vented, for safety reasons.

A further potential source of gas for parts of ECA is LNG. The only country in the region that imports LNG is Turkey, where it represented 22 percent of gas imports in 2011. Although other parts of ECA have not yet imported LNG directly, plans have been discussed for imports to Southeastern Europe, which has coastal locations suited to construction of an LNG receiving terminal. Life-cycle emissions from LNG are different from those for pipeline gas in that they must include the energy needed to cool gas at the liquefaction plant, to ship the LNG by tanker or by truck, and to regasify it at the receiving terminal. Some venting and leaks also occur in these processes. To these leaks, the emissions associated with gas production must be added, along with the emissions associated with transport of gas from the field to the LNG plant as well as during similar transportation after regasification.

To assess the amount of GHG emissions needed to deliver gas to Central and Eastern Europe, gas systems originating in Russia and Central Asia and gas supplies in the form of LNG from Algeria were simulated. Figure 6.2 shows the major sources of GHG emissions from these systems. As can be seen, the largest sources of GHGs come from the energy needed for operation rather than the venting and leakage of methane. In the case of pipeline transport, the dominant source is compressor fuel use. In the case of LNG, gas liquefaction is extremely energy intensive and contributes most of the emissions.

The efficiency of the pipeline system is an important determinant of emissions, especially when gas is transported over very long distance. For example, emissions for gas from Turkmenistan in Central Asia are much higher than those from Russia (even allowing for the additional distance involved). This is because the Russian emissions are based upon simulation of a new pipeline system from Yamal, while gas from Central Asia is assumed to travel through pipelines largely dating from the Soviet era, with very inefficient compressors.[7] The simulation of an existing Russian export pipeline shows emissions that are twice those from the new Yamal system.

A large proportion of production in ECA is used in the countries where it is produced. In 2010, 70 percent of total Russian and Central

FIGURE 6.2

Life-cycle GHG Emissions of Natural Gas Supply, as a Percentage of Field Production Volumes, in Europe and Central Asia

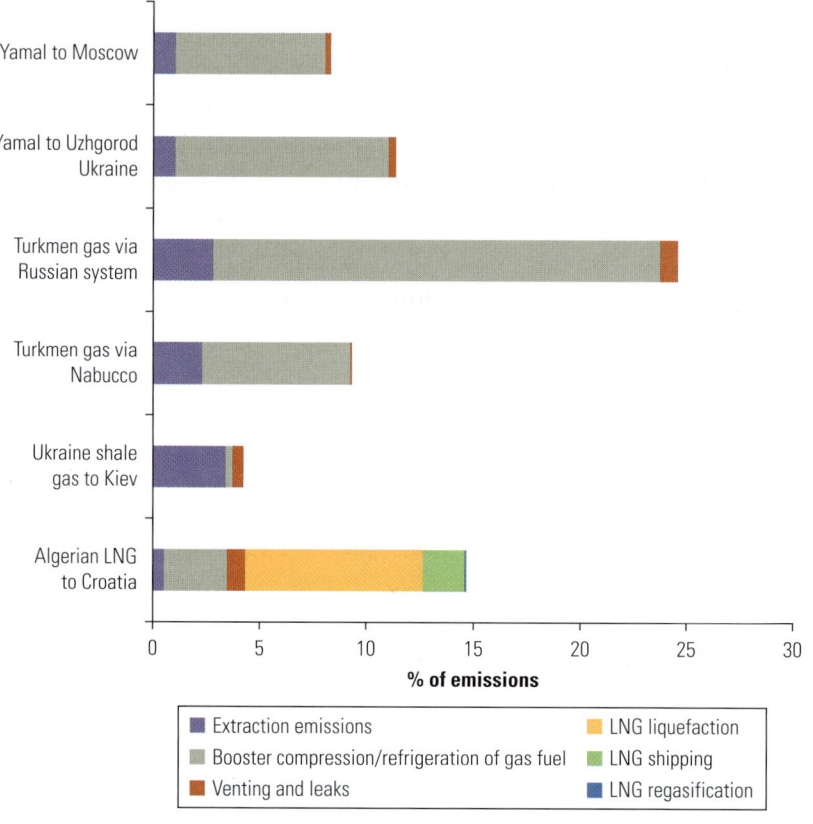

Source: Levitsky, Howorth, and Zhang 2012.

Note: Simulation shows the main sources of GHG emissions from the production, processing, and transmission of gas supplies originating in Russia and Central Asia and gas supplies in the form of LNG from Algeria. Assuming that 1 bcm in final use produces X tons of CO_2 equivalent of emissions and getting 1 bcm from field to use requires Y tons of CO_2 equivalent, then the percentage terms measured in the vertical axis equals Y/X. GHG = greenhouse gas; bcm = billion cubic meters; CO_2 = carbon dioxide; LNG = liquefied natural gas.

Asian production was used domestically. Given the pattern of energy consumption in Russia, much of the gas produced in Siberia will travel to central Russia. The length of a pipeline from Yamal to Moscow is 2,700 km, 40 percent shorter than the distance to Central and Eastern Europe. The life-cycle emissions for this route are correspondingly lower than for Russia's gas exports.

The life-cycle analysis described here considers gas up to the point where it is delivered from the high-pressure transmission line (the "city gate"). Significant GHG emissions can also arise from two further parts of the gas system. First, gas is often put into underground storage. This is done both to guarantee local security of supply and to allow gas supplies to consumers to adjust to meet changes in demand. Demand in ECA regions changes both seasonally and daily, and storage requirements are large relative to warmer regions, including most developing countries. Pumping gas into storage requires significant energy, and the whole process can generate leaks and venting. For example, in the case of the supply of gas to Moscow (as described above), operating storage could add around one-third to the total life-cycle emissions of gas.

Second, small users of gas, primarily in the residential and commercial sectors, receive gas through distribution systems. Unlike transmission, distribution uses little energy to transport gas but is more prone to leaks and to some operational venting because of the large number of pipeline connections involved. Estimates of the amount of gas that enters the distribution system but does not reach the end user vary greatly depending on the age and the quality of maintenance of the system. Although it is difficult to measure the overall supply balance of gas systems in practice, gas distribution systems are generally estimated to have loss rates ranging from 0.5 percent of total supplies, for the best system, to 2 percent or more for older systems that are less well maintained. Gas distribution systems in ECA are generally relatively old and may have had inadequate maintenance, and overall losses could be over 2 percent for many systems.[8]

Emissions from production, processing, and transport of natural gas produced and used within ECA could add between 8 percent and 25 percent to emissions from gas use. For LNG, this figure is 17 percent for gas from high-pressure systems, as used by power stations. Where gas is put through a low-pressure distribution system and storage is fully factored in, emissions upstream of final use could be around 18 percent of those from final consumption for Russian gas used in Central and Eastern Europe.[9] Although life-cycle GHG emissions add significantly to the emissions from gas in final use, they do not diminish the climate benefits of substituting gas for coal, the main alternative fossil fuel.

As a result of the size of GHG emissions from the ECA gas sector, there has been considerable interest in reducing them by upgrading gas infrastructure and improving operating practices. Three different types of interventions are available, as figure 6.3 illustrates. First, there are a number of relatively low-cost options to replace parts in compressor stations to reduce leaks considerably, of which the most significant is replacement of wet with dry seals on centrifugal compressors. Second, improved inspection and maintenance standards are also relatively low cost and can substantially reduce

FIGURE 6.3

Marginal Abatement Cost Curve for Emission Reductions from Natural Gas

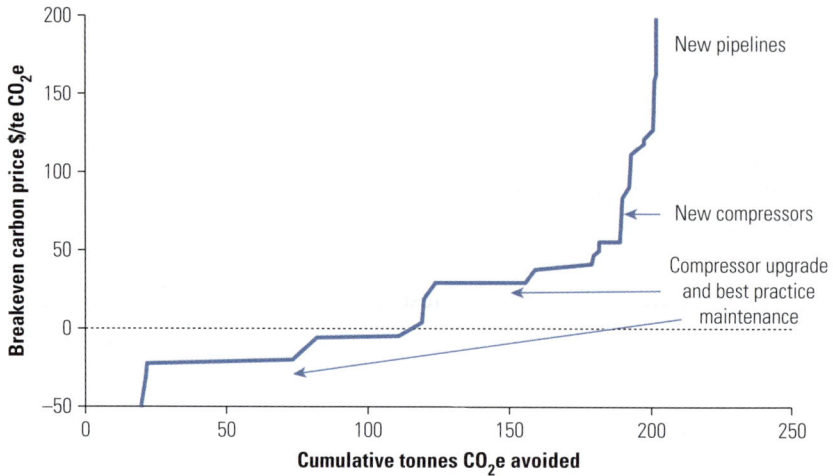

Source: Levitsky, Howorth, and Zhang 2012.
Note: A MAC curve is one way of illustrating the effect of technical emission-mitigation options on costs in terms of avoided emissions and cumulative emission reduction at a given break-even carbon price. CO$_2$e = carbon dioxide equivalent; t =tons; MAC = marginal abatement curve.

emissions. These two categories of low-cost interventions can reduce emissions by 20–30 percent. The third large category of emissions reductions involves replacing all old and outdated compressors. This action involves much larger investments, but it can reduce emissions by a further 30–40 percent in long-distance systems. Directed investments can thus reduce the emissions from large ECA transmission pipelines by some 50–70 percent. Given that these pipeline systems account for up to 10 percent of total GHG emissions in the region, such investments represent a large opportunity to curb emissions.

Shale Gas

Shale gas has been playing a rapidly growing role on the global gas stage, and parts of ECA could hold large resources. Shale gas is extracted from deep shale rocks using high-pressure injection of water and chemicals into wells to "fracture" the rock, which permits gas that is otherwise trapped within the rock to flow through the well to the surface. Over the past 10 years, shale gas has transformed the U.S. gas market from deficit to surplus; production has grown by a factor of 12 and now accounts for 25 percent of supplies, with this rapid growth expected to continue.

Many countries in ECA also contain shale formations that are likely capable of producing gas, but exploration is in the early stages and definite results will emerge from drilling over the next few years. According to an analysis of world resources by the U.S. Energy Information Administration (EIA 2011), Poland is one of the countries with the largest technically recoverable resources of shale gas, amounting to 5.3 trillion cubic meters (tcm), although this estimate has recently been questioned as too high.[10] If correct, this amount would be sufficient to cover 360 years of consumption at Poland's current rates. Ukraine also has major shale potential of about 1.2 tcm, equivalent to 23 years of domestic consumption. Successful development of these and other possible resources in Central and Eastern Europe would substantially change the pattern of regional gas production and trade flows.

Shale gas emissions are higher than conventional natural gas emissions at the preproduction phase because they involve more energy for drilling and fracturing and may also require more venting and flaring of gas during well completion. Water injected for fracturing comes back to the surface with gas in a solution. Current practices in the United States lead to this initial gas being vented or flared. Flaring converts the methane to CO_2, and this reduces the global warming impact of this phase of production by a factor of about nine because methane is a more damaging GHG in the short term. The volume of methane that needs to be flared is much higher than would usually arise in a conventional well. However, there is considerable controversy about how much more GHGs come from shale wells in the United States than from conventional wells. It has now been shown that best-practice operations for shale gas can lead to all or most of the early gas being collected, which greatly reduces emissions provided that regulations are well enforced (IEA 2012).

A further factor that affects the life-cycle GHG emissions from shale gas, when compared with conventional gas, is that shale wells produce gas at a very different rate over time than conventional wells. A conventional gas well's daily production may be constant for several years before declining by several percent per year. A shale well's daily production falls by 70 percent during the first year, after which it enters a slow decline.[11] For this reason, maintaining a given level of production from shale gas over a period of years requires drilling more wells than would be the case for conventional gas, and aggregate production emissions may also be correspondingly higher. For example, the number of wells needed to produce a given quantity of gas from Russia's highly productive wells will be a fraction of that needed for shale gas.

Shale gas production on a large scale from Poland and Ukraine would introduce a new element into the gas life cycle in these countries. The gas would supplement existing conventional gas production and would replace imports, mainly from Russia. The GHG emissions from transporting Russian gas to these markets would be avoided. To compare GHG emissions from supplying gas needs from local shale sources in Eastern Europe with emissions from Russian imports, a case was simulated in which 10 percent of Ukraine's imports from Russia were replaced by shale produced from the Dnieper-Donets Basin in Eastern Ukraine. It is assumed that shale gas production would use best practices to minimize GHG emissions from drilling.[12] Shale gas emissions were compared with emissions from the simulation for imports of Russian gas from Yamal. These estimates show that production and transport of local shale gas would produce approximately the same lifetime GHG emissions as gas imported from Russia through current pipeline systems. How best-practice shale production would compare with best-practice pipeline transportation is uncertain.

Shale gas production also has a much greater local environmental impact than conventional gas production. A range of factors are at work here: large demands for fresh water, the need to dispose of the wastewater after drilling, the larger footprint of well pads, the greater number of wells drilled, and the amount of traffic generated by all of these activities. In the case simulated for Ukraine to replace 10 percent of imports, about 260 wells would need to be drilled in the first year and 50 per year thereafter to maintain production.[13] The amount of fresh water used for fracturing would be around 200,000 cubic meters per day. In an area such as Ukraine, this water is easily available but is still significant; for example, it amounts to 23 percent of the daily flow rate of the Dnieper River, the largest in Ukraine. The wells drilled would directly cover an area of up to 145 square kilometers, and the activity would generate about 130 truck trips per day. This environmental impact needs to be carefully managed, with appropriate regulation and protection of local resources. Production of shale gas in Europe will benefit from the long and deep experience of the United States, and local environmental impacts can thus be contained to a lower level by application of best practices.

Natural Gas Outlook

Natural gas will be an important part of ECA's energy future for at least the next few decades and maybe beyond if CCS for natural gas becomes a realistic option. Modern, high-efficiency gas power plants are attractive economically at today's gas prices, although prices may of course rise in the future again. Small and flexible gas-fired units work well with intermittent renewable energy.[14] Efficient, small-scale combined heat and power (CHP) generation units fueled by natural gas already power some single-family homes in Western Europe with surplus power fed into the grid.

A "golden age of gas" may be possible (IEA 2011b). However, for climate goals to remain within reach, such an age would either need to involve CCS for natural gas or be relatively short lived. The International Energy Agency's (IEA) global natural gas scenario still leads to global warming of 3.5 degrees Celsius—beyond safe limits. The switch to a higher share of natural gas in ECA's energy mix therefore needs to be embedded in a broader low-carbon energy strategy that includes these elements:

- Reduce pipeline emissions by 50–70 percent to increase efficiency and realize climate benefits. Low-cost measures include replacement of wet seals with dry seals on centrifugal pipeline compressors and improved inspection and maintenance standards. A higher-cost option is to replace all or most parts of the compressors.

- Phase in natural gas power infrastructure gradually to maintain the option of switching to cleaner and more efficient technology as it becomes available or as gas prices rise in the future.

- Aggressively pursue energy efficiency gains, which also buy time or even avoid major capital investments.

- Ensure that gas systems are optimized to offset the intermittent nature of power from renewable sources, such as wind.

- Introduce some form of gradually increasing carbon charge as a natural check to fossil-fuel expansion. EU countries are already subject to such charges through the Emissions Trading System.

- Develop a sensible strategy for expanding the use of renewable energy that favors the most cost-competitive technologies, building on the continuing cost reductions achieved by research and development (R&D) and deployment growth in Western Europe and elsewhere.

Carbon Capture and Storage[15]

Within the European Union, Poland ranks third in total CO_2 emissions from large thermal power stations, with 139 million tons in 2009—well behind Germany and the United Kingdom, with 297 million and

171 million tons, respectively. However, Poland is also home to the largest CO_2-emitting power plant in Europe: PGE Elektrownia Bełchatów S.A. in the Łódź Voivodeship in south-central Poland. In 2009, the plant emitted almost 30 million tons of CO_2.[16] This is equivalent to about three-quarters of Poland's CO_2 emissions from the entire road transport sector (European Commission 2011).

The Bełchatów power plant is fueled by lignite (brown coal), which emits about 7 percent more CO_2 per unit of produced energy than hard coal and about 80 percent more than natural gas.[17] Coal is cheap and abundant in Poland, and it fuels well over 90 percent of electricity production. Coal mining employs more than 100,000 miners—although this is only about a quarter of the total employed in 1990 (Suwala 2010).

Although a large share of coal in the fuel mix is likely to persist for some time, to achieve national and EU climate goals, Poland, like several other ECA countries, will have to diversify its power mix and aim for a much larger share of cleaner and renewable energy. Poland could therefore benefit greatly from the successful development of carbon capture and storage technology. The Bełchatów plant will play a role in this process. In 2009, it was selected as one of six coal-fired power plants that received €180 million from the European Commission for CCS demonstration projects.

CCS involves capturing CO_2 released in burning fossil fuels before it reaches the atmosphere and permanently storing it underground or undersea. CCS is designed to operate on so-called point sources of emissions such as coal-fired power plants, where large amounts of CO_2 can be intercepted. Another requirement for CCS is a suitable CO_2 storage site, typically in stable geological formations such as depleted oil and gas fields or saline aquifers. Carbon dioxide capture can occur either postcombustion (after pulverized coal has been burned) or precombustion (in which carbon is separated out at an earlier stage). Despite its lower efficiency, most research currently focuses on post-combustion CO_2 capture because of the prospect of retrofitting existing coal power plants, most of which are pulverized coal plants and therefore use technology that would be suitable only for postcombustion capture.

The captured CO_2 is liquefied and transported by pipeline or tankers. Most scenarios envision a system of pipelines connecting large emission sources with suitable storage fields. CCS could also be applied to natural-gas power plants or large industrial emissions such as those from gas processing. More long-term scenarios envision CCS for biomass burning, enabling a net absorption of CO_2 from the atmosphere. This is known as bioenergy with CCS (BECCS). Other forms of capturing CO_2 directly from the atmosphere with subsequent sequestration are on the drawing board but remain a distant possibility.

Almost all parts of the CCS cycle have been applied in some form, so the technology is, in principle, well understood and proven. Large-scale applications that integrate all the processes involved in CCS have been scarce, however. At the moment, there are eight integrated commercial-scale CCS applications in Algeria, Canada, Norway, and the United States, although none is in power generation. Most applications are for enhanced oil recovery (EOR), in which gas is pumped into nearly depleted oil reservoirs to increase well pressure and produce more oil, improving the economics of CCS though not its contribution to climate action.[18] At the end of 2010, the Global CCS Institute identified a further 234 ongoing or planned projects across all sectors.

Despite the experience with many aspects of the technology, CCS faces a number of technical, economic, and social barriers before it can make a significant contribution to climate-change mitigation.

Technical Issues

As a mitigation strategy, CCS has a number of inherent inefficiencies. The carbon capture, transport, and storage process requires additional energy. Therefore, the amount of CO_2 produced and captured will always be significantly larger than the amount of emissions avoided. CCS will not be able to capture all emissions. Estimates are that by 2020, capture rates will be 68–87 percent—in exceptional cases, 95 percent but at increased cost.

However, the most challenging technical issue is safe storage. Determining the suitability of geological storage sites is a costly process. Because most of the operational projects to date have been for EOR, where the goal is to increase oil recovery rather than permanent storage, there has been relatively little monitoring. Even very small leakage rates can reverse the mitigation benefits.[19] More sudden, accidental releases would make climate benefits obsolete.

There are also concerns about health impacts if the release occurs near populated areas, although CO_2 generally dissipates quickly when released. These barriers and the time to develop the transport and storage infrastructure mean that CCS will not likely make a significant contribution to climate change mitigation until 2025 or 2030.

Economic Issues

Although electricity from renewable sources could end up costing the same as conventional sources, coal power with CCS will, by definition, always be more expensive than coal power without it. The IEA currently estimates that the cost of electricity with postcombustion CCS will be 63 percent higher. Additional capital costs include the CO_2 capture and transport equipment and the required larger plant capacity because the CCS process requires about 20 percent of the plant's output—though this may be reduced to 15 percent by 2015. Cost estimates will be unreliable until large-scale experience is available. However, because of this premium, CCS will always require some form of policy support such as a carbon tax or carbon finance.

The long period until large-scale deployment of CCS is likely also means that alternative clean power-generation technologies may catch up and become cost competitive before CCS is ready, especially if either resource prices or carbon charges rise significantly. For Western European countries with aggressive climate policies, some experts think that within 15 years, widespread use of renewables will make CCS in the power sector less relevant. However, this is unlikely to be the case in many ECA countries, let alone China or India, and CCS will still be desirable for industrial processes where CO_2 release is part of the manufacturing process.

Social Issues

As a result of concerns about the safety of storage, CCS has the attributes of other "not in my backyard" and liability issues such as high-voltage power lines or nuclear waste storage. Popular opposition has already derailed pilot projects in Western Europe. A study in Germany found that attitudes toward CCS varied greatly between the Ruhr area, where most of the carbon emissions originate (in favor), and potential storage sites in northern Germany (against) (Wuppertal Institute 2010).

Science needs to refine assessment methodologies, and companies that implement CCS must demonstrate the safety of storage—especially because current policies envisage that liability will transfer from the operator to the state after only 20 years.[20] Early information campaigns and stakeholder involvement will be essential for public acceptance. Furthermore, opposition could be reduced by allowing local communities to share in the benefits such as storage fees. Concern and impacts will be reduced by selecting storage sites in less densely populated areas or under the seabed as in existing Norwegian CCS projects.

CCS Outlook

Provided these barriers can be overcome, CCS can play a major role in climate change mitigation in the ECA region. However, countries vary in terms of the three main prerequisites for successful CCS implementation: climate change mitigation ambition, large point sources of emissions, and suitable storage sites (Kulichenko and Ereira 2011). The CCS outlook by country subgroup follows:

- *EU-member and candidate countries.* These countries have both the climate goals and the incentives to deploy CCS in principle. EU law already requires large coal-fired power plants to be ready for CO_2 capture, although this currently only involves setting aside land for capture equipment and demonstrating that retrofitting and transport to a suitable site are feasible (Graus et al. 2011). The EU also supports CCS using revenues from carbon trading and by funding pilot studies. A number of ECA's EU-member and candidate countries have explored CCS potential in more detail. Besides Poland, a recent World Bank study has investigated CCS storage options for the Southeastern Europe region; the study estimated transport and storage costs across a number of Balkan countries of about US$12 to US$18 per ton of CO_2 (Kulichenko and Ereira 2011).

- *Non-EU, non-Annex I countries.*[21] CCS has not yet been included as a mitigation tool under the Clean Development Mechanism (CDM), which supports emissions-reduction projects in developing countries. (See also box 6.3.) If CCS becomes eligible under CDM or another post-Kyoto Protocol market mechanism, countries such as Azerbaijan and Uzbekistan have CO_2 point sources and potential storage sites that could make CCS feasible. Azerbaijan, for instance, could use depleted oil or gas fields. However, little electricity is generated using coal in either country, so CCS would more likely be applied to natural-gas power plants or industrial processes. More study is needed to assess the suitability in other ECA countries that are eligible for funding through the CDM.

- *Non-EU, Annex I countries.* Russia, Kazakhstan, and Ukraine are the 5th-, 10th-, and 12th-largest coal producers in the world, respectively. Coal consequently plays a major role in electricity production in all three countries. In fact, Russia plans to build more coal power plants so it can export more natural gas. Under the current global climate regime, these countries have little incentive to aggressively reduce carbon emissions. The collapse of inefficient industries during transition lowered their emissions below 1990 levels, and they are not eligible for climate finance under the CDM. They are also not subject to EU regulations and are ineligible for EU financial incentives for CCS. With the exception of CCS for EOR, other emission reduction options may be more attractive for these countries.

This overview of CCS in the ECA region has two major implications. The first is that it would be wise for the region's countries to more systematically explore CCS as a climate change mitigation strategy and to develop the information base, regulatory framework, and institutional capacity to prepare for CCS. Currently, the region is lagging behind Western Europe in laying the groundwork for deployment once the technology is mature. All countries, not just the EU candidates, will likely face stricter emission limits in the future. Those that master the technology early could benefit from selling their expertise to others. Poland, for instance, is pursuing a number of CCS pilot studies. Besides Bełchatów, another utility is building a 288 MW coal-fired power plant equipped with precombustion CO_2 capture technology. In addition, the country is planning a "Zero Emission Power and Chemical Complex" in Silesia that will apply CCS to power and industrial processes. Poland has both the resources and incentives to become a world leader in CCS.

Second, however, the uncertain prospect of large-scale deployment of CCS technology should not distract from the urgent need to decarbonize energy systems. Over the next few decades, the cost of alternative energy generation may well match coal or gas with CCS or with a price on carbon, so investing in a large-scale, long-lived fossil-fuel power infrastructure that lasts many decades involves a risk. Perhaps most important, CCS does not address other environmental and health impacts from coal use. Besides CO_2, the Bełchatów plant, for instance, also emits significant amounts of mercury (1.6 tons in 2009), lead (316 kg), arsenic (258 kg), and sulphur oxides (50,700 tons). Safety issues in coal mining and the impact on natural landscapes—especially in open-pit or mountaintop removal mining—represent significant additional costs.

Nuclear Power

Nuclear power stations generate electricity without producing GHG emissions. Most climate change mitigation scenarios therefore expect nuclear power to contribute toward achieving climate goals while ensuring a sufficient supply of electricity. Under the IEA's new policies scenario, for instance, nuclear power's share of electricity production in Eastern Europe and Eurasia is expected to rise from about 10 percent to 13 percent by 2035 (IEA 2011a).

Eight ECA countries currently operate a total of 67 nuclear reactors, as shown in table 6.1 and map 6.2. They generated about 18 percent of Russia's electricity in 2011 and about half the electricity in the Slovak Republic and Ukraine. Twelve new reactors are under construction, and a further 45 are currently planned, including 13 in countries that do not yet or do not currently use nuclear power. (Lithuania closed its last nuclear reactor, which generated 70 percent of its electricity, at the end of 2009.) The total capacity of all ongoing and planned projects is 55 GW, about the same as current operating capacity. However, 60 percent of current plants are more than 25 years old, and 25 percent are older than 30 years. Therefore, many plants will need to be decommissioned in the coming decades.[22]

TABLE 6.1

Nuclear Power in Selected Europe and Central Asia Countries, 2011–12

Country	Nuclear electricity generation, 2011		Reactors operable			Reactors under construction		Reactors planned	
	billion kWh	Share of electricity (%)	No.	MW net	Median year	No.	MWe gross	No.	MWe gross
Armenia	2.4	33	1	376	1980	0	0	1	1,060
Bulgaria	15.3	33	2	1,906	1989	0	0	1	950
Czech Republic	26.7	33	6	3,764	1987	0	0	2	2,400
Hungary	14.7	43	4	1,880	1985	0	0	0	0
Kazakhstan	0	0	0	0	n.a.	0	0	2	600
Lithuania	0	0	0	0	n.a.	0	0	1	1,350
Poland	0	0	0	0	n.a.	0	0	6	6,000
Romania	10.8	19	2	1,310	2002	0	0	2	1,310
Russian Federation	162.0	18	33	24,164	1982	10	9,160	24	24,180
Slovak Republic	14.3	54	4	1,816	1992	2	880	0	0
Turkey	0	0	0	0	n.a.	0	0	4	4,800
Ukraine	84.9	47	15	13,168	1987	0	0	2	1,900
Total	**3,311**	**–**	**67**	**48,384**	**1986**	**12**	**10,040**	**45**	**44,550**

Source: World Nuclear Association (http://www.world-nuclear.org/info/reactors.html).

Note: Information as of April 2012, except energy production for 2011. "Median year" of "Reactors operable" refers to the year of reactor completion. kWh = kilowatt-hours; MW = megawatts; n.a. = not applicable; MWe = megawatts of electrical output.

MAP 6.2
Nuclear Power Plants in Europe and Central Asia, 2012

Map produced by ZOÏ Environment Network, June 2012

Nuclear power

Nuclear power plants 2012 (NPPs)

● Operational

◕ Reactor(s) under construction at existing NPP site

○ NPP under construction or planned

Installed capacity (after extension)

● > 3,000 MW

● > 1,000 MW

● < 1,000 MW

1 Ignalina: stopped, new station planned at the site after 2018

2 MAEK: stopped, new station will be constructed by 2016-2018

3 Bashkirskaya: construction stopped in the end of USSR, possible site for new station

4 Zarnowiec: construction stopped in 1990, possible site for new station

Source: Nuclear Power Stations of the World - GNV181, UNEP/DEWA/GRID-Geneva (http://www.grid.unep.ch)

Nuclear Power Issues

Although nuclear power will continue to contribute to the energy mix in several ECA countries, there are questions about the commercial feasibility of expansion plans and thus about the sector's contribution to climate action. The pros and cons of nuclear energy are well documented (for example, MIT 2009; Jacobson and Delucchi 2011). Apart from being a zero-emission-generation technology, nuclear power can be scaled up to provide very large amounts of power without consuming much land area. It uses relatively small quantities of uranium—an abundant fuel source. Kazakhstan is the world's largest producer, accounting for 33 percent of the global supply, and Russia, Uzbekistan, and Ukraine are also in the top ten.

On the other hand, nuclear power plants have raised concern about low-level radiation leaks with health risks that are hard to calculate (Beyea 2012). Major nuclear accidents in the Ukraine and more recently in Japan have demonstrated severe risks from catastrophic failures. Most nuclear countries have still not resolved the problem of safe storage of nuclear waste. Decommissioning of nuclear plants after their useful life span is complex, time consuming, and expensive. In addition, civilian nuclear operations pose the risk of proliferation for military purposes.

Large nuclear power plants may also not fit well into the power systems of the future—an important consideration because new nuclear plants will take many years to plan and construct and are expected to have a life span of half a century or more. Nuclear power plants currently provide continuous base load power, but in energy systems with a very high share of renewables, there will be a premium on flexible electricity sources that can be brought online or shut off quickly. Nuclear power plants take far longer than any other to shut down and restart, and to avoid disruptions during emergency shutdowns, they need very large backup capacity.

Nuclear Power Outlook

Although climate benefits have brought nuclear power much attention in recent years, the initial push for nuclear power more than 60 years ago was motivated mostly by military needs. Its deployment as a commercial power source was driven by presumed low costs. Reliable cost estimates for nuclear power are difficult to obtain, and those available are sometimes incomplete—for instance, omitting the full cost of nuclear waste storage or decommissioning. Estimates for the United States suggest that nuclear power is more expensive than electricity from coal or gas: market prices were about US$0.105 per kWh versus US$0.074 for coal and US$0.052 for gas in 2010 (Davis 2012).[23] It would require a CO_2 tax of more than US$35 per ton for nuclear power to compete with coal, and a tax of more than US$130 per ton to compete with gas. These cost estimates may also be low. Others who consider the crucial cost of capital, especially with construction time delays, find levelized costs as high as US$0.15–US$0.21 per kWh for nuclear power.[24] If damages from potential accidents were not implicitly covered by government guarantees—no private insurer can cover the risk of nuclear accidents—risk-adjusted insurance premiums would make nuclear power significantly more expensive.[25]

Furthermore, while operating costs of nuclear power plants have been relatively stable, construction costs have gone up, in part because of rising safety requirements. As documented for the United States and France—the two countries with the largest number of nuclear reactors—both construction times and costs have increased significantly over time (see figure 6.4 for France) (Davis 2012; Grübler 2010). Learning or experience curves show changes in costs as more and more generation units are installed. They are steeply downward sloping for most renewables (as the next section discusses); those for

FIGURE 6.4

Nuclear Reactor Construction Costs in France in Relation to Cumulative Capacity, 1977–99

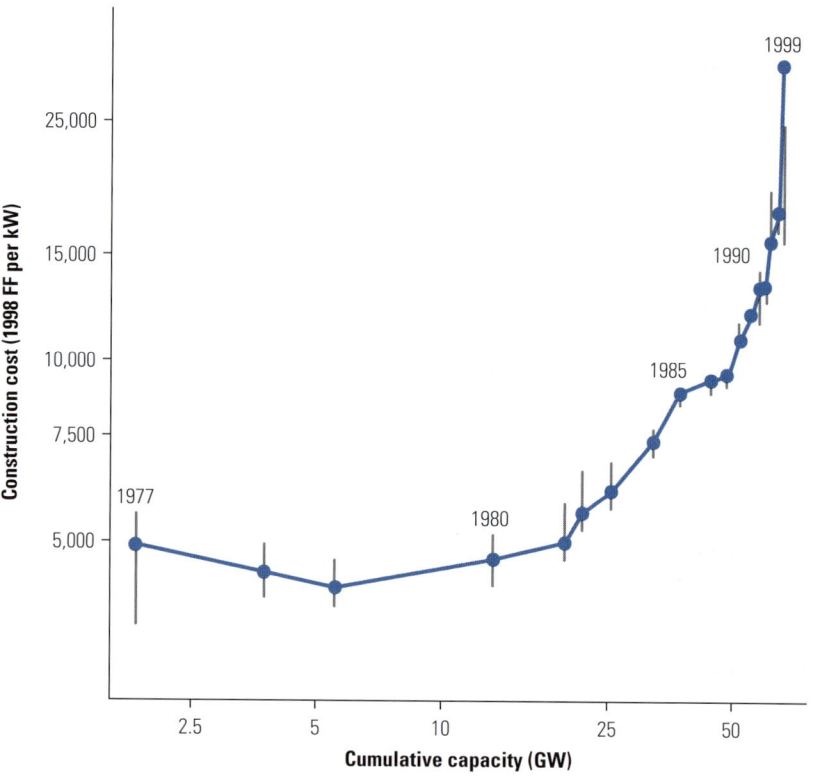

Source: Grübler 2010, table A12.
Note: The gray vertical lines indicate the min/max range of estimated construction costs. A total of 58 reactors were completed in this period. FF = French francs (1998); kW = kilowatt; GW = gigawatts.

nuclear power rise. Barring technical advances—such as break-throughs from China's ambitious nuclear program or from research into small, safe, and less expensive nuclear reactor units—recent cost trends suggest that nuclear power plants will not be able to compete in unregulated markets without significant government involvement. Countries may support nuclear power for noneconomic reasons such as energy security. However, combined with the short-term opportunities of a switch to natural gas and the medium- to long-term prospects for renewables and possibly CCS, nuclear power's role in climate change mitigation may not significantly increase.

Renewable Energy[26]

With the prospects for CCS and nuclear power uncertain and burning natural gas (without CCS) still releasing too much CO_2, a low-carbon energy future has to involve much larger use of renewable energy resources. As with the other cleaner-energy technologies, renewables are no panacea. How much and how quickly renewables can make a significant contribution to a country's energy mix depends most of all on technology costs and natural endowments. Recent years have seen significant cost reductions as the amount of energy produced through renewables expanded rapidly, especially in wind, solar, and some types of bioenergy. Their local competitiveness also depends on the costs of fossil-fuel alternatives, which may be artificially low because of subsidies or high because of energy or carbon taxes.

Resource capacity in most ECA countries appears sufficient for supplying all or most electricity needs with renewables, although currently not at any reasonable cost (for example, see Buys et al. 2009). Starting from generally small shares, a realistic medium-term goal for ECA would be the EU's 20 percent target, which some countries could achieve earlier than others. Five of the region's countries already exceed this level.

This section briefly reviews the current status of renewable energy, resource endowments in ECA, and technology costs. Some renewable energy options are already cost competitive, such as hydro, wind, or bioenergy, especially in places with high overall energy prices. Others require support. The section concludes with a discussion of the design of support policies, particularly feed-in tariffs. The most relevant types of renewable energy for power production are hydro, wind, solar, and geothermal. Some types of bioenergy also generate electricity, although it mostly produces biofuels for transport or heating. Bioenergy is discussed in more detail in chapter 10.

Renewable Energy Production and Renewable Resources

Globally, renewable energy investment trends appear to be going in the right direction. In just eight years, investments shot up: in 2011, US$263 billion was invested in renewables—up from US$33.7 billion in 2004. Of these investments, 95 percent were in Group of 20 (G-20) countries. Turkey accounted for US$1.2 billion of the investments, of which three-quarters were in wind energy (Pew Charitable Trusts 2012).[27] Over the 10 years from 2010 to 2020, renewables could be a US$2.3 trillion market if climate policies are strengthened—or US$1.7 trillion under current policies. With 83.5 GW in capacity added in 2011, globally installed renewable capacity is now at 565 GW. Although this is 50 percent more than installed nuclear power capacity, renewable generation will actually be less than its installed capacity figures suggest because renewables such as wind and solar are intermittent.

ECA falls well below the world average in renewable energy generation. In 2009, renewable resources contributed 4.7 percent of total primary energy supply in the region, or 649 million tons of oil equivalent compared with 10.4 percent in the EU-15 or 13 percent worldwide.[28] Hydropower and bioenergy each provide 2–3 percent of total energy supply in ECA, with very small fractions (well below 1 percent) coming from wind, solar, and other advanced technologies.

The distribution of renewable generation varies significantly among the region's countries, as shown in figure 6.5. Hydropower is the dominant renewable source in Albania, Georgia, the Kyrgyz Republic, and Tajikistan, accounting for 20–60 percent of total primary energy supply in 2009. The Baltic countries have a high share of biomass production, with 30 percent in Latvia, 15 percent in Estonia, and 10 percent in Lithuania. Turkey and Georgia lead the region in the use of geothermal resources, ranking 10th and 11th globally, respectively. Wind, solar, and other more-advanced renewables are still almost unnoticeable in ECA's energy mix, although they increased significantly in Turkey and the EU-10 countries over the past decade, as figure 6.6 illustrates.[29]

The potential to generate power from wind and solar sources is unevenly distributed across the region, as map 6.3 shows. Among ECA countries, Estonia, Latvia, Lithuania, and areas of the Czech Republic and Romania have higher potential for producing electricity from onshore wind power, while Southeastern Europe and Turkey have good conditions for solar. Russia, the largest country in the world, has enormous technical potential for bioenergy, wind, solar, and geothermal. At 170 terawatt-hours (TWh), it is also the

FIGURE 6.5

Shares of Renewable Energy in Total Primary Energy Supply, Europe and Central Asia Countries, 2009

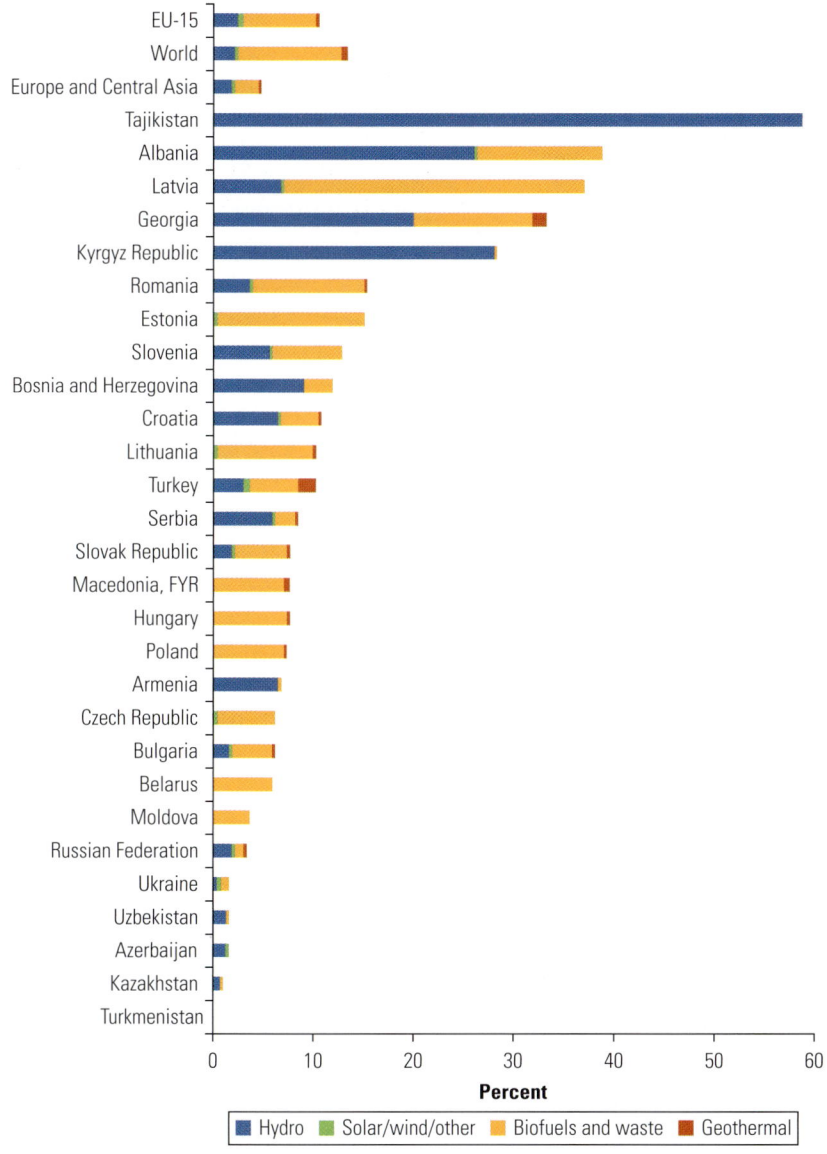

Source: Calculations based on IEA 2011c.
Note: EU = European Union. The EU-15 countries include Austria, Belgium, Denmark, Finland, France, Germany, Greece, Ireland, Italy, Luxembourg, the Netherlands, Portugal, Spain, Sweden, and the United Kingdom.

fifth-largest hydropower producer in the world, using 18 percent of its potential. (See also box 6.1 on hydropower.)

Identifying the geographic distribution of these resources is critically important for the profitability of investments in renewables because choosing the best location strongly affects how much

FIGURE 6.6

Total Energy Generation from Solar and Wind, Europe and Central Asia, Selected Regions, 2000–09

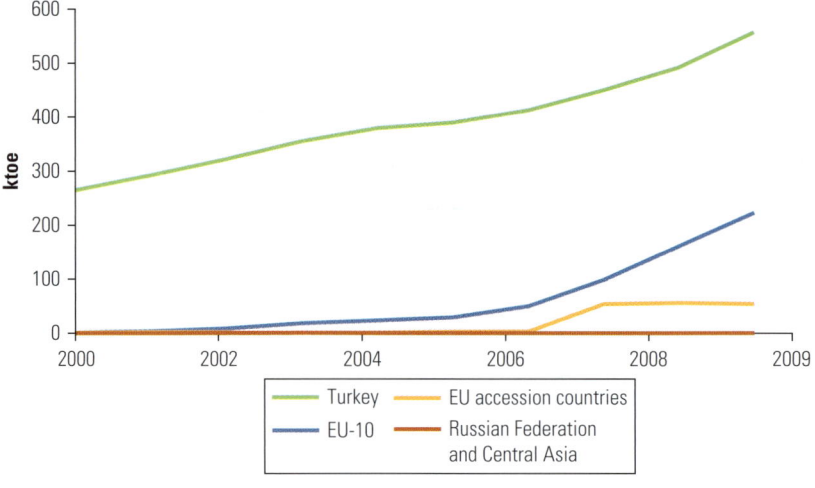

Source: IEA 2011c.
Note: The EU-10 countries include Bulgaria, the Czech Republic, Estonia, Hungary, Latvia, Lithuania, Poland, Romania, the Slovak Republic, and Slovenia. ktoe = kilotons of oil equivalent; EU = European Union.

energy can be generated by an intermittent resource—also known as capacity utilization. Boosting capacity utilization by just a few percentage points can influence rates of return as much as a carbon credit or a significant change in construction costs (World Bank 2010a). Some ECA countries have had renewable resource assessments, typically resulting in wind or solar atlases. A study for the Commonwealth of Independent States (CIS) countries used data from 3,600 surface meteorological stations and 150 upper air stations to generate a detailed wind resources map. It estimates technical potential for wind energy at 14,000 TWh per year (Nikolaev et al. 2010).[30]

Costs of Renewables

Renewables, especially more advanced technologies, are generally still more expensive than fossil fuels. Energy cost comparisons are notoriously unreliable because local construction and fuel costs vary greatly, as do the methods for estimating full life-cycle costs—especially for large, capital-intensive plants. Incorporating the damages to health and the environment that were avoided through the use of renewables changes relative costs. A technology's suitability for specific tasks such as covering peak demand also changes

MAP 6.3

Renewable Energy Resources in Europe and Central Asia, 2012

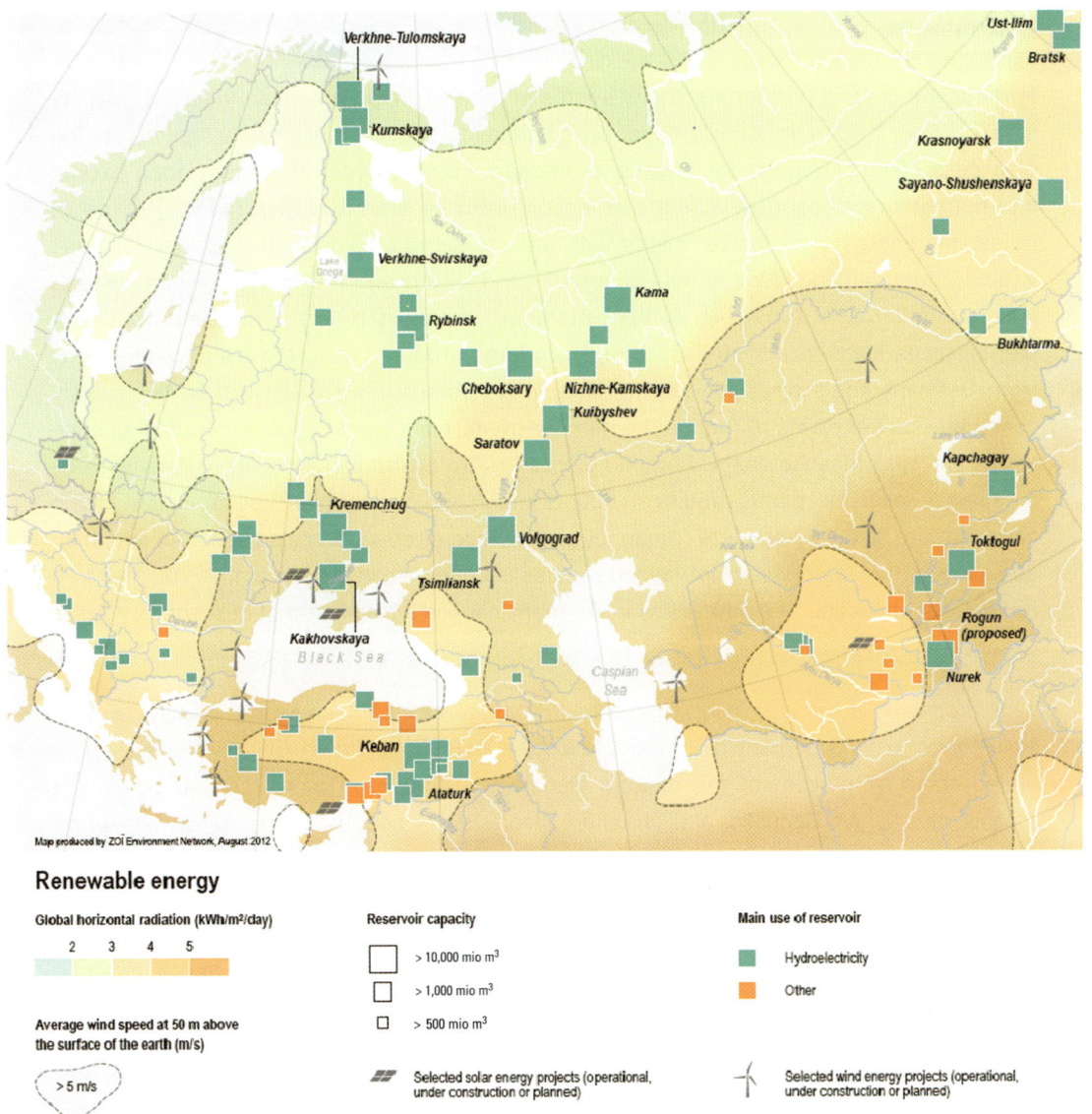

Renewable energy

Global horizontal radiation (kWh/m²/day)

2   3   4   5

Average wind speed at 50 m above the surface of the earth (m/s)

> 5 m/s

Reservoir capacity

☐ > 10,000 mio m³

☐ > 1,000 mio m³

☐ > 500 mio m³

▦ Selected solar energy projects (operational, under construction or planned)

Main use of reservoir

🟩 Hydroelectricity

🟧 Other

⟟ Selected wind energy projects (operational, under construction or planned)

Source: NASA Surface meteorology and Solar Energy (SSE) Release 5 Data Set (Jan. 2005) [windspeed; edited] and Release 6.0 Data Set (Jan 2008) [insolation] (http://eosweb.larc.nasa.gov/sse); Global Reservoir and Dam (GRanD) database, Version 1.1 (http://www.gwsp.org/85.html and http://sedac.ciesin.columbia.edu/pfs/grand.html).

competitiveness. A large share of renewables will also usually require complementary investments for grid integration of new power generation units and extension of transmission lines if the best renewable resource locations are far away from major demand, which raises their costs. Finally, much depends on the overall price of electricity. Where fossil fuels are unsubsidized or

BOX 6.1

Hydropower

Hydropower is an important electricity source in the ECA region, with an installed capacity of 107 GW, or 21 percent of total capacity. It is also important for load balancing: the storing of excess electrical power for release as needed. Further, hydropower is a good backup intermittent power source—a role that becomes more critical with a growing share of renewable energy. Hydropower plants with storage serve as giant batteries that "charge" when system demand is lower than generation and feed power into the grid when demand is high. The 716 MW Zarnowiec pumped-storage hydro plant in Northern Poland, for example, balances loads in a system that receives large amounts of intermittent wind energy from Germany and increasingly from Polish wind turbines. It also allows thermal—mostly coal-fired—power plants to reduce the number of start-ups and shut-downs, reducing operations and maintenance costs. Pumped-storage hydro plants can start up within five minutes, while fossil-fuel plants require an hour or more, and nuclear plants several days. Hydro can therefore also help stabilize frequency and voltage after sudden load shifts. In less than 30 seconds, the Zarnowiec plant helped reestablish grid stability in November 2006 when a routine disconnection of a power line in Germany caused a blackout that cascaded as far as Greece and Portugal.

Hydropower is the dominant source of low- or zero-carbon renewable energy, responsible for almost 96 percent of renewable power in ECA, as figure B6.1.1 shows. It generated about 17 percent of the region's electricity in 2009 and more than 90 percent in Albania, the Kyrgyz Republic, and Tajikistan. Nearly every country in the region has some hydropower capacity. However, the distribution is uneven. Russia and Turkey account for almost 60 percent of generation capacity, as shown in figure B6.1.2. There is large potential for hydropower energy expansion. The overall technical potential is estimated at more than 2,622 TWh per year in ECA, only 12 percent of which has been exploited so far, as shown in figure B6.1.3.

For all their benefits, hydropower projects can face environmental, social, and economic challenges and risks.

Large hydropower projects can affect aquatic habitats and threaten ecological balance. Even smaller run-of-river projects—designed to minimize the need for a reservoir—could cause stream-flow variations. When multiple projects are built along the same site, the cumulative impacts can be significant. Building access to a dam area imposes environmental impacts that are both direct (for example, on vegetation) and indirect (for example, air pollution). From the

continued

BOX 6.1 *continued*

FIGURE B6.1.1

Hydropower vs. Total Renewable Power Capacity in Europe and Central Asia, 1990–2010

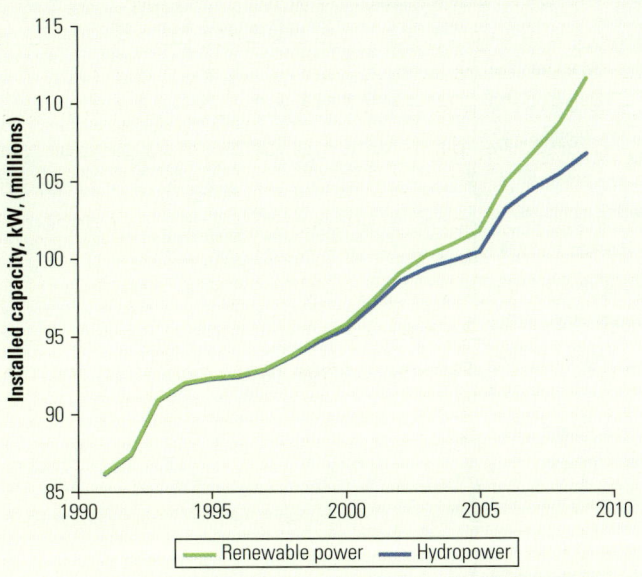

Source: EIA n.d.
Note: kW = kilowatt.

FIGURE B6.1.2

Distribution of Hydropower Capacity, by Subregion, in Europe and Central Asia, 2009

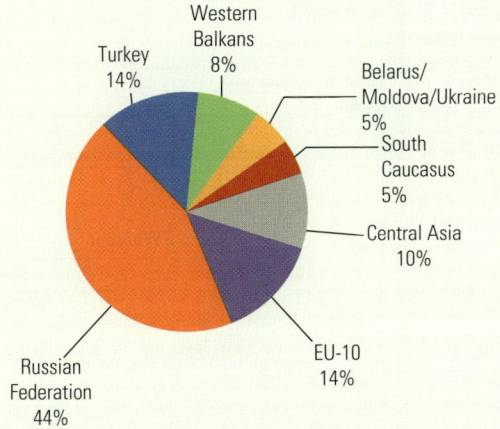

Source: EIA n.d.
Note: EU-10 = Bulgaria, the Czech Republic, Estonia, Hungary, Latvia, Lithuania, Poland, Romania. the Slovak Republic, and Slovenia.

continued

BOX 6.1 *continued*

FIGURE B6.1.3
Hydropower Development Potential and Production, Europe and Central Asia Countries, 2009

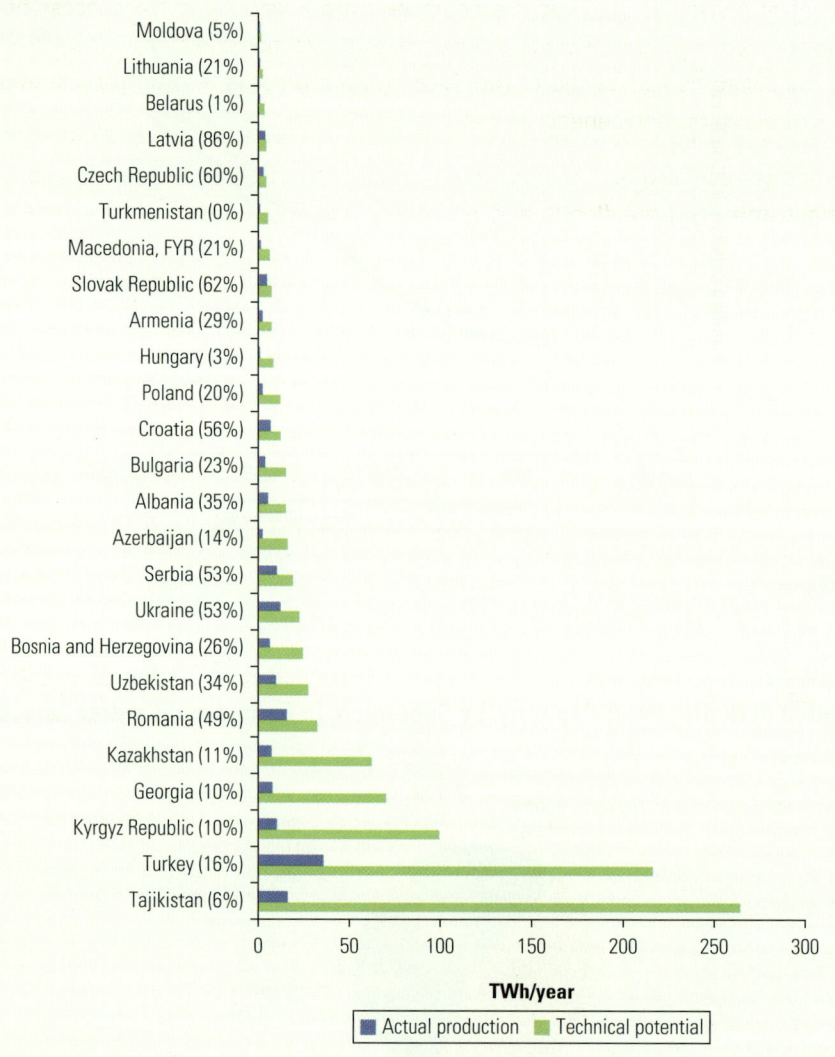

TWh/year

■ Actual production   ■ Technical potential

Source: WEC 2010; EIA n.d.
Note: The percentage number indicates the ratio of production to potential. Production bar shows 2009 actual generation. Russia's production (162 TWh) and potential (1,670 TWh) are not shown. TWh = terawatt-hours.

climate change perspective, hydroelectric plants can produce CO_2 and methane from the decomposition of biomass in the reservoirs—although this is a minor issue in ECA. The construction of dams can also change GHG fluxes because of land use change. A comprehensive accounting and assessment of a dam's contribution to GHGs should take these effects into account.

continued

BOX 6.1 *continued*

Hydropower development can encourage rural development by creating jobs and income by enabling rural electrification and by providing flood control and irrigation services. On the other hand, hydropower development often requires resettlement, which results in additional social and livelihood consequences. Public acceptance plays a key role in the success or failure of such a project. Local citizens, including those stakeholders most affected by the project, need to be fully consulted as part of the project development process. Project objectives should be clearly communicated and beneficial for all parties involved.

Hydropower can also be a tool for climate adaptation. Dammed reservoirs provide a buffer against impacts from floods and droughts. However, climate change and its consequences for water resources also increase the uncertainty over hydropower generation. Variability in rainfall has already had significant impacts on the stability of electricity supply in Southeastern Europe and Central Asia. Understanding the impacts of potential changes in climate and hydrology is essential for project design and for the operation of new and existing power facilities. Furthermore, hydropower development involves competition between energy and food production for water and land use as well as benefits sharing between downstream and upstream communities. This has sometimes created conflicts in the development of new dams. It is important to develop integrated land and water management strategies to ensure optimal allocation of land and water, as well as mechanisms to ensure shared benefits.

These potential environmental, social, and economic costs need to be carefully assessed and managed to develop hydropower in a sustainable manner. In the short run, ECA should focus on rehabilitating aging plants. Rehabilitation is a cost-effective way of expanding capacity with minimal risks.

Sources: IEA 2010; Eurelectric 2011; World Bank 2009.

---

subject to environmental taxes, renewables are more competitive. In Turkey, for example, a tight supply-demand balance results in high wholesale electricity prices (around US$0.10 per kWh). All of this implies that simple levelized cost comparisons are incomplete descriptions of the relative competitiveness of different generation options, as box 6.2 discusses further.

Figure 6.7 shows a comprehensive recent cost comparison from the Intergovernmental Panel on Climate Change. Renewable power from wind and small hydro is often already cost competitive in the wholesale market. Although many types of renewable energy can be cost competitive under good conditions—especially if co-benefits

BOX 6.2

Comparing Costs of Electricity Generation Technologies

Cost assessments of different electricity generation technologies usually compare their level-ized costs—or the price for energy from a given source after all of the related costs for it have been factored in (see, for example, figure 6.7). These are the total life-cycle costs per unit of electricity output, including capital and operating costs. This comparison works well for "dis-patchable" base load power plants (typically coal or nuclear, sometimes natural gas) that can produce power around the clock. However, renewable generating technologies such as wind or solar are "intermittent": they can only feed power into the grid when the resource (wind or sun-shine) is available. They can be highly variable, although predicting generation potential hours or even a day in advance is steadily becoming more accurate.

Levelized costs do not take into account the fact that in many electricity markets, the price of electricity varies over the course of a year and even during a single day. The difference between off-peak and peak electricity prices can be several orders of magnitude because demand varies strongly, the capacity to store electricity is still very limited, and power systems always need to balance supply and demand exactly. A generating technology that produces power at peak times will thus have more favorable economics than one available only during off-peak times. For instance, a dispatchable generator and an intermittent generator may have identical level-ized costs, but if the intermittent resource mostly produces during peak hours when prices are high, it can be more profitable than the dispatchable technology. If it mostly produces off-peak, it would be less competitive. Levelized cost comparisons therefore often undervalue solar power (which is available during the day when prices are high) and sometimes overvalue wind (if it is more available during off-peak times). Competitive power markets—and policy makers considering support for renewables—need to evaluate the economics of all power generation technologies based on the expected market value of supplied electricity and their life-cycle costs, which together determine profitability.

Sources: Joskow 2011; Borggrefe and Neuhoff 2011.

are considered—energy from fossil fuels net of their social costs is generally cheaper.

Costs tend to fall with installed capacity as learning and standard-ization help improve efficiency and refine technology. In the past 15 years, wind generation costs have fallen by more than 50 percent. This is mostly related to learning associated with increased market experience (Neij 2008). Figure 6.8 shows how wind and solar technologies have consistently fallen in price with the expansion of production—in contrast to the trends in nuclear power plant

FIGURE 6.7

Levelized Cost of Energy, by Source, 2011

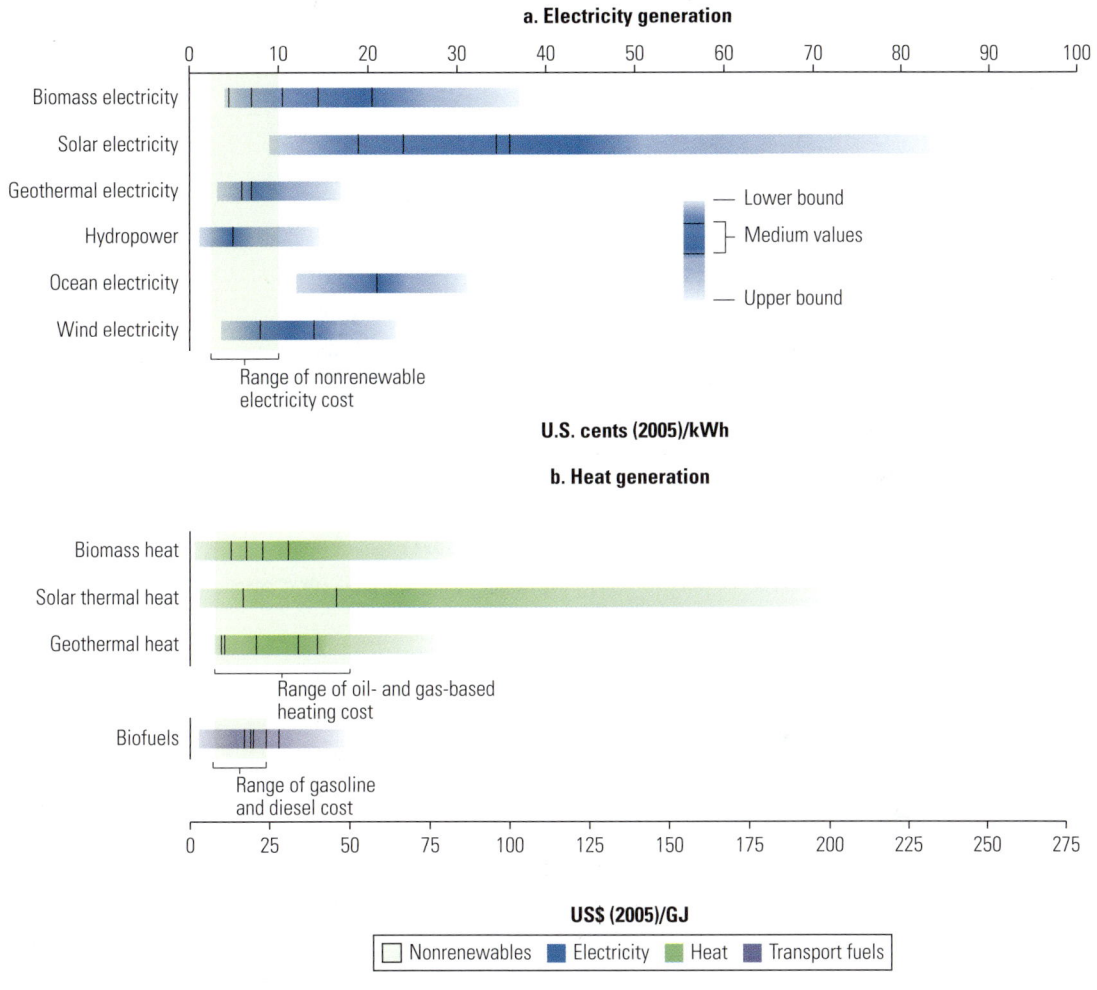

Source: IPCC 2011.
Note: kWh = kilowatt-hour; GJ = gigajoule.

construction discussed in the previous section. In many cases, wind is already competitive with fossil fuels. Solar photovoltaics (PV) could match the cost of other energy sources in high-energy-price countries within the next few years. Further cost reductions for renewables are likely but difficult to predict. The recent rapid decline in the cost of solar PV shown in figure 6.7, for instance, is due to massive expansion of production capacity in China. The cost of silicon, a key input for producing solar panels, has fallen 93 percent—from US$475 per kilogram three years ago, when there were supply shortages, to US$33 per kilogram now.

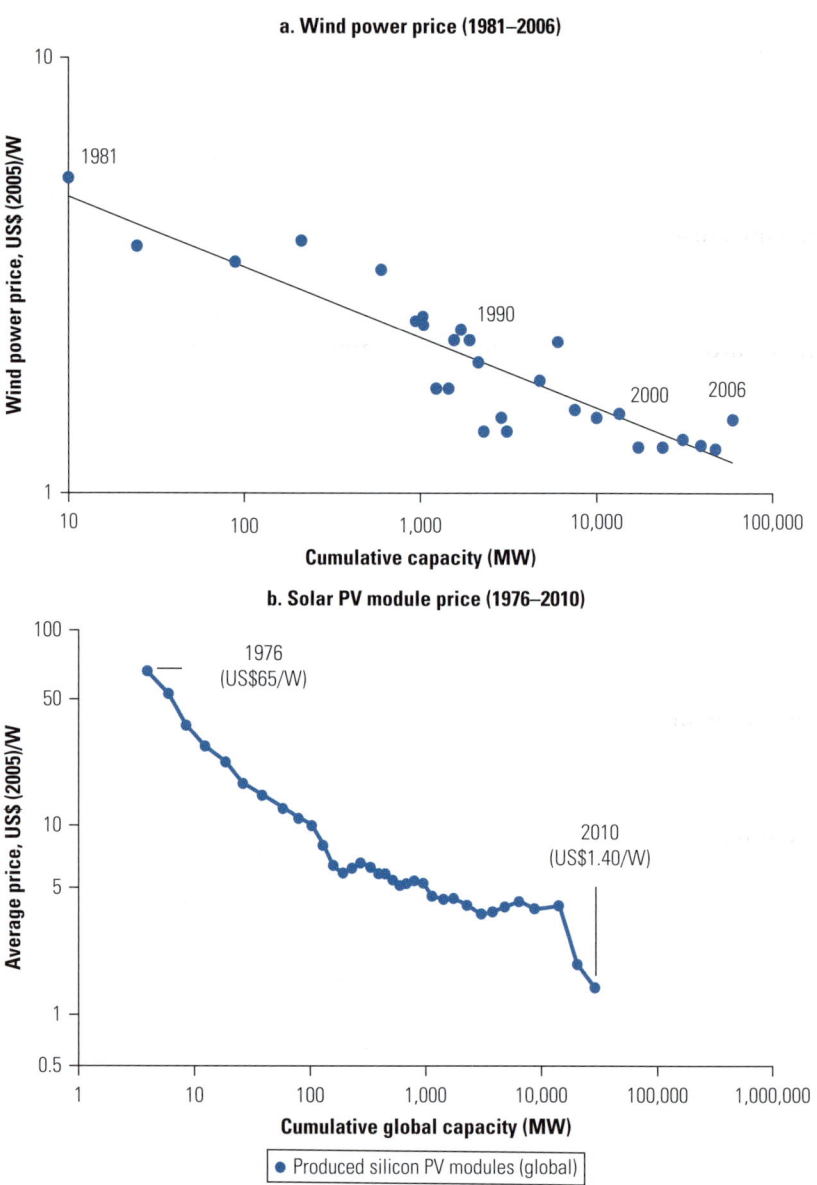

FIGURE 6.8

Learning Curves of Renewable Technologies in Relation to Price

a. Wind power price (1981–2006)

b. Solar PV module price (1976–2010)

Source: IPCC 2011.
Note: W= watt; MW = megawatts; PV = photovoltaic.

Renewables Incentive Policies

The expansion of renewable power that drove these cost reductions has so far predominantly been policy induced rather than market driven. Fairly generous support mechanisms in Western Europe and parts of the United States created a market and thus incentives for research and development and for scaling up production. The implicit or explicit subsidies either come out of government budgets or are collected by utilities that charge all customers extra to cover the cost of renewable energy. These policies for promoting renewable energy have raised some controversy in high-income countries but are generally accepted because the added cost represents a small share of household expenditures.[31] The question is whether they are also justified in poor or middle-income countries. In other words, should households with far lower incomes than those in Western Europe, for instance, pay extra for energy to reduce global emissions and to bring technology costs down?

Various incentive programs promote renewable energy deployment around the world. These programs include direct subsidies or tax incentives; mandates that a certain share of energy production must be renewable (so-called renewable portfolio standards); and competitive bidding for government-sponsored renewable energy contracts. Table 6.2 summarizes the main design features of the most important renewable energy support policies and compares their pros and cons. Table 6.3 shows which policies are used, to varying degrees, in ECA countries for which information is available.

All of the region's countries for which information is available have some form of renewable energy support policy. Feed-in tariffs are the most common. These are long-term contracts for clean energy producers that compensate them based on their costs of production (solar power producers are paid more, wind power producers, less). Renewable portfolio standards, fiscal incentives, and public loans or grants are further support mechanisms used by some EU-10 countries that introduced comprehensive policy packages. Renewable energy auctions, in which the utility buys a specified amount of renewable power from the lowest bidder, have so far not been used in ECA. Table 6.3 also shows current renewable energy shares and the 2020 targets for EU members.

Feed-in tariffs are by far the most popular support program worldwide. Nearly all ECA countries that have adopted renewable incentive programs have chosen feed-in tariff mechanisms, with Poland and Romania the only two opting for tradable quotas. Under a feed-in tariff scheme, governments set prices for different types of renewable power to compensate producers for the higher cost of producing clean energy. Utilities are required to purchase power from renewable electricity producers at this price, but they can spread the additional cost across their entire customer base or receive compensation from the government to cover the incremental costs.

Feed-in tariff policies have been effective in accelerating the deployment of renewable resources, but setting the "right" level of

TABLE 6.2

Comparison of Major Renewable Energy Policies

	Design feature	Advantage	Disadvantage	Examples
Feed-in tariffs	Give renewable energy installations long-term contracts to sell renewable generation to the electricity grid at a guaranteed price (normally with a price premium above the electricity market price).	The certainty of renewable electricity prices enhances the attractiveness of investment. Renewable prices are less susceptible to market manipulation or political influence on specific projects.	Overly generous feed-in tariffs may encourage inefficient production and increase government or household fiscal burden.	Globally, feed-in tariffs are now used in more than 65 countries, including 13 in ECA.[a]
Renewable portfolio standards (tradable green certificates)	Creates a market for renewable energy certificates, which are awarded to renewable producers based on their renewable energy output. Electricity suppliers must purchase certificates or otherwise supply renewable energy for a certain percentage of their total end-use delivery.	The fixed number of certificates avoids the risk of overinvestment (quantity certainty).	Investors may require a higher expected premium to compensate for the price uncertainty of tradable certificates. There is no price differentiation among technologies, and standards tend to benefit nearly commercial technologies.	Belgium, Italy, Poland, Romania, Sweden, and the United Kingdom
Renewable auction mechanisms	Utilities are required to conduct auctions for the supply of a fixed amount of renewable capacity. The cheapest proposals are awarded contracts first, followed by the next cheapest proposals, and so on, until each utility meets its quota for auction.	A specific capacity target is likely to be met with the least costly and most viable projects.	Auction mechanisms tend to favor larger-scale projects, putting small firms and locally distributed generation at a disadvantage. A robust market of large players may be needed to ensure a competitive auction.	India, Ireland, and South Africa

Source: Fischer and Preonas 2012.

a. In order of introduction since 1999: Slovenia, Latvia, the Czech Republic, Lithuania, Estonia, Hungary, the Slovak Republic, Turkey, Albania, Bulgaria, FYR Macedonia, Ukraine, and Serbia. For details, see http://www.globalfeedintariffscom/.

support is difficult. A feed-in tariff has to provide sufficient incentives to achieve an overall quantity of renewable generation; on the other hand, it should not be so generous as to allow poorly performing renewable energy investments to survive solely based on heavy subsidies. Furthermore, if the implicit subsidy for renewables is paid by energy consumers, high feed-in tariffs may impose an adverse impact on growth and affordability. This is a particular concern to developing countries, where households devote a larger portion of income to energy and are more vulnerable to rising tariffs (see chapter 5).

This trade-off in setting feed-in tariffs is evident in Ukraine (Trypolska 2012). The country's feed-in tariff for utility-scale solar

TABLE 6.3

Renewable Energy Support Policies in Europe and Central Asia, 2011

	Regulatory policies					Fiscal incentives				Public financing		RE shares (%)	
	Feed-in tariff (including premium payment)	Electric utility quota obligation/ RPS	Net metering	Biofuels obligation/ mandate	Tradable REC	Capital subsidy, grant, or rebate	Investment or production tax credits	Reductions in sales, energy, CO_2, VAT, or other taxes	Energy production payment	Public investment, loans, or grants	Public competitive bidding	Renewable share in 2009/2010[a]	EU Directive target 2020
Armenia												6.7	n.a.
Belarus								●		●		6.0	n.a.
Bosnia and Herzegovina						●					●	12.1	n.a.
Bulgaria								●		●		6.3	16
Croatia												11.0	n.a.
Czech Republic						●	●	●				6.8[a]	13
Estonia								●	●			14.4[a]	25
Hungary						●		●				7.9[a]	13
Kazakhstan												1.1	n.a.
Kyrgyz Republic								●		●		28.4	n.a.
Latvia										●		37.2	40
Lithuania												10.4	23
Macedonia, FYR										●		11.3	n.a.
Moldova								●		●		3.5	n.a.
Poland						●		●		●	●	7.6[a]	15
Romania						●						15.4	24
Russian Federation												3.4	n.a.
Serbia												8.4	n.a.
Slovak Republic						●	●					6.9[a]	14
Slovenia								●			●	13.1[a]	25
Turkey												11.0[a]	n.a.
Ukraine												1.6	n.a.

Sources: REN21 2011; IEA 2011a.

Note: No information available for unlisted ECA countries. RE = renewable energy; RPS = renewable portfolio standards; REC = renewable energy credit; CO_2 = carbon dioxide; VAT = value added tax; n.a. = not applicable.

a. Indicates renewable share in 2010.

projects is €0.46 per kWh, the highest in Europe and more than twice Germany's in 2012. As a consequence, the largest PV plant in Europe, at 100 MW capacity, opened recently at Perovo on the Crimean Peninsula. The country is targeting 1 gigawatt of solar capacity by 2015 as part of a strategy to diversify from Russian gas imports that also includes construction of an LNG terminal. There may be some justification for higher tariffs to compensate for higher project risk in less-mature markets for renewable energy. However, there is also a real risk that tariffs are set higher than necessary and provide excessive windfall profits for investors.

Global experiences show that higher feed-in tariffs do not necessarily yield greater levels of renewable deployment. A number of noneconomic barriers can reduce their effectiveness, such as administrative hurdles (including planning delays and restrictions, lack of coordination between different authorities, and long lead times in obtaining authorizations), grid access problems, electricity market design issues, lack of social acceptance, and disputes over how the costs are distributed (see box 6.3) (IEA 2008).

An analysis of 35 European countries suggests that guaranteed grid access, the length of guaranteed feed-in tariffs, and the structure of the electricity market are more important than actual feed-in tariff levels in determining the effectiveness of renewables policy (Zhang 2012). In fact, high subsidies may have driven up investment costs by allowing inefficient investment in low-wind-speed sites. The success

---

BOX 6.3

How Will the Costs of Feed-In Tariffs Be Distributed?

Consumer Costs

The price impacts of a feed-in tariff in an ECA country are actually quite hard to predict, being so intertwined with the electricity market regulatory structure as well as the tariff's design. Assume that the burden of the premium is imposed on the electricity sector, as opposed to being financed explicitly from public coffers; then the question is how the system passes along costs. If prices are set by public authorities with nonmarket goals in mind (for example, price stability or below-cost energy access rather than cost recovery or profit maximization), then the burden of additional shortfalls will likely be passed back to the government. However, the trend in ECA countries is toward greater liberalization and, at a minimum, pricing for cost recovery.

continued

BOX 6.3 *continued*

If retail prices are set based on average costs, the feed-in tariff costs will be transferred to utility customers. If, on the other hand, wholesale electricity prices are determined competitively rather than by the government, then what matters is the effect of introducing renewables on the cost of producing the last unit of electricity—the marginal generation cost—in the market to meet demand. It is possible that a modest feed-in tariff could lower marginal generation costs and consequently lower the wholesale rates, which would mitigate the tariff's effects on consumers. For example, the variable cost of wind production is generally lower than fossil-fuel-based power generation. When wind is strong during peak demand hours, wind generation can replace gas-fired power plants and reduce the marginal cost of power supply.

Fiscal Costs

The fiscal costs of feed-in tariff policies could be high for either of two reasons. First, when government owns a large share of the assets in the generation sector, subsidies to renewable generation are paid by taxpayers. Second, governments may choose to subsidize end-use electricity consumption to offset price increases for low-income consumers. Governments may also have responsibilities for making the necessary infrastructure investments to support distributed and intermittent generation. Thus, public officials designing feed-in tariffs must also consider the trade-offs with other priorities for public finances, including health, education, other infrastructure investments, and so on.

CDM, JI, and Technology Transfer

One opportunity available to ECA countries that are not EU members is the potential for selling credits through Kyoto Protocol compliance vehicles, including the Clean Development Mechanism (CDM) and Joint Implementation (JI). These initiatives can help fund renewable energy projects in developing countries in exchange for credits that count toward meeting buying countries' emissions targets.[a] As yet, ECA countries have not been major contributors of these certified emission reductions (CERs), representing only around 1 percent of cumulative CDM credits. However, a recent review found favorable institutional conditions and some economic potential for CDM and JI implementation in most of the region's countries. Although landfill gas and methane capture projects typically generate the most CERs, significant potential for renewable energy projects was identified in Armenia, Azerbaijan, and FYR Macedonia. Governments in these countries thus have the opportunity for additional external financing. However, proper institutions and streamlined, transparent administrative processes must be in place to attract and manage CDM projects.

Source: Fischer and Preonas 2012.

a. JI is relevant for countries (like Belarus and Turkey) listed in Annex B of the Kyoto Protocol, which sets binding emissions targets. CDM is designed for developing countries without firm targets.

of countries with the highest level of wind deployment stems from high investment stability guaranteed by a long-term guaranteed feed-in tariff, an appropriate framework that allows competition in the renewable energy market, low administrative and regulatory barriers, and favorable grid access conditions. The average feed-in tariffs in the countries with largest deployment were often lower than the average in Europe and also lower than those in countries applying tradable quota systems, as shown in figure 6.9. In contrast, the highest levels of feed-in tariffs in ECA are seen in Croatia and the former Yugoslav Republic of Macedonia. Yet neither of these countries achieved high levels of wind power development, likely due to significant noneconomic barriers.[32]

To improve the effectiveness and cost-efficiency of renewable incentive programs in ECA, some lessons can be drawn from global experiences with feed-in tariffs (Fischer and Preonas 2012):

- The chosen mechanism should emphasize simplicity, transparency, and compatibility with the national circumstances, including the degree of competition in renewable energy markets and the capacity of government institutions. A feed-in tariff works well if the price cannot be manipulated and all successful competitors have guaranteed access. Other mechanisms, including tradable certificates, auctions, and certain kinds of grants, are

FIGURE 6.9

Average Annual Generation of Onshore Wind Energy as a Share of Power Generation and Relative to Average Feed-In Tariffs and Tradable Certificate Prices in Selected Europe and Central Asia Countries, 1991–2010

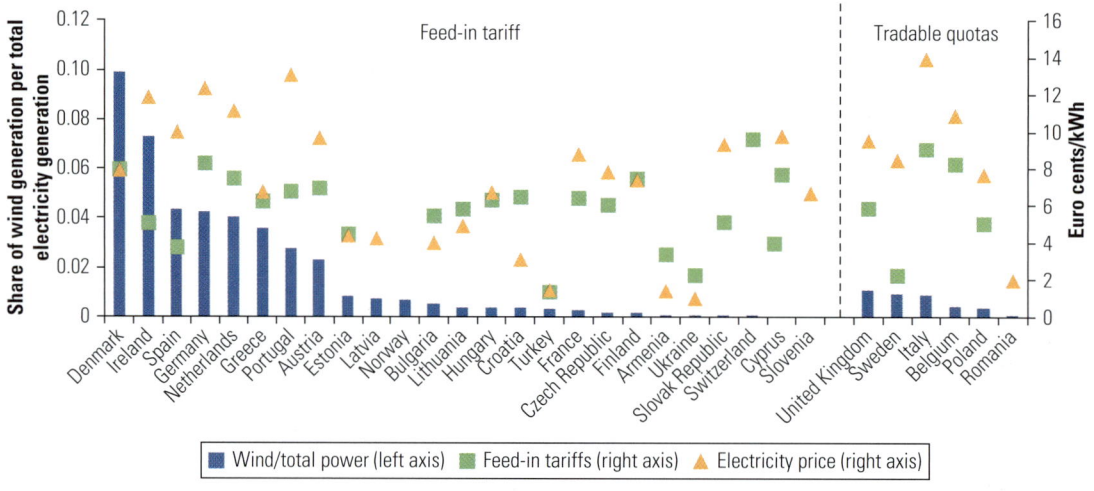

Soruce: Zhang 2012.

Note: Annual wind installation is normalized by per capita GDP to reflect the country scale effect on renewable development. Feed-in tariffs are in 2000 euros deflated by country consumer price index. FiT = feed-in tariff; kWh = kilowatt-hour.

more susceptible to market manipulation or political influence if robust competition and transparency are not assured. Feed-in tariffs should be as simply designed as possible to avoid tailoring to specific favored projects.

- Feed-in tariffs will not be effective without institutional foundations. Serbia, for instance, introduced a feed-in tariff for wind energy of €0.095 in 2009 (Komarov et al. 2012). However, complex legal procedures for permits, a lack of standardization, and difficult grid connections have slowed investments. Policies supporting renewables need to be accompanied by general energy market reforms, grid access requirements, net metering regulations, training and certification of installers, monitoring capacity, permitting and transmission line siting processes, and the removal of other administrative barriers; all of these activities also improve the efficiency of the electricity sector generally.

- Because renewable generation will be more distributed, complementary investments need to ensure grid access and more sophisticated grid management than is necessary with fewer, large generators. Coping with intermittency may require investments in electricity storage (where options are so far still limited) or additional flexible generation capacity (in the short term, mostly natural-gas power plants).

- Limiting the scope of feed-in tariffs to the locally most appropriate technologies keeps costs low. Attention should be focused on the few renewable energy technologies with the greatest resource availability in a given country, the lowest costs, and the greatest potential to lower those costs through learning. Locations and technologies that are also likely to improve energy access or displace local pollutants such as sulfur and particulate matter may also be given greater weight.

Notes

1. However, some realize that, in practice, the solution is not so simple: "We know enough to be very clear that a simple approach confined to price equals marginal social cost will not be good enough" (Stern 2010, 275).
2. For more information, see EcoGrid EU: http://www.eu-ecogrid.net.
3. This section has been drafted by Michael Levitsky and is based on Levitsky, Howorth, and Zhang (2012).
4. After it has been processed, natural gas consists mainly of methane.
5. Global Warming Potential is the amount by which a substance increases temperatures over a certain time frame, compared with CO_2. Because

methane's life in the atmosphere is only about 12 years, compared with over 100 years for CO_2, methane's impact is relatively greater in the short term. For a list of Global Warming Potentials, by species and chemical formula, see http://unfccc.int/ghg_data/items/3825.php.

6. Some natural gas is produced together with oil ("associated gas"). This raises similar issues of coproduction. Associated gas has not been included as an element in the calculations of GHG emissions in the estimates presented here.

7. In practice, gas from Central Asia (Kazakhstan, Turkmenistan, and Uzbekistan) is exported to Russia, and this allows a similar quantity of Russian gas to be exported. The whole route from Turkmenistan to Central and Eastern Europe is simulated to provide a "maximum" for the life-cycle emissions of gas trade within the region.

8. Despite the importance of GHG emissions from distribution systems, data about these are scarce, and accurate measurements are difficult over large systems.

9. This is estimated as follows: Russian gas export system 13 percent, storage system 4 percent, and distribution 1 percent.

10. A more recent study from the Polish Geological Institute puts reserves at 0.35–0.77 tcm; this demonstrates the highly uncertain and early state of understanding of shale gas outside the United States.

11. This estimate is based upon the behavior of wells in the Marcellus shale in the northeastern United States. The United States is the only country where shale gas data are available. Productivity of actual wells drilled in Poland and Ukraine is likely to be very different, although the basic pattern of rapid initial decline will be the same.

12. It is assumed here that future policy in Eastern and Central Europe will regulate for best practices in shale drilling as a means of gaining public acceptance of this industry.

13. Development of shale gas on a large scale in Europe will call for much larger resources or services, particularly drilling rigs, than are currently available. Production is thus likely to be ramped up over a period of many years.

14. To offset intermittency from renewables, gas systems need to have additional flexibility, which may require additional investments in storage and pipelines.

15. This section is based on Kulichenko and Ereira (2011).

16. Bełchatów accounts for 2.5 percent of all CO_2 emissions from the 982 facilities in the European Pollutant Release and Transfer Register. The register includes facilities from EU member states plus Iceland, Liechtenstein, Norway, Serbia, and Switzerland. For more about the register, see http://prtr.ec.europa.eu.

17. Lignite emits 101.2 $kgCO_2$ per gigajoule (GJ); hard coal emits 94.6; and natural gas emits 56.1 (Quaschning 2011).

18. EOR's use as a mitigation option is, of course, questionable because for every ton of CO_2 injected, the processing and use of the produced oil will release four tons of CO_2 (Wuppertal Institute 2010).

19. Shaffer (2010), for instance, suggests that leakage rates must remain below 1 percent per 1,000 years to retain mitigation benefits. Dooley (2010) disputes this estimate.

20. As suggested in the EU CCS Directive (2009/31/EC).

21. Parties to the UNFCCC are classified as (a) Annex I countries, (industrialized countries and economies in transition), and (b) non-Annex I countries (developing countries).

22. The design life span of nuclear power plants is 30 to 40 years, but many reactors' licenses have been extended by 10 to 20 years.

23. These estimates are for the United States. The cost of gas power is higher in Europe and Asia.

24. Lovins, Sheikh, and Markevich (2009) surveyed several detailed estimates. At a cost of US$5,500 per kW produced by a nuclear plant, capital accounts for 75 percent of total nuclear costs in Europe and America (*Economist* 2012).

25. A study by insurance sector experts suggests insurance costs per kWh of at least another US$0.18 and potentially much more for nuclear power plants in Germany (Versicherungsforen Leipzig 2011).

26. This section draws on Fischer and Preonas (2012) and Zhang (2012).

27. Included are all biomass, geothermal, and wind generation projects of more than 1 MW; all hydropower projects between 0.5 and 50 MW; all commercial solar projects of more than 0.3 MW; all marine energy projects; and all biofuel projects with a capacity of 1 million liters or more per year. This leaves out a large number of smaller-scale systems.

28. The EU-15 countries include Austria, Belgium, Denmark, Finland, France, Germany, Greece, Ireland, Italy, Luxembourg, the Netherlands, Portugal, Spain, Sweden, and the United Kingdom.

29. The EU-10 countries include Bulgaria, the Czech Republic, Estonia, Hungary, Latvia, Lithuania, Poland, Romania, the Slovak Republic, and Slovenia.

30. The CIS countries are Armenia, Azerbaijan, Belarus, Georgia, Kazakhstan, the Kyrgyz Republic, Moldova, the Russian Federation, Tajikistan, Turkmenistan, Ukraine, and Uzbekistan.

31. For instance, according to the German Environment Ministry, in 2011 this extra charge added €0.035 per kWh to cover the feed-in tariff in Germany. Electricity prices for German households had gone up from €0.143 in 2000 to €0.216 in 2008. Many blamed the feed-in tariff. However, only about 5 percent of the charges and 7–17 percent of annual increases were due to feed-in tariff legislation. Market structure in the electricity sector (lack of competition among large suppliers) contributed far more.

32. These patterns hold under alternative definitions of deployment such as generation/wind potential or MW/per capita GDP.

References

Aldy, Joseph E., Alan J. Krupnick, Richard G. Newell, Ian W. H. Parry, and William A. Pizer. 2010. "Designing Climate Mitigation Policies." *Journal of Economic Literature* 48 (4): 903–34.

Beyea, Jan. 2012. "Special Issue on the Risks of Exposure to Low-Level Radiation." *Bulletin of the Atomic Scientists* 68 (3).

Borggrefe, Frieder, and Karsten Neuhoff. 2011. "Balancing and Intraday Market Design: Options for Wind Integration." DIW Berlin Discussion Paper 1162, Climate Policy Initiative, DIW Berlin (German Institute for Economic Research), Berlin.

BP (British Petroleum). 2011. *Statistical Review of World Energy*. London: BP. http://www.bp.com/statisticalreview.

Buys, P., U. Deichmann, C. Meisner, T. T. That, and D. Wheeler. 2009. "Country Stakes in Climate Change Negotiations: Two Dimensions of Vulnerability." *Climate Policy* 9 (3): 288–305.

Davis, Lucas W. 2012. "Prospects for Nuclear Power." *Journal of Economic Perspectives* 26 (1): 49–66.

Dooley, J. J. 2010. "Response to 'Long-Term Effectiveness and Consequences of Carbon Dioxide Sequestration'." PNNL-19547, Pacific Northwest National Laboratory, U.S. Department of Energy, Richland, WA.

Economist. 2012.

EIA (Energy Information Administration). 2011. *World Shale Gas Resources: An Initial Assessment of 14 Regions Outside of the United States*. Washington, DC: EIA, U.S. Department of Energy.

———. n.d. International Energy Statistics (database). EIA, U.S. Department of Energy, Washington, DC. http://www.eia.gov/cfapps/ipdbproject/IEDIndex3.cfm.

Eurelectric. 2011. "Hydro in Europe: Powering Renewables." Report for the Eurelectric Renewables Action Plan (RESAP), Electricity for Europe, Brussels.

European Commission. 2011. "Mobility and Transport." In *EU Transport in Figures*. Brussels: European Commission.

Fischer, C., and L. Preonas. 2012. "Feed-In Tariffs for Renewable Energy: Effectiveness and Social Impacts." Background paper, Europe and Central Asia Region, World Bank, Washington, DC.

Graus, Wina, Mauro Roglieri, Piotr Jaworski, Luca Alberio, and Ernst Worrell. 2011. "The Promise of Carbon Capture and Storage: Evaluating the Capture-Readiness of New EU Fossil Fuel Power Plants." *Climate Policy* 11 (1): 789–812.

Grübler, Arnulf. 2010. "The Costs of the French Nuclear Scale-Up: A Case of Negative Learning by Doing." *Energy Policy* 38 (9): 5174–88.

IEA (International Energy Agency). 2008. "Developing Renewables: Principles for Effective Policies." Report, IEA, Paris.

———. 2010. *Energy Technology Perspectives*. Paris: IEA.

———. 2011a. *World Energy Outlook 2010*. Paris: Organisation for Economic Co-operation and Development and IEA.

———. 2011b. "Are We Entering a Golden Age of Gas?" World Energy Outlook Special Report, Organisation for Economic Co-operation and Development and IEA, Paris.

———. 2011c. World Energy Statistics and Balances (online database). http://www.iea.org/stats/index.asp.

———. 2012. "Golden Rules for a Golden Age." World Energy Outlook Special Report on Unconventional Gas, Organisation for Economic Co-operation and Development and IEA, Paris.

IPCC (Intergovernmental Panel on Climate Change). 2011. "Special Report on Renewable Energy Sources and Climate Change Mitigation." Summary report of a comprehensive assessment for IPCC Working Group III, IPCC, Abu Dhabi.

Jacobson, Mark Z., and Mark A. Delucchi. 2011. "Providing All Global Energy with Wind, Water, and Solar Power, Part I: Technologies, Energy Resources, Quantities and Areas of Infrastructure, and Materials." *Energy Policy* 39 (3): 1154–69.

Joskow, Paul L. 2011. "Comparing the Costs of Intermittent and Dispatchable Electricity Generating Technologies." *American Economic Review* 101 (3): 238–41.

Komarov, Dragan, Slobodan Stupar, Aleksandar Simonovic, and Marija Stanojevic. 2012. "Prospects of Wind Energy Sector Development in Serbia with Relevant Regulatory Framework Overview." *Renewable and Sustainable Energy Reviews* 16 (5): 2618–30.

Kulichenko, Natalia, and Eleanor Charlotte Ereira. 2011. "Prospects for Carbon Capture and Storage in the ECA Region." Background paper, Europe and Central Asia Region, World Bank, Washington, DC.

Levitsky, Michael, Gary Howorth, and Fan Zhang. 2012. "Life-Cycle Greenhouse Gas Emissions for Natural Gas Supply in Europe and Central Asia." Background paper, Europe and Central Asia Region, World Bank, Washington, DC.

Lovins, Amory B., Imran Sheikh, and Alex Markevich. 2009. "Nuclear Power: Climate Fix or Folly?" Technical report summary, Rocky Mountain Institute, Snowmass, CO.

MIT (Massachusetts Institute of Technology). 2009. "Update of the MIT 2003 'Future of Nuclear Power.'" Study update, MIT Energy Initiative, Cambridge, MA.

Neij, Lena. 2008. "Cost Development of Future Technologies for Power Generation: A Study Based on Experience Curves and Complementary Bottom-Up Assessments." *Energy Policy* 36 (6): 2200–11.

Nikolaev, V. G., S. V. Ganaga, K. I. Kudriashov, R. Walter, P. Willems, and A. Sankovsky. 2010. "Prospects of Development of Renewable Power Sources in Russian Federation." Results of TACIS project, Europe Aid/116951/C/SV/RU, Atmograph, Moscow.

Nitsch, Joachim, Thomas Pregger, Yvonne Scholz, Tobias Naegler, Dominik Heide, Diego Luca de Tena, Franz Trieb, Kristina Nienhaus, Norman Gerhardt, Tobias Trost, Amany von Oehsen, Rainer Schwinn, Carsten Pape, Henning Hahn, Manuel Wickert, Michael Sterner, and Bernd Wenzel. 2012. "Long-Term Scenarios and Strategies for the Deployment of Renewable Energies in Germany in View of European and Global Developments." Deutsches Zentrum für Luft- und Raumfahrt (DLR), Fraunhofer Institut für Windenergie und Energiesystemtechnik (IWES), Ingenieurbüro für neue Energien (IFNE), Stuttgart Teltow.

Pacala, S., and R. Socolow. 2004. "Stabilization Wedges: Solving the Climate Problem for the Next 50 Years with Current Technologies." *Science* 305 (5686): 968–72.

Pew Charitable Trusts. 2012. "Who's Winning the Clean Energy Race? Growth, Competition, and Opportunity in the World's Largest Economies." G-20 Clean Energy Factbook, The Pew Charitable Trusts, Washington, DC.

Quaschning, Volker. 2011. *Regenerative Energiesystemse* [Renewable Energy Systems]. Munich: Hanser Verlag.

REN21 (Renewable Energy Policy Network for the 21st Century). 2011. "Renewables 2010 Global Status Report." Annual publication, REN21, Paris.

Shaffer, Gary. 2010. "Long-Term Effectiveness and Consequences of Carbon Dioxide Sequestration." *Nature Geoscience* 3: 464–67. DOI: 10.1038 /NGEO896.

Stern, Nicholas H. 2010. "Imperfections in the Economics of Public Policy, Imperfections in Markets, and Climate Change." *Journal of the European Economic Association* 8 (2–3): 253–88.

Suwala, Wojciech. 2010. "Lessons Learned from the Restructuring of Poland's Coal-Mining Industry." Study for the Global Subsidies Initiative (GSI) of the International Institute for Sustainable Development (IISD), Geneva.

Trypolska, Galyna. 2012. "Feed-In Tariff in Ukraine: The Only Driver of Renewables' Industry Growth?" *Energy Policy* 45 (June): 645–53.

Umweltbundesamt (German Federal Environment Agency). 2010. "Energy Target 2050: 100% Renewable Electricity Supply." Study report, Umweltbundesamt, Dessau, Germany.

Versicherungsforen Leipzig. 2011. "Calculation of a Risk-Adjusted Insurance Premium to Cover the Liability Risks Resulting from the Operation of Nuclear Plants." Study report, Versicherungsforen Leipzig, http://www.versicherungsforen.net.

WEC (World Energy Council). 2010. "Survey of Energy Resources." London.

World Bank. 2009. "Directions in Hydropower: Scaling Up for Development." Water Working Notes 49017, World Bank, Washington, DC.

———. 2010a. *Climate Change and the World Bank Group—Phase II: The Challenge of Low-Carbon Growth*. Independent Evaluations Group (IEG) Study Series. Washington, DC: World Bank.

———. 2010b. *Lights Out? The Outlook for Energy in Eastern Europe and the Former Soviet Union*. Washington, DC: World Bank.

Wuppertal Institute. 2010. "Comparison of Renewable Energy Technologies with Carbon Dioxide Capture and Storage (CCS)." Final report to the Federal Ministry for the Environment, Nature Conservation and Nuclear Safety (BMU), Wuppertal, Germany.

Zhang, F. 2012. "How Fit Are Feed-In-Tariff Policies? Evidence from the European Wind Market." Background paper, Europe and Central Asia Region, World Bank, Washington, DC.

MAP 7.1

CO$_2$ Emissions and Industrial Energy Intensity in Europe and Central Asia

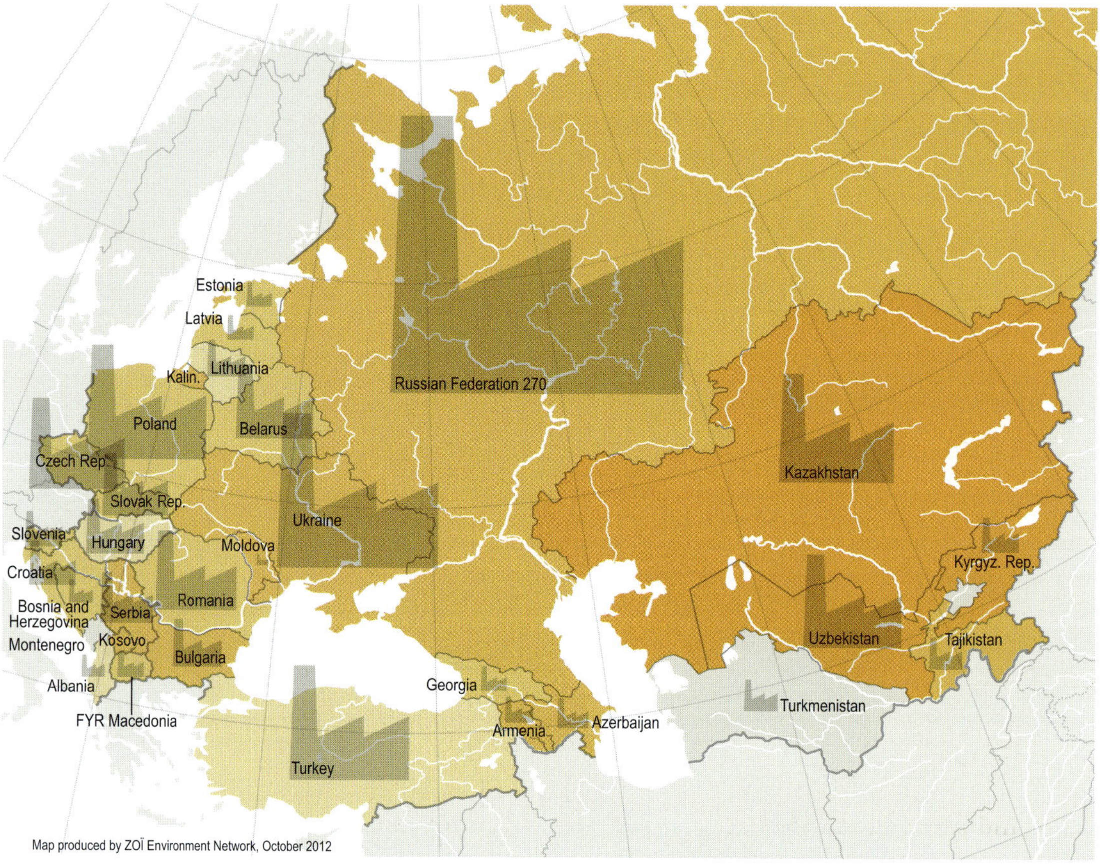

Estonia
Latvia
Lithuania
Kalin.
Poland
Belarus
Czech Rep.
Slovak Rep.
Slovenia
Hungary
Moldova
Croatia
Bosnia and Herzegovina
Serbia
Romania
Montenegro
Kosovo
Bulgaria
Albania
FYR Macedonia
Turkey
Ukraine
Russian Federation 270
Georgia
Armenia
Azerbaijan
Kazakhstan
Kyrgyz. Rep.
Uzbekistan
Tajikistan
Turkmenistan

Map produced by ZOÏ Environment Network, October 2012

Production

Carbon dioxide emissions from manufacturing, 2009 (million tonnes)

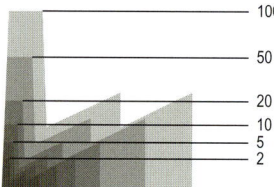

100
50
20
10
5
2

Energy intensity of industry, 2008 (in kilogram oil equivalent per 2005 PPP dollars)

0.1   0.2   0.4   no data

Source: World Energy Council (WEC) (http://wec-indicators.enerdata.eu); The World Bank (http://data.worldbank.org).
Note: PPP = purchasing power parity.

CHAPTER 7

Production

Main Messages

- Industrial production accounts for 28 percent of Europe and Central Asia's (ECA) energy use and a third of its carbon dioxide (CO_2) emissions. Industrial energy efficiency has increased by 30 percent since transition—largely through genuine efficiency improvements rather than changes in what is being produced.

- Industrial energy intensity remains high. In several ECA countries, it is several times higher than in Western Europe. The region accounts for 5 percent of global manufacturing output but 9 percent of industrial energy use. By achieving global best practice, energy savings of 50 percent are possible in energy-intensive sectors and 10–20 percent across all sectors.

- Accelerating efficiency gains in the industrial sector will require price instruments, regulations, and investments, especially in information provision. Gains can also come

from voluntary agreements in sectors with a few large
players such as cement production or from private sector
initiatives where innovative companies pursue efficiency
gains to raise competitiveness.

Industry was the cornerstone of the centrally planned economies in
Europe and Central Asia (ECA).[1] Today, it still plays an important
role, measured by both its economic contribution and its environmental impact. Industry is the largest energy user, accounting for
28 percent of total energy use, and the second-largest carbon emitter,
responsible for 34 percent of total carbon dioxide (CO_2) emissions in
the region in 2009. Recent value added by manufacturing totaled
US$1 trillion in 2005 purchasing power parity (PPP) dollars, or
17 percent of overall gross domestic product (GDP) in the region.[2] In
countries such as Belarus and Turkmenistan, the share is as high as
30 percent and 25 percent, respectively. The industrial sectors offer
perhaps the largest opportunities for energy savings in ECA, increasing competitiveness and reducing emissions. That does not mean,
however, that such savings will come automatically. As two prominent examples show, efficiency-oriented entrepreneurial management yields large energy savings, but sometimes a nudge from policy
makers helps set the process in motion.

DuPont is the world's third-largest chemical company and also one
of the world's largest industrial energy consumers: its annual energy
bill is over US$1.1 billion. DuPont's corporate leaders realize that
energy management is essential to maintaining competitiveness.
Between 1990 and 2010, DuPont reduced its energy use by 18 percent
despite a 40 percent growth in production. This improvement saved
US$6 billion in energy use. DuPont achieved these reductions with
projects that required little or no out-of-pocket spending and funded
capital projects with a blended rate of return of 65 percent.[3] Savings
came from a wide range of projects: repairs and improvements to steam
traps, correcting metering problems with purchased energy, and
upgrading boilers with new equipment design. DuPont also built large,
efficient heat and power cogeneration plants. More important, there
was a deep change of culture inside the company. In 2008, it launched
the "Bold Energy Plan" that called for all plants to accelerate energy
efficiency improvements. DuPont has now set a new target: a 5 percent
(US$50 million) annual decrease in energy use and a 65 percent reduc-
tion in greenhouse gas (GHG) emissions below 1990 levels by 2020.

In China's 11th Five Year Plan in 2006, the government adopted the slogan "Jieneng Jianpai"—"Save Energy! Cut Emissions!"—as a pillar for economic development. The plan was to reduce energy consumption and to create a "new industrial system": competitive new industries based on low-carbon technologies that will make China a leader in green technology. The goal was to reduce energy intensity by 20 percent between 2005 and 2010. This ambitious objective was pursued in many ways. One of the key initiatives was the "Top-1,000 Energy-Consuming Enterprises Program," which aimed to cut energy consumption by 70 million tons of oil equivalent (mtoe) among China's 1,000 largest energy users by 2010. Each company signed conservation agreements with local governments, which in turn signed agreements with the central government. In return for accepting binding energy targets, companies received capacity building, such as training workshops on energy benchmarking and audits. The program is reported to have saved some 14 mtoe in its first year from increased attention to energy management, including the appointment of energy managers, the closure of inefficient production processes, and the implementation of retrofit projects such as the renovation of fans and pumps. Almost overnight, the program has become a model for other companies. A new "Top-10,000" program is currently under preparation.

ECA's Industrial Energy Intensities

ECA had one of the least efficient industrial sectors in the world because of the region's historical emphasis on heavy industries and artificially low energy prices during the Soviet era. Since the transition period started two decades ago, the sector's energy productivity output per unit of (energy inputs) has consistently improved—by almost 30 percent by 2009. The bulk of the improvement came from enhancements in efficiency rather than changes in the mix of industrial activities. Both economic growth and energy price reform have driven the change. Economic growth makes it possible to build new plants and close down old ones, while energy price reform ensured that enterprises had proper incentives to minimize energy costs.

Despite big strides in reducing energy intensity, substantial opportunities remain to further improve efficiency and reduce CO_2 emissions. Many ECA countries currently use several times more energy per unit of output than their Western partners, as shown in figure 7.1. The region as a whole produces 5 percent of manufacturing output in the world but consumes 9 percent of energy. If all countries

FIGURE 7.1

Energy Consumption from Manufacturing, Selected Europe and Central Asia Countries Relative to World and Western European Averages, 2010

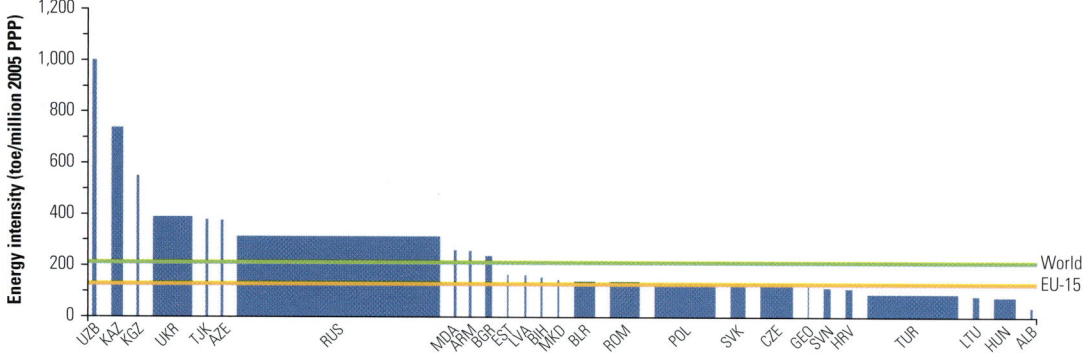

Source: Data from IEA 2011; World Bank, *World Development Indicators.*
Note: The x-axis is proportional to value added of manufacturing at constant PPP in 2005 international prices. toe = tons of oil equivalent; PPP = purchasing power parity; EU = European Union.

adopt best-practice technologies, the energy savings potential is estimated to be 50 percent of current energy use in the energy-intensive sectors (iron and steel, cement, and pulp and paper) and 10–20 percent across the board. Given the old age of many of the region's industrial plants, upgrading to energy-efficient technology is likely to be cost effective if undertaken as part of the natural cycle of plant replacement and offers an excellent opportunity for achieving energy savings and associated GHG emission reductions in the near to medium term.

Most industry sectors consist of a relatively small number of large players with similar manufacturing processes. This helps achieve large gains quickly. In the International Energy Agency (IEA) "BLUE Map" scenarios—in which global warming stays between 2 degrees Celsius (°C) and 3°C—more than a quarter of all energy efficiency gains come from the industrial sector, largely by changing the pattern of industrial energy use (IEA 2011). By 2050, this scenario anticipates CO_2 emission reductions of more than 30 percent in the industrial sector of transition economies. This goal is unlikely to be achieved without government action. As shown in figure 7.2 in the baseline scenario (assuming currently implemented or planned policies), total industrial CO_2 emissions in ECA would rise 62–70 percent in the low- and high-demand scenarios in 2050 compared with 2006 levels. Continued energy price reform and help for businesses to overcome financing and information barriers could instead yield large energy savings and avoid about 1 gigaton of CO_2 emissions per year.

FIGURE 7.2

Total Industrial CO$_2$ Emissions in Transition Economies under Baseline and IEA BLUE Scenarios, 2006 and 2050

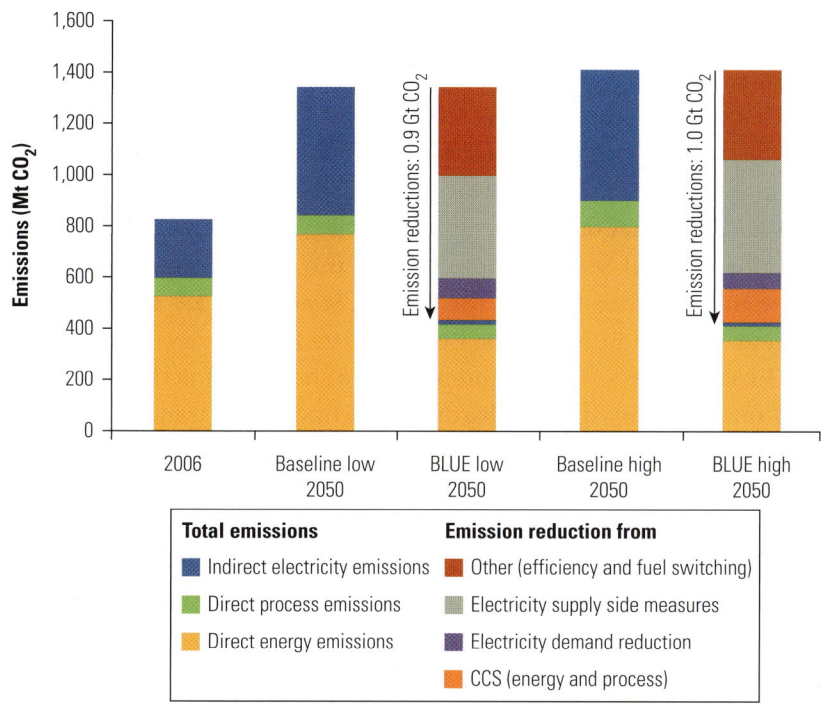

Source: IEA 2011.

Note: The IEA Baseline scenario assumes that governments introduce no new energy and climate policies. In contrast, the BLUE Map scenario sets the goal of halving global energy-related CO$_2$ emissions by 2050 (compared with 2005 levels) and examines the least-cost means of achieving that goal through the deployment of existing and new low-carbon technologies. "Low" and "high" refer to low- and high-demand scenarios. Demand is assumed to be 15–30 percent lower in the low-demand cases than in the high-demand cases in 2050, depending on the industrial sectors. CO$_2$ = carbon dioxide; MtCO$_2$ = millions of tons of carbon dioxide; GtCO$_2$ = gigatons of carbon dioxide; IEA = International Energy Agency; CCS = carbon capture and storage.

A Region Catching Up

In the Soviet era, leaders in ECA emphasized rapid industrialization, particularly of heavy industry, to foster economic growth. Much of the production was linked to the military-industrial complex, including the vast armaments-producing factories, with their demands for steel, chemicals, and energy. In 1988, the Soviet Union produced 17 times more steel per dollar of GDP than the United States (with GDP measured in PPP terms) (Sachs 1995). Meanwhile, the Eastern European countries developed large industrial sectors to process Soviet raw materials and then reexported them to the Soviet Union in semifinished or finished form. In 1987, for example, the industrial sector in Poland produced 52 percent of GDP, compared with just 23 percent in Organisation for Economic Co-operation and Development (OECD) countries.

This heavy industry was also vastly inefficient. To facilitate industrial production and social welfare, central planners kept energy
prices well below cost-recovery levels. Of all non-OECD subsidies to
fossil fuels (including electricity) in 1991, totaling US$270–US$330
billion, roughly two-thirds were in the former Soviet Union and
Eastern Europe (Myers and Kent 2001). Subsidies distorted prices
and led to extraordinarily wasteful use of energy. In 1990, the former
Soviet Union's energy intensity—the ratio of primary energy consumption to GDP—was 70 percent higher than in the United States
and 2.5 times that of Western Europe. The gap was especially evident
in the industrial sector, where energy use per unit of output was
three times higher than in the United States and 3.5 times higher
than in Western Europe (U.S. Congress 1993).

The early years of transition brought a sharp contraction in industrial output following the disruption of traditional trade and financial links and an end to centrally planned production (Havrylyshyn,
Izvorski, and Rooden 1998). Between 1991 and 1993, industrial
value added fell on average by 30 percent. In parallel, industrial
energy consumption dropped on average by 27 percent across
all countries, as shown in figure 7.3.[4] Most transition countries

FIGURE 7.3

Decoupling of Energy Consumption from Manufacturing Output in Europe and Central Asia, 1990–2008

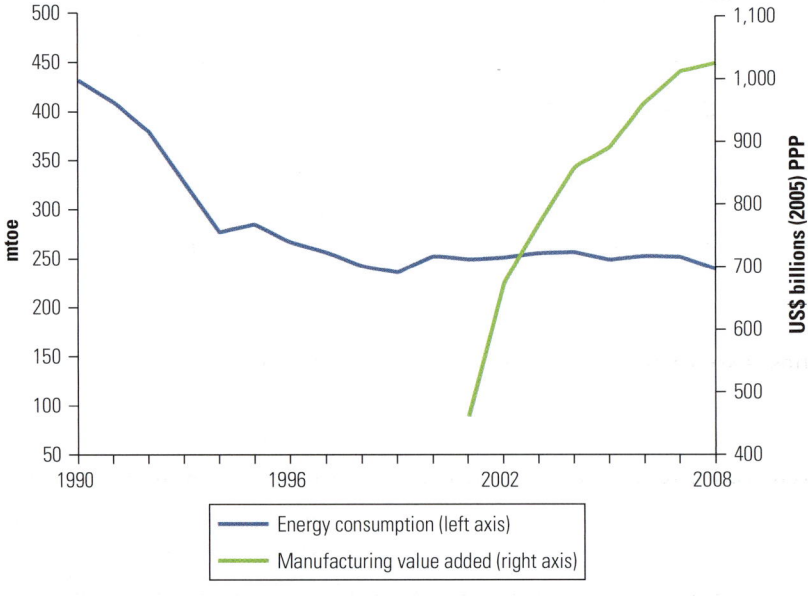

Source: Data from IEA 2011; World Bank's *World Development Indicators.*
Note: Manufacturing value-added data for Russia and Serbia are not available before 2002. Energy consumption data are
not available for Montenegro and Kosovo. mtoe = million tons of oil equivalent; PPP = purchasing power parity.

continued the comprehensive reform programs, including the liberalization of energy prices. Price supports for fossil fuels in the Russian Federation and Eastern Europe were greatly reduced. In Poland, for example, industrial coal prices quintupled in January 1990 (Myers and Kent 2001). By 1992, electricity and gas prices for industry in the Czech Republic, Hungary, and Poland were close to levels in Western Europe (UNEP 2003). Collections have also improved dramatically since 1992, in part because of privatization and metering. Bringing laws into agreement with the European Union's (EU) further accelerated the process of eliminating remaining subsidies. Even in Russia, gasoline and diesel prices for industry (which received two-thirds of energy subsidies in 1994) have increased to world market levels, although gas prices still remain below border price levels (World Bank 2008).

The market-friendly reform had a far-reaching impact on the transformation of the energy economies. When the sharp initial decline gave way to gradual economic recovery, energy consumption continued to decline and then stabilized, resulting in an absolute decoupling of industrial production and energy consumption, as figure 7.3 illustrates. Although manufacturing output has more than doubled between 2002 and 2008, annual industrial energy use has stabilized at around 250 million tons of oil equivalent (mtoe). During the same period, industrial energy intensity declined by 28 percent, from 369 tons of oil equivalent (toe) per US$1 million (PPP) in 2002 to 266 toe per US$1 million dollars (PPP) in 2008, also shown in Figure 7.3.[5] If energy intensity had remained at its 2002 level, energy demand would have been 776 mtoe higher, the equivalent of 10 years' total energy consumption in Ukraine today.

This rapid decline in energy intensity is largely a story of catch-up. The region as a whole is converging to EU-15 levels, and the difference in energy productivity between ECA and EU-15 countries has almost halved over the past decade.[6] Meanwhile, the gap between the most- and least-efficient countries within the region has also shrunk at an average rate of 4.7 percent per year since 2002, as shown in figures 7.4 and 7.5. However, not all countries are doing equally well. While most parts of ECA have sharply reduced energy intensity (faster than the EU-15 and world averages), the Central Asian countries have been lagging.[7] The average energy intensity in Central Asia is still seven times higher than that of Turkey, as shown in figure 7.6. From 2000 to 2009, Kazakhstan, the Kyrgyz Republic, and Tajikistan all experienced a decline in energy productivity.

FIGURE 7.4

Convergence in Industrial Energy Intensity of Europe and Central Asia Relative to EU-15, 2002–09

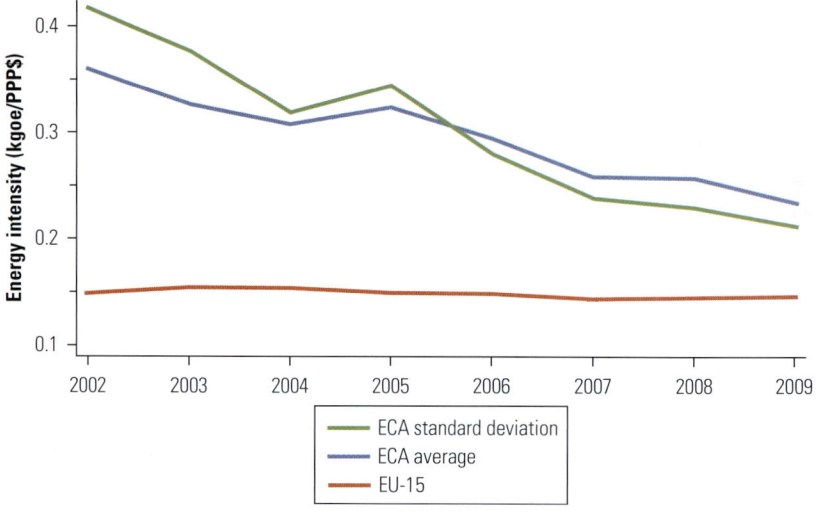

Source: Data from IEA 2011; World Bank's *World Development Indicators.*
Note: EU-15 countries include Austria, Belgium, Denmark, Finland, France, Germany, Greece, Ireland, Italy, Luxembourg, the Netherlands, Portugal, Spain, Sweden, and the United Kingdom. PPP = purchasing power parity; kgoe = kilograms of oil equivalent; EU = European Union.

FIGURE 7.5

Convergence in Industrial Energy Intensity within Europe and Central Asia, 1996–2009

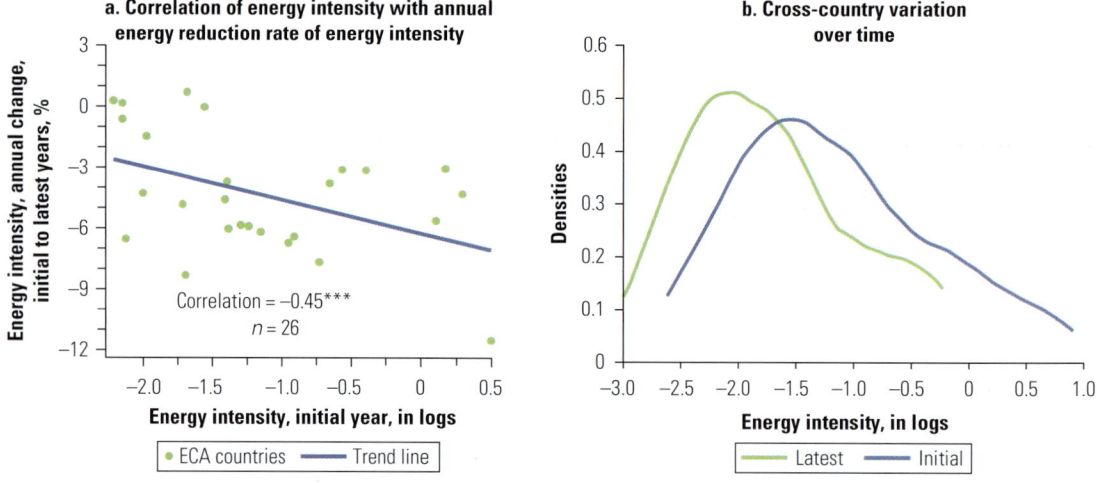

Source: Data from IEA 2011; World Bank's *World Development Indicators.*
Note: The initial and latest years are, for most countries, 1996 and 2009, respectively. The initial year for the Russian Federation is 2002, while the latest year for Azerbaijan, Estonia, Hungary, the Kyrgyz Republic, and Slovenia is 2008.

FIGURE 7.6

Industrial Energy Intensity in Europe and Central Asia, by Subregion, 1996–2009

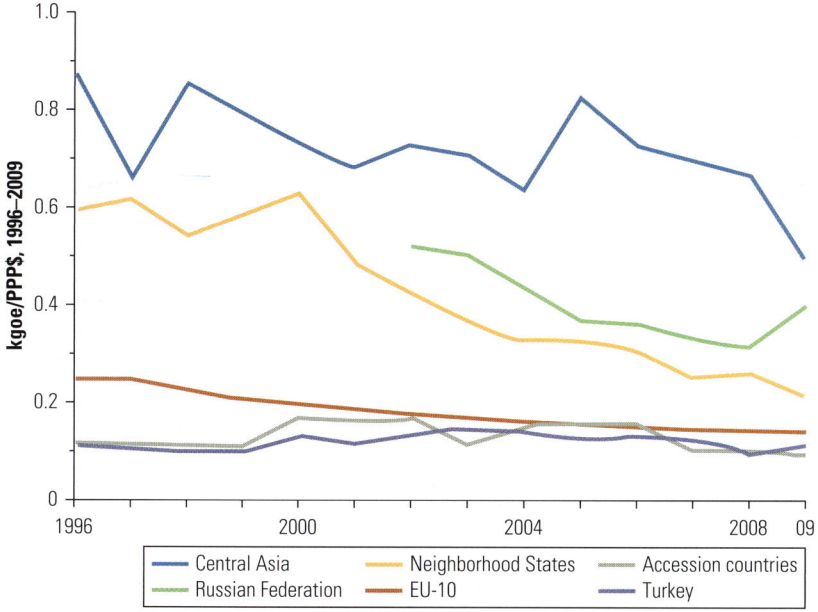

Source: Data from IEA 2011; World Bank's *World Development Indicators*.

Note: EU-10 countries include Bulgaria, the Czech Republic, Estonia, Hungary, Latvia, Lithuania, Poland, Romania, the Slovak Republic, and Slovenia. EU accession countries in this analysis include Albania, Bosnia and Herzegovina, Croatia, the former Yugoslav Republic of Macedonia, and Serbia. Neighborhood states include Armenia, Azerbaijan, Belarus, Georgia, Moldova, and Ukraine. Central Asian nations include Kazakhstan, the Kyrgyz Republic, Tajikistan, Turkmenistan, and Uzbekistan. Energy consumption data for Kosovo and Montenegro are not available, and these two countries are not included in the analysis. kgoe = kilograms of oil equivalent; PPP = purchasing power parity.

Structural or Efficiency Change

Industrial energy intensity can drop for two main reasons: manufacturing shifts toward different products that require less energy to make (the structural effect), or increased efficiency of manufacturing processes so they require less energy to produce the same products (the efficiency effect). Changes in what is being produced—the *industrial production mix*—can have a major impact on overall energy intensity. For example, the machinery industry may grow much faster than the steel industry over time, and because using steel to make machines is substantially less energy intensive than making steel itself, these differential growth rates make the manufacturing sector less energy intensive.

Although both factors contributed to the ECA's impressive reduction in energy intensity, the bulk of these gains was due to efficiency improvements, whereas structural change had less influence—and even contributed to increasing energy intensity in some countries.

Energy-Intensive Industries

Some industries require significantly more energy than others to produce a dollar's worth of output. The top four energy-using sectors in ECA consume more than half of the total industrial energy use but produce only 28 percent of value added (gross economic output less the value of purchased inputs). The four producing sectors are iron and steel (accounting for 25 percent of industrial energy use);[8] chemicals, including plastics, rubber, fertilizers, and pesticides (19 percent);[9] nonmetallic minerals such as glass, ceramic, and cement (11 percent); and paper and pulp (5 percent). Most industrial value added is created not by these basic materials industries but by manufacturing firms that process their output at lower energy intensity, as shown in figure 7.7.[10] Although the structural composition of industry varies from country to country, the energy-intensive sectors generally command most of a country's industrial energy consumption, as figure 7.8 illustrates.

Decomposition of Structural and Efficiency Effects

Efficiency improvements (rather than shifts from energy-intensive to less-intensive manufacturing activities) contributed to most of the improvements in energy productivity in Albania, Azerbaijan, Hungary, Lithuania, the former Yugoslav Republic of Macedonia, Moldova, Poland, Romania, and Russia (Zhang 2012). For example, in Moldova, total energy intensity of the manufacturing sector in 2008 was 85 percent of its intensity level in 2004. The index measuring the impact of industrial production mix on energy intensity (the structural index) was 97 percent of its level in 2004, while the index measuring energy consumption per unit of production (the efficiency index) was 88 percent of its 2004 level. In other words, had the energy efficiency stayed the same, total energy intensity would have decreased by just 3 percent between 2004 and 2008. In Russia, the 32 percent reduction in energy intensity is almost entirely due to improvements in energy efficiency between 2001 and 2008, as shown in figure 7.9, panel b. In Azerbaijan, Lithuania, and Poland, shifts in the mix of economic activities offset gains in energy efficiency by 7 percent, 26 percent, and 12 percent, respectively.

Another way to disentangle the impact of structural change and efficiency improvement on overall productivity is to analyze the trend of energy intensity and manufacturing value added.[11] Efficiency has

FIGURE 7.7

Industry Energy Intensity vs. Value Added, by Subsector, in Europe and Central Asia, 2009

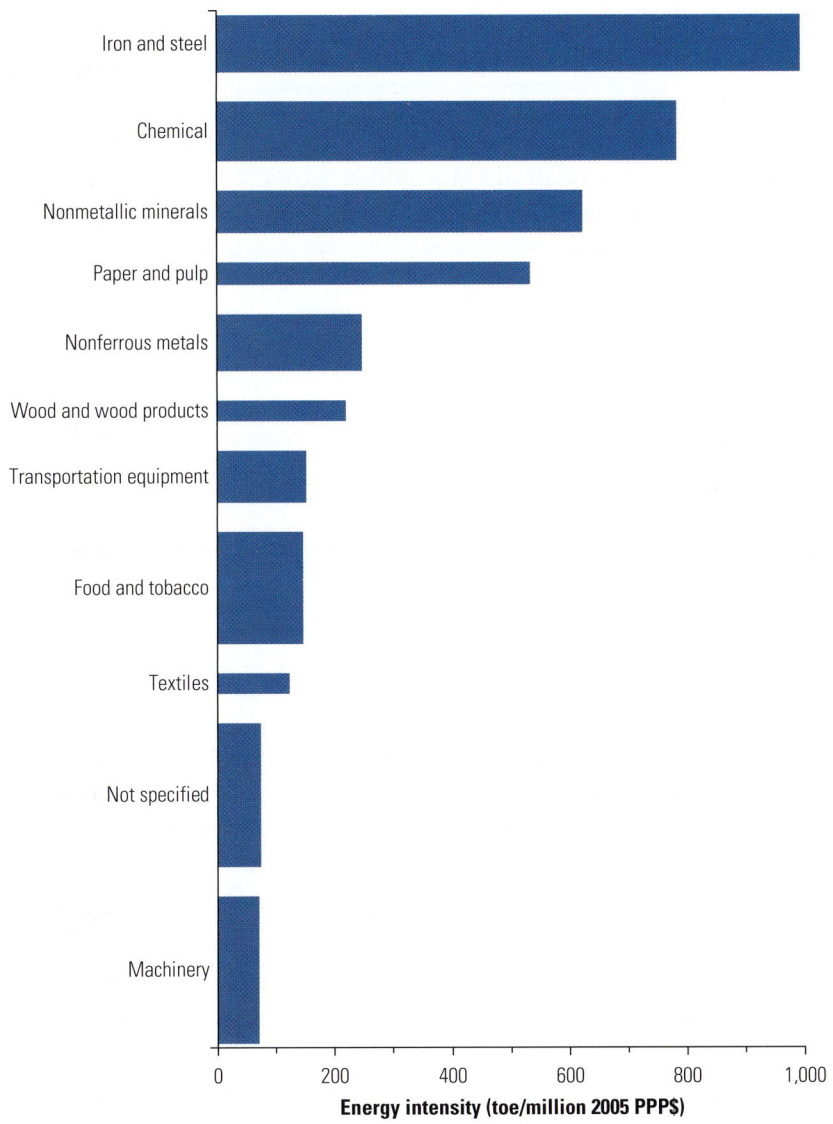

Sources: Calculations based on IEA 2011; UNIDO 2011.

Note: The figure depicts industry energy intensity by sector versus value added. The width of the bars is proportional to the total energy use of each subsector. The figure is based on aggregated energy consumption and value-added data of Albania, Azerbaijan, the Czech Republic, Hungary, Lithuania, FYR Macedonia, Moldova, Poland, and Russia. Those countries represent 77 percent of industrial energy use and 76 percent of industrial value added in ECA. Other countries of the region are not included because of limited data availability. Non-specified industry corresponds to ISIC Rev.4 22, 31 and 42 (manufacture of furniture) and any other manufacturing that is not included in the other categories. When countries cannot provide a complete industrial breakdown for all fuels, the non-specified industry category is used. toe = tons of oil equivalent; PPP = purchasing power parity.

FIGURE 7.8

Industrial Sectors Consuming Largest Shares of Energy in Europe and Central Asia Countries, 2009

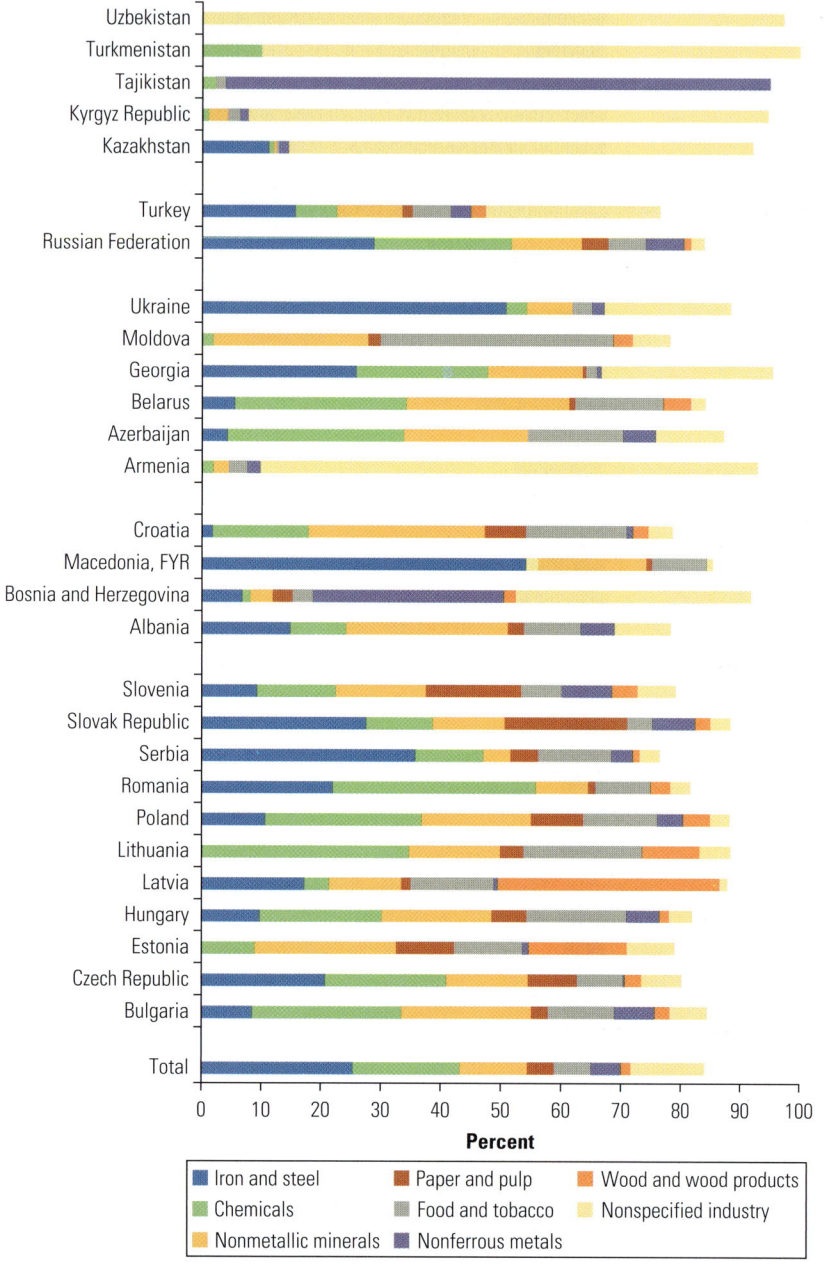

Source: Calculations based on IEA 2011.

Note: "Nonspecified industry" corresponds to ISIC Rev.4 22, 31, and 42 (manufacture of furniture) and any other manufacturing that is not included in the other categories. When countries cannot provide a complete industrial breakdown for all fuels, the "nonspecified industry" category is used. To calculate the regional-level industrial energy share (the bottom bar in the figure labeled "Total"), countries for which the nonspecified industry constitutes more than 15 percent of the total energy use are excluded (Armenia, Bosnia and Herzegovina, Georgia, Kazakhstan, the Kyrgyz Republic, Turkey, Turkmenistan, Ukraine, and Uzbekistan).

FIGURE 7.9

Relationship of Changes in Economic Structure, Energy Efficiency, and Energy Intensity in Manufacturing Sectors of Moldova and Russia, 2001–08

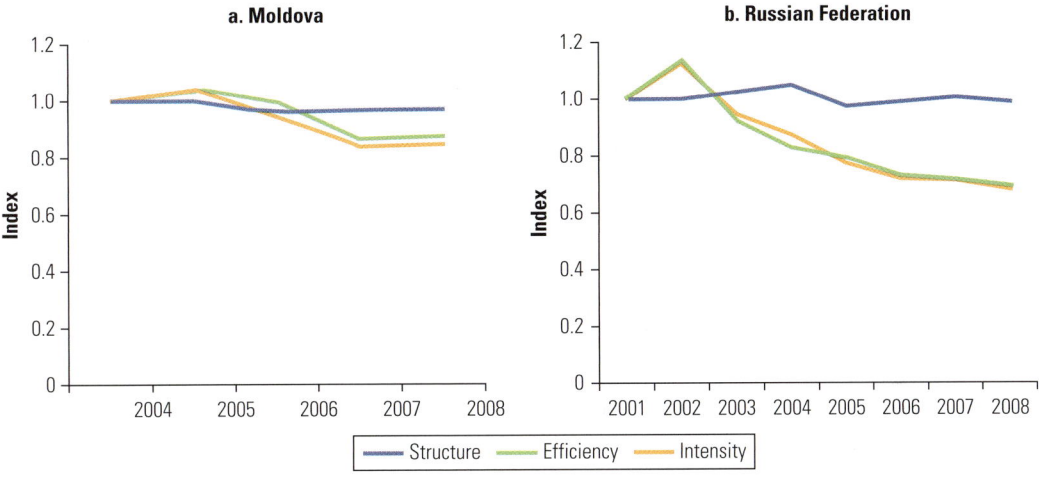

Source: Calculations based on IEA 2011; UNIDO 2011.
Note: The structure index measures the composition of industrial activities. The efficiency index measures the energy efficiency of industrial production. The intensity index measures the overall energy intensity of industrial production.

improved in all heavy industries, especially for iron and steel, chemicals, and nonferrous metals (like zinc, copper, and aluminum) in most EU countries—including Bulgaria, the Czech Republic, Estonia, Latvia, the Slovak Republic, and Slovenia. In addition, the value-added share of heavy industries has been relatively stable. The machinery sector experienced strong economic growth in the Czech Republic, Estonia, and Turkey, while food and textiles had a declining economic trend. Because food and textile production are more energy intensive than machinery, economic shifts may have contributed slightly to the decrease in overall energy intensity in these countries.

Trade Specialization and Carbon Leakage

Efficiency improvements had a stronger positive impact than changes in the economy on energy productivity for a number of reasons. First, production increased enormously for some of the energy-intensive industries in ECA countries during the commodity boom in the 2000s. For example, Russian crude steel production increased by more than 22 percent between 2000 and 2007.[12] Cement output increased by 25 percent between 1998 and 2007 in EU countries (European Commission n.d.). To facilitate output growth, countries have increased the use of existing production capacity—for example, as they have in Hungary (Odyssee 2009). Alternatively, they built new manufacturing plants that are more efficient than old ones.

Second, EU enlargement offers access to a larger market and opportunities for specialization. Within the EU, some new member states including Bulgaria, Estonia, Hungary, Latvia, and Romania have ECA's most specialized industrial structures. For example, Hungary is highly specialized in refined petroleum products, while Bulgaria shows high specialization in mining and quarrying (European Commission 2011). Specialization brings economies of scale, which increases efficiency in energy use. Countries may have also reinforced their position in industries of higher energy intensity through specialization.

Finally, EU energy polices, such as the Emissions Trading System, could encourage relocation of energy-intensive industries to ECA countries that have fewer carbon restrictions. Industries such as cement and aluminum that are emission intensive and cannot simply add the increased energy costs to product prices are most likely to relocate. Because the region's industrial energy efficiency is lower on average than in EU countries, outsourcing of emission intensive products leads to concerns about "carbon leakage" (see also chapter 4). That would mean that an increase of energy-intensive manufacturing products in non-EU countries will offset the decrease in EU manufacturing sector emissions. Many studies have simulated the potential leakage rates in specific industries. Based on different assumptions regarding CO_2 prices, these studies find carbon leakage rates could range from the very low to significant at 30 percent or more (Reinaud 2008).[13]

Determinants of Industrial Energy Intensity

The convergence in energy productivity and changes in the composition of economic activities and energy efficiency are driven by several factors, including income and price effects. As income grows, consumer demand and investment for manufactured goods expand. Countries turn over their capital stock—equipment and factories—to meet increasing demand, and efficiency rises. When energy prices are high, there is a greater incentive for firms to invest in efficiency. When it is cheap, there is little incentive for industry to reduce energy use or shift to less energy-intensive sectors.

Both income growth and rising energy prices helped lower overall industrial energy intensity in ECA (see figure 7.10), mostly through changes in energy efficiency rather than changes in the mix of manufacturing activities (see figure 7.11).[14] Controlling for other influences, a 1 percent increase in per capita GDP is on average associated with a 1.3 percent reduction in energy intensity (Zhang 2012). A 10 percent increase in an electricity price is associated with a 2.3 percent reduction in energy intensity.

FIGURE 7.10

Inverse Relationship between Energy Intensity and Income or Energy Price in Selected Countries, Europe and Central Asia, 1998–2009

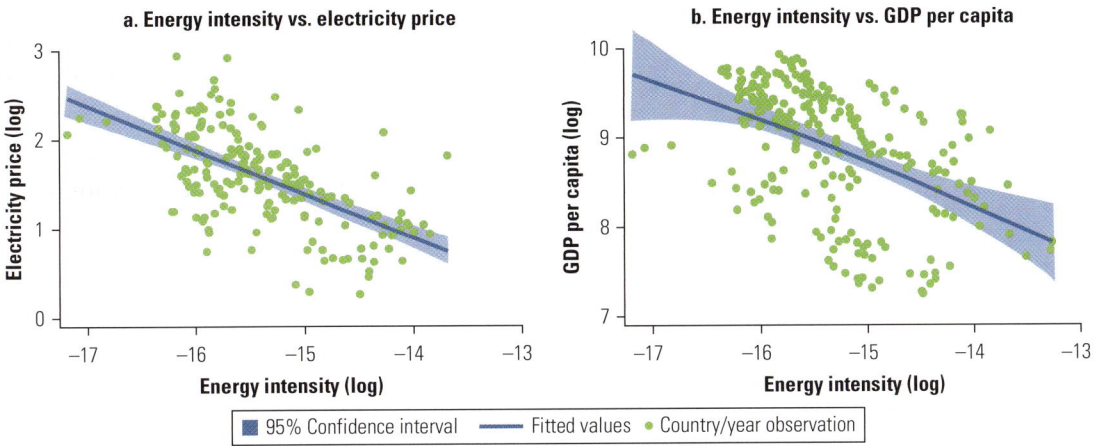

a. Energy intensity vs. electricity price

b. Energy intensity vs. GDP per capita

■ 95% Confidence interval ━━ Fitted values ● Country/year observation

Source: Calculations based on ERRA n.d.; IEA 2011; World Bank's *World Development Indicators.*
Note: Industrial electricity prices are measured in constant 2000 U.S. dollars. Per capita GDP is measured in constant 2005 PPP dollars. The charts are based on the following countries where industrial energy prices are available: Albania, Armenia, Azerbaijan, Bulgaria, Croatia, Estonia, Georgia, Hungary, Kazakhstan, the Kyrgyz Republic, Latvia, Lithuania, Poland, Romania, Turkey, and Ukraine. PPP = purchasing power parity.

FIGURE 7.11

Effects of Price and Income Changes on Energy Efficiency, Energy Intensity, and Industrial Structural Composition in Europe and Central Asia, 1998–2009

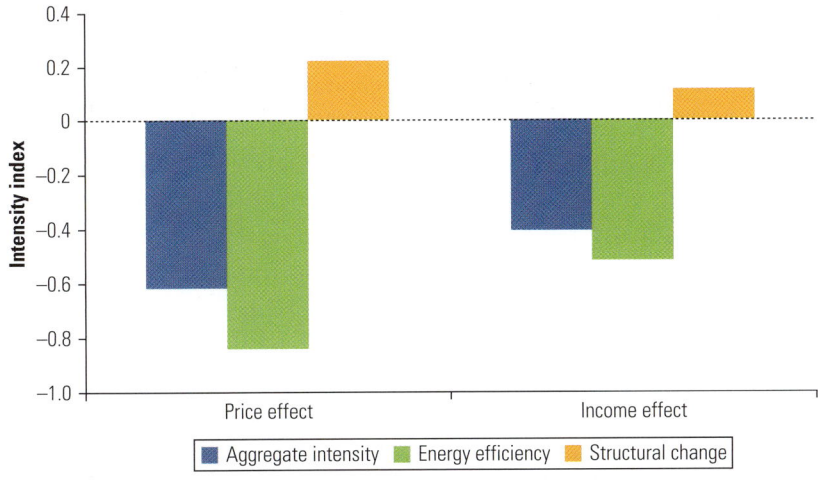

■ Aggregate intensity ■ Energy efficiency ■ Structural change

Source: Calculations based on Zhang 2012.
Note: The graph shows the sensitivity of aggregate energy intensity, energy efficiency, and structural indexes to price and income changes. A 1 percent increase in electricity prices is associated with a 0.41 percentage point decrease in aggregate energy intensity and a 0.52 percentage point improvement in energy efficiency, but a 0.22 percentage point increase in the structural index. A 1 percent increase in per capita GDP is associated with a 0.6 percentage point decrease in energy intensity and a 0.8 percentage point improvement in energy efficiency, but a 0.1 percentage point increase in the structural index. The structural index increases with price and income, partially offsetting the efficiency gains. See Zhang (2012) for regression results.

Before the global financial crisis, GDP grew on average by 7 percent annually in ECA countries. This economic growth led to an increase in the demand for manufacturing commodities such as iron and steel. To satisfy increased demand for commodities, countries added new, more efficient factories, thereby reducing the share of smaller, less efficient plants. In addition, industries undergoing rapid growth were less capital-restricted and were more receptive to efforts to improve their competitive edge through increased management of energy costs. For example, because Romania experienced strong growth in demand, industrialists rapidly introduced modern technologies into outmoded or newly built facilities; increased the production of commodities (especially those for sale in foreign markets); and invested profits into further upgrading and expansion of industrial production. By contrast, in Central Asia, where production levels have stalled, manufacturers have failed to upgrade to more-efficient technologies.

Rising energy prices contribute to energy productivity by encouraging the adoption of energy-saving technologies—especially in sectors where energy constitutes a significant portion of their total costs. These efforts include the introduction of more energy-efficient machines, production processes, or materials. For example, producing clinker from raw material is the main energy-consuming process in a cement factory. Using other synthetic material to substitute for clinker can significantly reduce energy use in cement production.

The cost and availability of capital significantly shapes energy intensity in industry. When capital and energy are interchangeable, a more capital-intensive manufacturing sector will consume less energy. In much of ECA, however, capital and energy are complements. A larger share of capital-intensive heavy industries (such as iron and steel, chemicals, and so on) is associated with higher energy intensity. This effect indicates that the legacy of heavy industry continues to mark the region because a negative relationship between capital and energy inputs would indicate a transition to high-tech industries such as electronics.

Finally, trade brings information and technology spillovers that improve energy efficiency (see chapter 4). Trade also exposes enterprises to higher levels of competition in international markets and creates pressure to cut costs and save energy (EBRD 2010). On the other hand, trade could increase differences in energy intensity between countries by stimulating international specialization (Grossman and Helpman 1991). Energy intensity increases slightly with higher shares of manufacturing exports in ECA countries, implying that trade induces specialization in more energy-intensive industries. Many of the region's countries are net carbon exporters, as the average carbon intensity of exports in many of the countries is higher than the world average, as figure 7.12 shows. In fact, Ukraine, Russia, and Kazakhstan are the first-, third-, and fourth-most carbon- (and energy-) intensive exporters in the world (Davis and Caldera 2010). This finding is not surprising. Energy- and emission-intensive goods tend to be produced in countries with abundant energy resources, low energy prices, and more lenient emissions rules. This effect dominates other, potentially positive effects of trade on energy productivity, such as increased competition.

One implication of this result is that domestic manufacturers whose products have enjoyed high energy subsidies would lose competitiveness if ECA countries integrated more closely into global markets where they are more bound by international norms. With World Trade Organization (WTO) membership, industrial producers in Russia, for instance, "should understand one thing: the time of national markets is over. There will be no more comfortable niches," Vladimir Putin stated in a blog post (*Financial Times* 2012). With greater market opening, industrial energy efficiency must improve to international standards.

FIGURE 7.12
Carbon Intensity of Trade, Selected Countries, Relative to World and EU-15 Averages, 2004

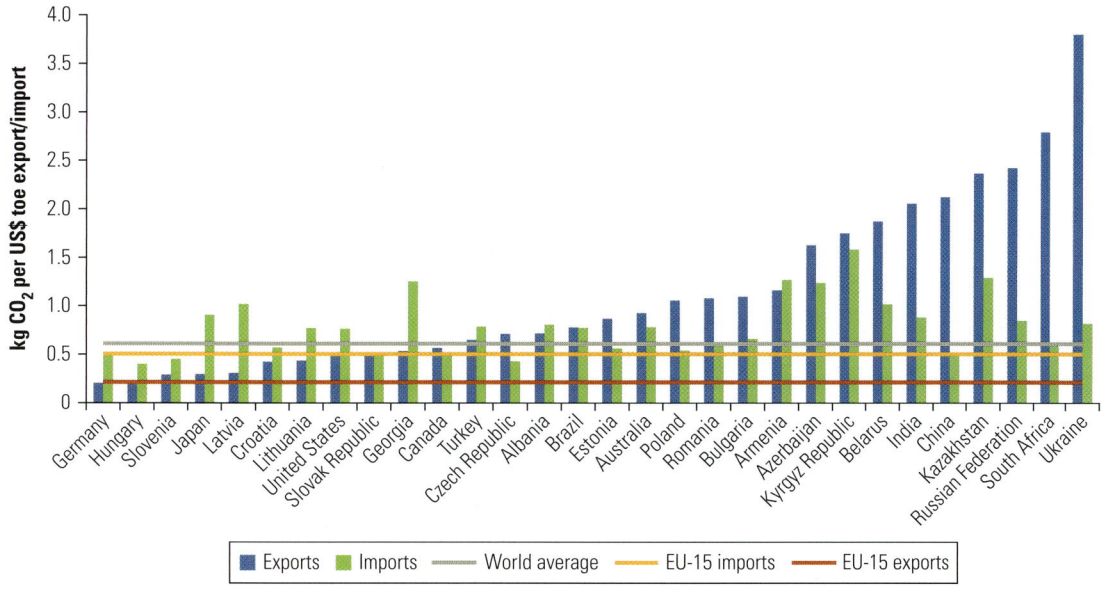

Source: EBRD 2010.
Note: EU-15 countries include Austria, Belgium, Denmark, Finland, France, Germany, Greece, Ireland, Italy, Luxembourg, the Netherlands, Portugal, Spain, Sweden, and the United Kingdom. kgCO$_2$ = kilograms of carbon dioxide; toe = tons of oil equivalent.

Industrial Energy Efficiency: Where Are the Low-Hanging Fruits?

Although ECA countries have made significant strides in reducing industrial energy intensity, substantial opportunities to further improve efficiency and reduce CO_2 emissions remain. The energy savings potential of the three most energy-intensive sectors—iron and steel, cement, and paper and pulp—is estimated to be 50 percent of current energy use with best-practice technologies, already proved to be cost effective.[15] In addition, generic technologies that could improve steam and motor systems could further deliver 10–20 percent of energy savings across the board, as shown in table 7.1.

Iron and Steel

In ECA, a region with a strong steel manufacturing tradition, steel has maintained its important role in production but also in terms of its environmental impact. This sector is the largest industrial energy user (consuming 61,288 mtoe in 2009) and the largest industrial source of CO_2 emissions (at about 35 percent of total manufacturing emissions). Steel production is highly energy intensive and relies on coal as the main energy source. Russia, Ukraine, and Turkey are the third-, seventh-, and ninth-largest steel producers in the world, respectively. The ECA region accounted for 12 percent of total world steel production in 2010.[16]

Variations in production technology contribute to considerable variation in countries' energy efficiency of steel production, as figure 7.13 illustrates. Russia uses almost twice as much energy to produce a ton of steel as the world average. After Russia, Bulgaria, Ukraine, and Croatia are also among the world's least-efficient steel producers.

TABLE 7.1

Potential Energy Savings from Adoption of Best-Practice Commercial Technologies in Europe and Central Asia

Sector	Energy savings potential	Share of current energy use (%)
Iron and steel	15.5 GJ/t	56
Cement	2.8 ~ 3.2 GJ/t	50
Paper and pulp	3.7 ~ 7 GJ/t	50
Steam systems	0.31 EJ/yr	10
Motors	0.22 EJ/yr	20

Sources: Calculations based on UNIDO 2010; IEA 2007b; IEA 2009; World Bank 2008; Worrell et al. 2008.
Note: GJ/t = gigajoules per ton; EJ/yr = exajoules per year.

FIGURE 7.13

Energy Intensity of Steel Production in Selected Europe and Central Asia Countries Relative to Other Regional Averages, 2008

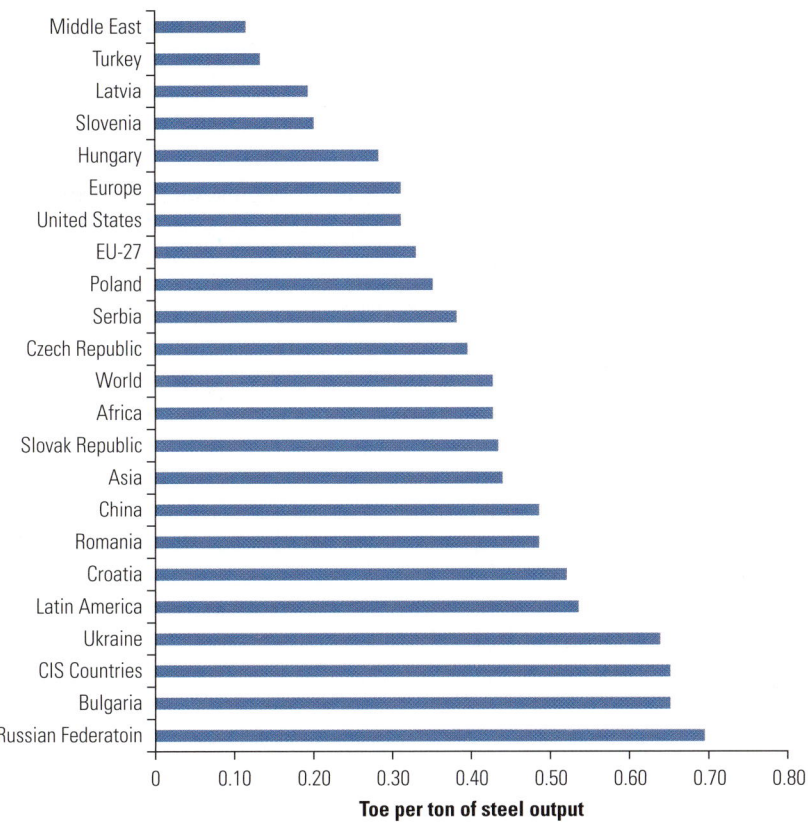

Toe per ton of steel output

Source: Global energy and CO_2 data from Enerdata 2011.
Note: EU-27 countries include all current EU members. CIS countries include Azerbaijan, Armenia, Belarus, Georgia, Kazakhstan, the Kyrgyz Republic, Moldova, the Russian Federation, Tajikistan, Turkmenistan, Ukraine, and Uzbekistan. Energy intensity is the energy used per ton of steel produced. CO_2 = carbon dioxide; toe = tons of oil equivalent; EU = European Union; CIS = Commonwealth of Independent States.

Across the world, two major processes are applied for the production of crude steel: (a) the blast furnace or basic oxygen furnace (BF/BOF) process, which uses iron ore and scrap; and (b) the electric arc furnace process (EAF), which uses direct reduced iron, scrap, and cast iron. The EAF process is much less energy intensive (4–6 gigajoules [GJ] per ton) than the BF/BOF process (13–14 GJ per ton) because there is no need to convert iron ore into iron, and it removes the need for the ore preparation steps (IEA 2009). However, the steel industry in ECA (except in Turkey) relies heavily on the more energy intensive BF/BOF processes, as shown in table 7.2.[17] Only 1.2 percent of all steel in the world is produced through other processes such as open-hearth furnaces, which are outdated and extremely

TABLE 7.2

Steel Production and Technology in Europe and Central Asia Countries Relative to Selected Other Countries and Regions, 2010

Country or region	Production (Mt/year)	Share in world (%)	Electric steel (%)	BOF steel (%)	Open hearth furnace steel (%)
Russian Federation	67.0	4.7	26.9	63.3	9.8
Ukraine	33.6	2.4	4.5	69.3	26.2
Turkey	29.0	2.1	71.7	28.3	0
Poland	8.0	0.6	50.0	50.0	0
Czech Republic	5.2	0.4	8.1	91.9	0
Slovak Republic	4.6	0.3	7.3	92.7	0
Kazakhstan	4.3	0.3	n.a.	n.a.	n.a.
Romania	3.9	0.3	46.5	53.5	0
Hungary	1.7	0.1	5.4	94.6	0
China	626.7	44.3	9.8	88.7	1.5
India	66.8	4.7	60.4	39.6	0
United States	80.6	5.7	61.3	38.7	0
Asia	881.2	62.3	19.9	80.0	0.1
EU-27	172.9	12.2	41.9	57.7	0.4
Latin America	43.8	3.1	34.5	65.5	0
World	1,414	100.0	29.0	69.8	1.2

Source: IISI 2011.

Note: The nine ECA countries listed in the table represent more than 97 percent of total steel production in the region. EU-27 countries include all current EU members. Mt = millions of tons; BOF = basic oxygen furnace (process); n.a. = not applicable; EU = European Union.

inefficient. Gone in most of the world, this process is still used in Russia and Ukraine.[18]

Other factors influencing energy efficiency in steel production are economies of scale, the level of waste-energy recovery, the quality of iron ore, operational know-how, and quality control.[19] So even using the same production technology, there is significant difference in energy efficiency between plants. For example, electric arc furnaces in former Soviet Union (FSU) countries use on average 630 kWh per ton versus 400 kWh per ton in OECD countries (IEA 2007b). Continuous casting, which uses two to three times less energy, is used in only 17 percent of FSU castings, and around 20 percent in the EU-10, compared with 53 percent in the United States and 90 percent in Japan (UNIDO 2008). Overall, economies in transition have the highest energy intensity and also the highest potential for energy savings (15 GJ per ton of crude steel production), as shown in figure 7.14.

FIGURE 7.14

Energy Intensity and Energy Savings Potential for the Iron and Steel Industry, Selected Countries and Regions, 2005

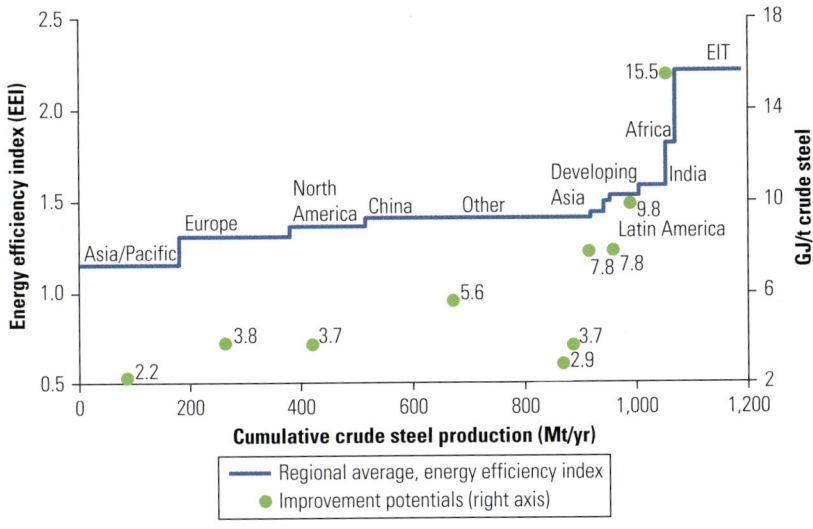

Source: UNIDO 2010.
Note: Regional average energy savings potential is denoted by the purple dots with respect to the y-axis on the right side. The energy efficiency index divides actual energy use of steel plants by the energy use of the best practice technology. EIT = economies in transition; GJ/t = gigajoules per ton; Mt/yr = millions of tons per year.

In the past decades, the region has made great progress in modernizing steel production (as in Romania; see box 7.1). The industry can further improve its energy efficiency by closing open-hearth furnaces, switching from oxygen to electric processes, increasing continuous casting, improving quality control and process management, and adopting other energy-efficient technologies.[20] The sector can also make more use of recycled materials to reduce its energy consumption because the remelting of scrap typically requires about 40 percent less energy than the production of iron and steel from iron ore. Finally, diminishing the role of the iron and steel industry in the economy is a structural strategy to improve economywide efficiency. This can be accomplished by reducing metals use, either absolutely or by substituting with other lighter-weight materials such as aluminum.

Cement

In growing economies from Asia to Eastern Europe, cement is the glue of progress. The main ingredient of concrete, cement is essential for constructing buildings and laying roads. Cement is also the single biggest material source of carbon emissions in the world. It requires less energy per ton of product than steel or aluminum, but

BOX 7.1

Energy Efficiency Investment in Romania's Steel Industry

Ductil Steel is one of Romania's main producers of steel billets, wire, and wire products. To increase its production capacity by 2.5 times without increasing energy costs, the plant replaced an old furnace with a state-of-the-art electric arc furnace and introduced continuous melting technology. The new system is expected to lower energy use per ton of liquid steel by 44 percent, cutting energy costs by €15 million per year. The GHG emissions are expected to fall by around 140,000 tons per year.

BETA Buzau is a Romanian company making steel and nonferrous specialized equipment, such as pressure vessels, tanks, and heat exchangers. The company implemented a suite of 12 energy efficiency investments that yielded 67 percent energy savings on average. For example, by replacing a press and drawing technology that removed the need for heating the metal, the company achieved 91 percent (3.7 gigawatt-hours [GWh]) energy savings per year with an investment of €50,000. By improving the insulation of two industrial buildings, the company saved another 1.1 GWh per year with an investment cost of €400,000.

Source: Case studies, EU/EBRD Energy Efficiency Financing Facility (http://www.eeff.ro/).

the volume of cement production is much larger, with an estimated 2.6 billion tons produced in 2007 (IEA 2009). Russia and Turkey are the sixth- and ninth-ranked cement manufacturers in the world, accounting for 2.1 percent and 1.9 percent of global output, respectively.

Cement is produced from a feedstock of limestone, clay, and sand. The raw material is fed into a rotating kiln and exposed to intense heat to form "clinker." Clinker is then ground with other additives to create cement. Most of the energy consumption occurs during the fabrication of clinker using either a "wet" or "dry" process, depending on the water content of the raw material feedstock. The dry process using preheaters consumes 20–30 percent less energy than the wet process because it avoids the need for water evaporation (IEA 2007b). Kiln type also affects energy consumption: the rotary kilns are more efficient than vertical shaft kilns.

Figure 7.15 compares the average energy consumption per ton of clinker between regions. In the Commonwealth of Independent States (CIS) countries, the thermal energy needed per ton of clinker is the highest in the world at 6.1 GJ per ton of clinker in 2009 (as shown in figure 7.15), compared with 3.3 GJ per ton in Asia and 4.1 GJ per ton in North America.[21] Former Soviet Union countries

FIGURE 7.15

Thermal Energy Consumption in Production of Clinker, Selected Countries and Regions, 1990–2009

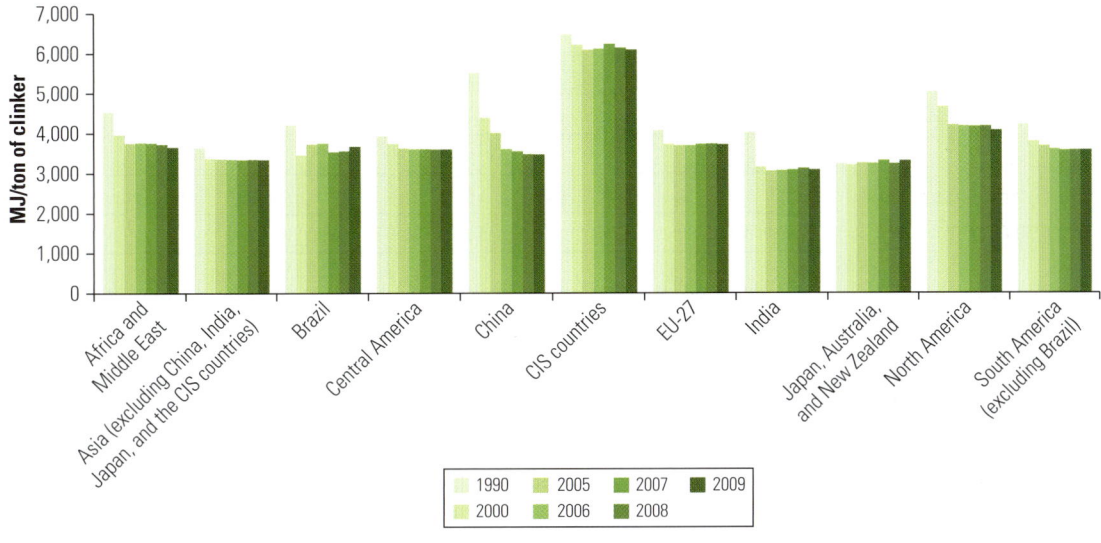

Source: CSI 2009.

Note: EU-27 countries include all current EU members. CIS countries include Armenia, Azerbaijan, Belarus, Georgia, Kazakhstan, the Kyrgyz Republic, Moldova, the Russian Federation, Tajikistan, Turkmenistan, Ukraine, and Uzbekistan. MJ = megajoules; CIS = Commonwealth of Independent States; EU = European Union.

(which in the IEA data include the CIS as well as the Baltic countries of Estonia, Latvia, and Lithuania) have the highest share of less-efficient wet process kilns—used for 78 percent of production compared with 3 percent in China and 18 percent in the United States, as table 7.3 shows. Replacing production technology could significantly improve the industry's energy efficiency in ECA. Garadagh, the largest cement producer in Azerbaijian, reduced its energy consumption by about 50 percent by turning from a water-intensive wet system to a more energy-efficient dry system in 2009.

The high energy intensity of ECA's cement production also stems from the age of the capital stock. Almost 80 percent of the plants were built before the 1980s, requiring more than 6.0 GJ per ton level (see figure 7.16). In Western Europe, new plants represent almost 80 percent of the capital stock. Replacing or retrofitting these plants could halve energy inputs to 2.9–3.3 GJ per ton of clinker (UNIDO 2010). In addition to phasing out wet kilns, increasing the use of alternative fuels (such as waste plastic) and clinker substitutes (reducing the amount of clinker produced) also improve energy efficiency.

TABLE 7.3

Cement Production Processes, Selected Countries and Regions, 2002–06

Percent

Country or region	Process type			
	Dry	Semi-dry	Wet	Vertical
Former Soviet Union[a]	12	3	78	7
Europe	92	5	4	0
China	50	0	3	47
United States	82	0	18	0
India	50	9	25	16
Latin America	67	9	23	1

Source: IEA 2007b.

a. "Former Soviet Union" countries include Commonwealth of Independent States (CIS) countries as well as the Baltic states of Estonia, Latvia, and Lithuania. CIS countries include Azerbaijan, Armenia, Belarus, Georgia, Kazakhstan, the Kyrgyz Republic, Moldova, the Russian Federation, Tajikistan, Turkmenistan, Ukraine, and Uzbekistan.

FIGURE 7.16

Energy Efficiency of Clinker Production, CIS Countries Relative to EU-27, 2009

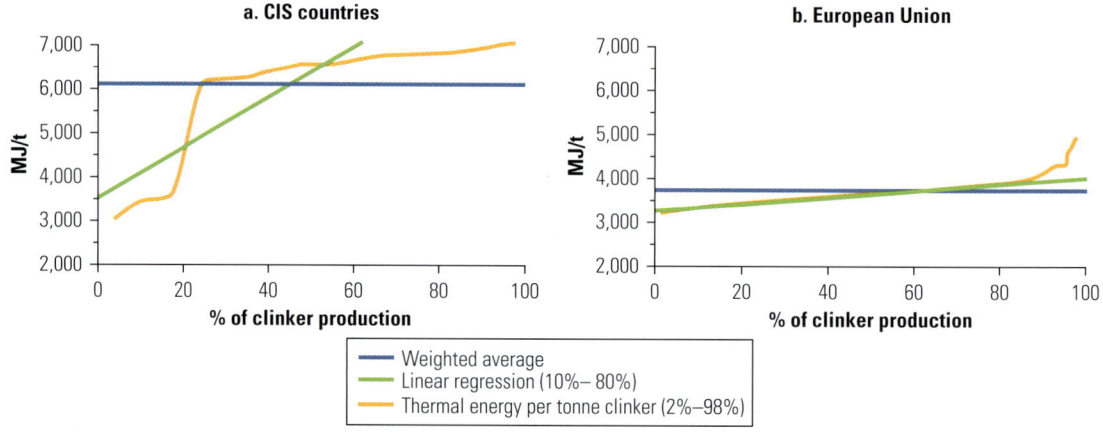

Source: CSI 2009.

Note: The figure compares the distribution of energy consumption for clinker production in CIS and EU-27 countries (at company level). Plants producing at around 3,500 MJ per t of clinker or less are new, built since the 1980s. The figure therefore reflects the age profile of cement plants in the two regions. EU-27 countries include all current members of the European Union. CIS = Commonwealth of Independent States. CIS countries include Armenia, Azerbaijan, Belarus, Georgia, Kazakhstan, the Kyrgyz Republic, Moldova, the Russian Federation, Tajikistan, Turkmenistan, Ukraine, and Uzbekistan. MJ/t = megajoules per ton.

Paper and Pulp

Paper production is an extremely energy-intensive process. Every pound of wood pulp used to make a paper product also requires 100 pounds of water. During the production process, this water must then be removed mechanically or by evaporation, which makes the industry the fourth-most energy intensive in ECA. The sector consumed 9,825 mtoe of energy in 2009, or 4 percent of total industrial energy consumption. Russia is one of the world's leading producers of paper and pulp, accounting for 2 percent of world production (IEA 2007b). Poland and the Czech Republic are two of the largest paper producers among the newer EU member states, accounting for 0.7 percent and 0.3 percent of global output, respectively (UNIDO 2010; Enerdata 2011).

Most energy used in paper production is for mechanical pulping and paper drying using heat and electricity. Economies in transition, particularly Russia, have the largest energy savings potential by adopting best-practice technologies, as figure 7.17 indicates. The region currently also has the highest energy intensity in the world, as shown in table 7.4. Within the EU, the Czech Republic and Poland are among the least efficient, using almost twice as much energy per ton of paper as the average level of the EU-15 (Enerdata 2011).

Russia has some of the oldest stock of paper and pulp mills in the world, with many over 30 years old, as shown in figure 7.18. Energy efficiency gains from using the latest technology differ depending on whether the mills are greenfield mills or retrofits. Retrofitting can be more costly and less effective than greenfield investment, but upgrading equipment and processes still yields large energy savings (see box 7.2).

FIGURE 7.17

Energy Efficiency Benchmarks for the Pulp and Paper Industry, Selected Countries and Regions, 2006

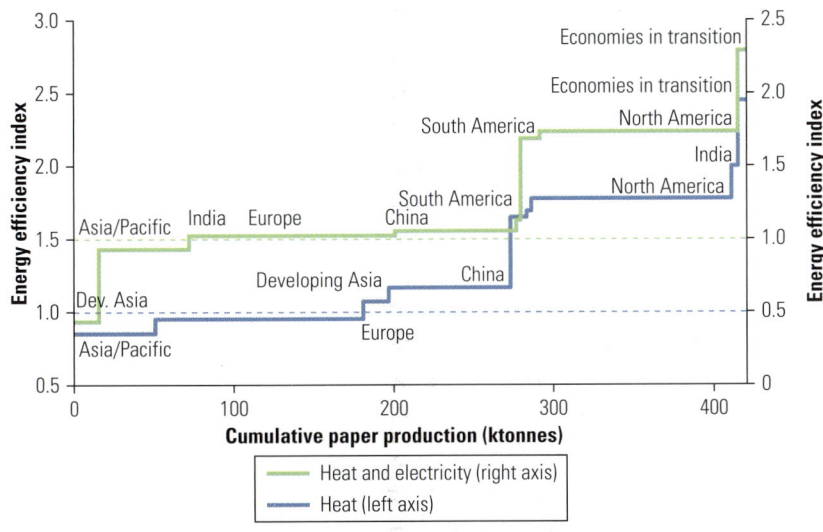

Source: UNIDO 2010.

TABLE 7.4

Energy Savings Potential in the Pulp and Paper Sector Using Best Available Technology, Selected Countries, 2006

Region or country	Improvement potential (GJ/t)
OECD Asia	0.2–0.5
OECD Europe	0.6–2.0
OECD North America	5.2–7.0
Brazil	2.4
China	0.9
Russian Federation	11.6

Source: IEA 2007b.
Note: OECD = Organisation for Economic Co-operation and Development. GJ/t = gigajoules per ton.

FIGURE 7.18

Age Distribution of Paper and Pulp Mills, Selected Countries, 2009

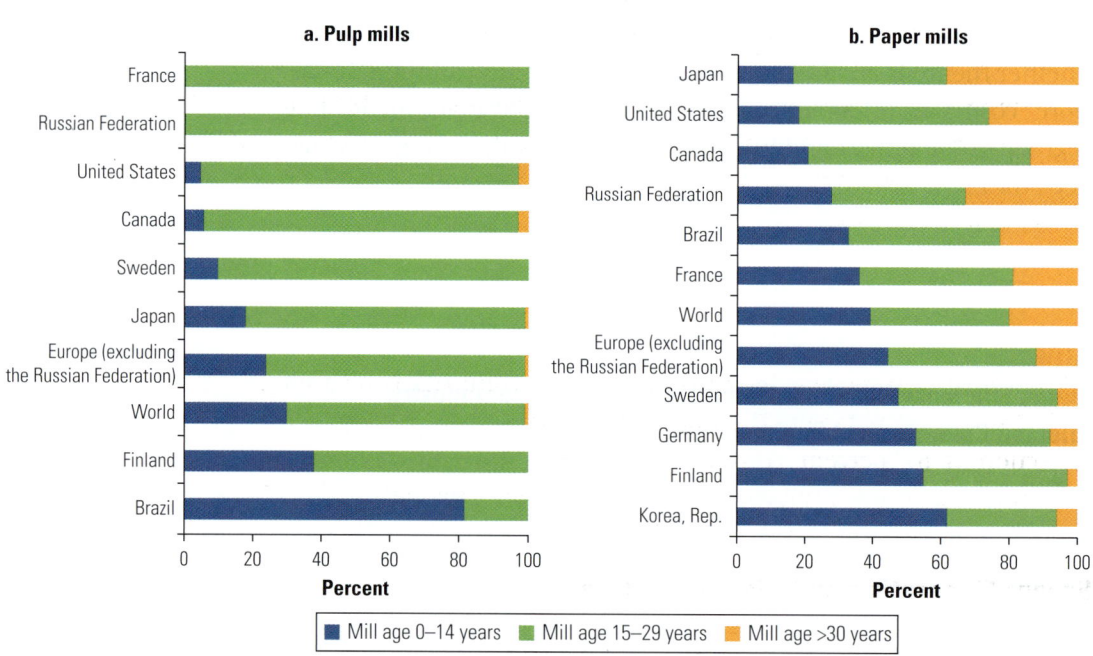

Source: IEA 2009.

Increased paper recycling and use of recovered paper can also substantially reduce energy consumption in the sector. Pulp production is more energy intensive than paper production. Paper produced from recycled paper uses 10 GJ less energy per ton than the production of raw pulp.[22] Each 1 percent increase in paper recycling saves 39.2 petajoules (PJ) of energy (IEA 2007b). Current levels of recovered paper production is 30 percent in Russia, and varies from

BOX 7.2

Efficiency Investment in Arkhangelsk Pulp and Paper Mill in Russia

Arkhangelsk Pulp and Paper Mill (APPM) is one of the leading pulp and paper producers in Russia and Europe. The company specializes in manufacturing paperboard and market pulp, fiberboard, paper, and paper stationery products. To strengthen its leadership in the industry, the company adopted an energy efficiency strategy in 2002. In 2003, the company became the first Russian enterprise to accept voluntary GHG emission reduction obligations and pledge to keep emissions under 2.6 million tons of CO_2 equivalent until 2012. This represented a 12 percent reduction from its 1990 level, while pulp production was set to increase by 8.5 percent during the same period.

During 1994–2006, despite doubling of production, APPM's consumption of energy has increased by just 25 percent, while consumption of fossil fuel (coal and black oil) has remained flat and biofuel consumption has increased twofold. The energy use per ton of pulp decreased by 35 percent from 2.24 toe in 1994 to 1.45 toe in 2006. The fossil-fuel intensity dropped by almost half, from 1.7 toe per ton to 0.9 toe per ton. To achieve this result, APPM updated its processing and production equipment and its lye recovery system. Moreover, APPM installed two new fluidized-bed boilers that allow efficient use of bark, wood waste, and even deposits from treatment facilities (up to 30 percent) without using black oil for flame stabilization.

Source: Yulkin 2005, 2010.

10 percent to 50 percent in many non-OECD countries compared with 70 percent in Japan. The upper technical limit to waste paper collection is 81 percent (IEA 2011). More effective policies that encourage paper recycling could realize large energy savings.

Steam Generation and Electric Motors

There are numerous opportunities for significant energy savings through the use of simple, low-cost retrofit technologies, such as fixing pipes, plugging leaks, insulating buildings, and improving the lighting systems. Below we review two cross-cutting options: improvement of steam generation and electric motors. In many cases, investments in these technologies pay for themselves in less than one year. Realizing these options also requires shifts in corporate culture that enable energy-focused management in the context of total corporate sustainability or social responsibility commitment. Companies that have made the shift in organizational culture report many benefits in cost savings, productivity, and operational efficiency. Where companies do not pursue efficiency opportunities on their own, the state can provide incentives, as in the case of China's "Top 1,000" program described previously.

Despite the diversity of products and processes of industrial operation, most industrial energy use has just two purposes: heating materials (typically through steam) and running motors. According to the IEA, steam and motor systems account for 15 percent and 38 percent, respectively, of global final manufacturing energy use (IEA 2007a).

Simple measures such as insulating tanks and pipes, repairing steam leaks, installing and maintaining steam traps, and operating boilers at optimal temperatures and pressures can deliver big energy savings. For example, increased insulation and steam traps each alone can reduce energy use by 5 percent (IEA 2006). The use of insulated pipelines and steam traps in non-OECD countries is 50 percent lower than in OECD countries (IEA 2009). Adding electronic temperature controls and installing improved boilers involve higher up-front cost but have short payback times. Arkhangelsk Pulp and Paper Mill decreased fuel consumption by 35 percent by upgrading to a high-efficiency boiler (Yulkin 2010). New chemical catalysts and process routes can reduce the need for steam.

Combined heat and power (CHP) systems use surplus heat from power generation or generate power as a by-product of industrial processes. CHP systems can yield fuel-use reduction of up to 35 percent. They lower transmission and distribution losses because they generate heat and electricity on site—ideal for industries that need both steam and electricity, such as chemicals, pulp and paper, aluminum, metallurgy, food, textile, and minerals. Estimated energy savings from existing industrial CHP systems in Russia and Turkey are 384 PJ and 75 PJ, respectively (IEA 2007b). A few ECA countries, such as the Czech Republic and the Slovak Republic, have a relatively high share of industrial CHP, but many could greatly expand its use, as shown in figure 7.19.

Electric motors consume more than 60 percent of industrial electricity (IEA 2009). These motors are used for pumps, fans, compressors, materials processing (such as grinding), and materials movement (cranes, elevators, and so on). New high-efficiency motors are about 85–95 percent more efficient than standard-efficiency motors. Replacing outdated air compressors with higher-efficiency models could reduce energy costs of the largest ferroalloy manufacturer in Russia by around 40 percent (RUSEFF 2011). These motors typically cost about 20 percent more than standard motors but can pay back the investment rapidly.

FIGURE 7.19

CHP Share of Industrial Power Generation, Selected Europe and Central Asia Countries and Others, 2008

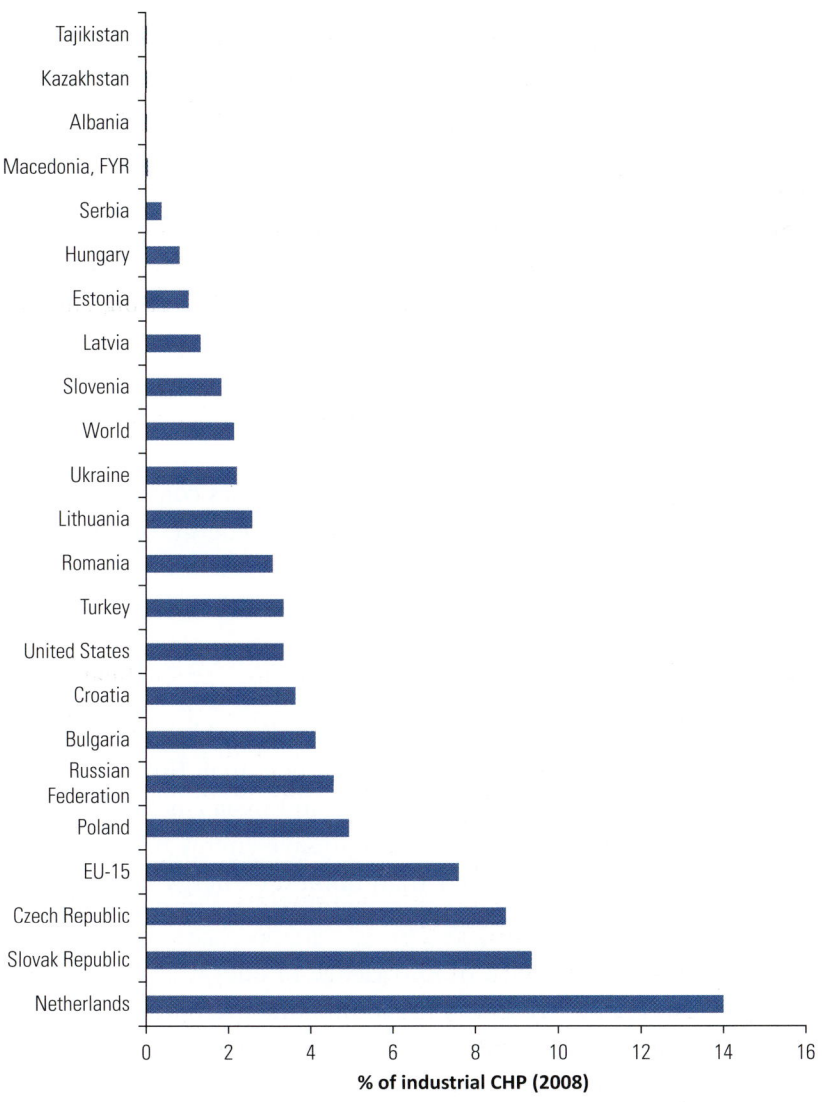

Source: Global energy and CO_2 data from Enerdata 2011.
Note: EU-15 countries include Austria, Belgium, Denmark, Finland, France, Germany, Greece, Ireland, Italy, Luxembourg, the Netherlands, Portugal, Spain, Sweden, and the United Kingdom. CHP = combined heat and power (system); EU = European Union; CO_2 = carbon dioxide.

Overall, a combination of energy-efficient technologies and improved maintenance practices is expected to save around 10 percent of total steam energy use and 20 percent of total motor energy use in the four largest manufacturing countries in ECA: Poland, Russia, Turkey, and Ukraine, as shown in table 7.5.

TABLE 7.5

Steam and Motor System Energy Savings Potential in Selected Europe and Central Asia Countries, 2006

exajoules per year

Country	Manufacturing fossil-fuel-based electricity use	Steam system energy use	Steam systems savings potential	Motor systems energy use	Motor systems savings potential
Poland	0.70	0.25	0.02	0.09	0.02
Russian Federation	5.32	2.13	0.21	0.72	0.14
Turkey	0.84	0.34	0.03	0.13	0.03
Ukraine	1.40	0.49	0.05	0.14	0.03

Source: IEA 2007a.

Closing the Efficiency Gap

Improving energy efficiency lowers costs and raises competitiveness, particularly in an environment of fluctuating energy prices. However, as with household energy efficiency, firms might underinvest in lowering energy use for several reasons (see also chapter 2):

- *Price barriers.* Low energy prices or uncertainties related to energy prices can lead to high hurdle rates for energy efficiency investments. Russia is sometimes called the "Saudi Arabia of energy conservation" because of its energy savings potential. However, some of the most energy-intensive industries in Russia (such as iron and steel and cement) lack the incentive to save energy because product prices are growing faster than domestic energy tariffs, especially for natural gas (IEA 2011).

- *Information barriers.* Lack of knowledge of energy-efficient technologies or the limited ability to monitor and evaluate energy consumption is another barrier. An International Finance Corporation (IFC) survey of 1,350 industrial enterprises in Armenia, Azerbaijan, Belarus, Georgia, Russia, and Ukraine found that companies have consistently underestimated potential energy savings by a wide margin—deviating an average of 40 percent from international best practice. They have limited information on their own energy use and of improved technology and processes (World Bank 2010a).

- *Capital barriers.* Insufficient long-term capital to finance energy-efficient modernization is a problem, especially for small and medium-size enterprises. This shortage makes it difficult to finance capital-intensive investments such as the transition from wet to dry kilns in cement manufacturing, or from the blast furnace and coke plant to direct iron ore reduction in steelmaking.

Industrial energy efficiency is an important economic and climate action priority. EU countries have implemented or considered around 260 industrial energy efficiency policies (Intelligent Energy 2009). Many CIS countries, such as Russia and Ukraine, have also passed national energy efficiency legislation and introduced national action plans (World Bank 2010a). A comprehensive energy efficiency law in Belarus brought substantial reductions in energy use. There is no silver-bullet policy. The best solution will be an "integrated approach" that combines fiscal incentives with regulations and investments.

Price Incentives: Energy Price Reform, Taxes, and Subsidies

Despite recent progress in energy price reforms, subsidies still persist in many ECA countries, as seen in a sample of countries in table 7.6. Because saved energy is increasingly seen as an energy source, the cost of energy efficiency measures will be compared with the cost of energy production. Removing energy price subsidies makes this a fair comparison.

The broader benefits of energy efficiency will also often justify more direct support, such as reduced taxes on energy-efficient equipment, accelerated depreciation, and tax credits. For example, in France, investments in energy efficiency are awarded with lease credits. These credits can be used to finance associated costs such as equipment, construction, land, and transport (Bernstein et al. 2007). Tradable energy efficiency certificates ("white certificates") are another emerging policy instrument. Under a white-certificate scheme, energy suppliers, distributers, and end users must fulfill

TABLE 7.6

Fossil-Fuel Consumption Subsidies, Selected Europe and Central Asia Countries, 2010

Country	Average subsidization rate (%)	Subsidy (US$ per person)	Total subsidy (% of GDP)
Azerbaijan	22.1	90.4	1.5
Kazakhstan	29.3	269.2	3.1
Russian Federation	22.6	274.3	2.7
Turkmenistan	65.1	994.9	19.3
Ukraine	25.7	168.7	5.6
Uzbekistan	57.1	433.7	30.5

Source: IEA 2011.

certain energy savings targets. A white certificate is issued to certify that a certain reduction of energy consumption has been achieved. White certificates are tradable so those who have overcomplied can sell their unused certificates to those who have not met their obligations. In 2009, Poland became the first ECA country to introduce a white certificate scheme. It is expected to achieve energy savings of about US$4.2 billion by 2020.

Regulations: Standards, Voluntary Energy-Saving Agreements, and the Importance of Data Collection

Economies of scale matter in many energy-intensive industrial sectors. A company that achieves economies of scale lowers the average production cost through increased production because fixed costs are shared over an increased number of goods. These industries are therefore typically dominated by relatively few large firms, which reduces the coordination costs of regulation and encourages negotiated or voluntary agreements. One frequent difficulty in developing efficient regulation is scarcity of data at both the firm and country level.

Governments have imposed technology specifications in cases (such as motors and boilers) where equipment shares large commonalities. Another form of direct regulation involves energy management standards. An energy management standard requires a facility to develop an energy management plan, which may include appointment of energy managers, mandatory periodic audits, and the implementation of energy efficiency plans. Governments can also introduce targets and agreements at the plant, firm, or sector levels that, without specifying technologies and processes, encourage firms to identify and implement appropriate technical action. These agreements range from completely voluntary, to voluntary with the threat of future taxes or regulation if shown to be ineffective, and voluntary but associated with an energy or carbon tax (Price 2005).

An IEA evaluation shows that regulating equipment performance standards has a high potential to save energy but also requires a high level of technical support so that policies are designed to leave little flexibility for interpretation (Tanaka 2009). In contrast, energy management standards are seen to be relatively effective under all criteria. Experiences with negotiated voluntary agreements are mixed. The most effective agreements are those that set realistic long-term targets (typically 5–10 years); require facility or company-level implementation plans for reaching the targets; require annual monitoring and reporting of progress toward the targets; and include a real threat of increased government regulation of energy or GHG taxes if targets are not achieved. Government support, such as information sharing, financial assistance, and awards and recognition is an essential part of an effective voluntary program (Bernstein et al. 2007).

Voluntary agreements for industrial energy efficiency improvement and reduction of energy-related GHG emissions have been used by several governments since the early 1990s, such as in Brazil, Denmark, and the United States. Energy savings achieved after the implementation of long-term agreements are estimated to be 10–20 percent in EU countries (Intelligent Energy 2009). In Lithuania, three energy-intensive companies in the cement industry have signed voluntary agreements with the government and achieved electricity savings of more than 10 GWh during 2007–09 (Balezentis, Balezentis, and Streimikiene 2011). Currently, the Czech Republic and Romania are also preparing to adopt target-setting voluntary agreements (Odyssee 2009).

The lack of energy-use data is one of the stumbling blocks to developing rational energy efficiency strategies and policies. At the facility level, metering is important for monitoring and evaluating the implementation of energy efficiency measures. Auditing is important for identifying where energy can be saved and for prioritizing energy savings opportunities. However, fewer than 20 percent of the 1,350 companies that participated in the IFC survey had installed meters at the division level. Even fewer enterprises had conducted energy audits (World Bank 2010b). In recent years, many countries have recognized the importance of energy audits. For example, in Poland, enterprises received financing of energy audits through government energy efficiency programs. Meanwhile, in Bulgaria and Romania, all enterprises whose annual energy consumption surpasses certain thresholds have to carry out an energy audit every year (Odyssee 2009).

At the macro level, important data gaps also remain, especially in Central Asian countries. For example, as shown previously in figure 7.8, 97 percent of all industrial energy use is reported as "nonspecified" in Uzbekistan (the shares are 90 percent, 88 percent, and 78 percent in Turkmenistan, the Kyrgyz Republic, and Kazakhstan, respectively). The lack of monitoring poses a major problem for industrial energy efficiency and climate policy making. Collection of reliable, timely, comparable, and detailed data that go well beyond those currently available should be a priority in these countries to provide the basis for rational policy making.

Investment: Access to Capital and Information Programs

In the long term, major increases in energy efficiency will come from replacement of technically outdated industrial facilities with new facilities that use modern technologies and practices. The capital requirements for rebuilding industrial facilities will be enormous. Liquidity constraints particularly affect small and medium enterprises (SMEs), which lack the scale and capacity to overcome relatively high transaction costs and technical complexities in project identification.

The public sector can help leverage private capital for energy efficiency. Developing an industry that profits by bundling small projects overcomes the difficulty that small-scale energy efficiency projects have in attracting financiers. This industry consists of energy auditors, construction firms, equipment suppliers, and energy service companies (ESCOs). Improving their capacity and demonstrating the viability of such a business is the key to success. In China, for example, the government worked with the World Bank for more than 10 years to develop an ESCO market. There are now hundreds of ESCOs in China doing billions of dollars of energy efficiency projects each year (almost all in industry) (World Bank 2010a). In Turkey, a recent World Bank project was specifically designed to finance energy efficiency of SMEs. The project not only supports the development of ESCOs but also encourages the provision of standardized loan products to further reduce transaction costs (see box 7.3). Chapter 2 provides more details on energy efficiency financing mechanisms.

Finally, an opportunity available to ECA countries that are not EU members is to take advantage of the Clean Development Mechanism (CDM) and Joint Implementation (JI) mechanisms under the Kyoto Protocol to finance energy efficiency.[23] So far, the region's countries have not been major contributors of certified emission reductions (CERs), representing only around 1 percent of cumulative CDM credits (Fischer and Preonas 2012). Multilateral organizations such as the World Bank and IFC have provided a direct line of credit or worked with financial intermediaries to provide long-term financing and build capacity of local banks to support industrial energy efficiency in various countries across the region. Many countries have also benefited from the EU's funding for energy efficiency and European Bank for Reconstruction and Development (EBRD) Energy Efficiency Financing Facility.

The role and potential contribution of public awareness campaigning to promote rational use of energy is widely recognized in the OECD countries. However, the full potential of communicative

BOX 7.3

Turkey SME Energy Efficiency Project

The industrial sector is the largest energy user in Turkey, accounting for about 39 percent of total final energy consumption. More than 99 percent of industries are classified as small and medium enterprises (SMEs), having fewer than 250 employees. Despite recent legislation for energy efficiency, various audit and incentive schemes, and energy price reforms, energy efficiency (EE) has remained relatively untapped in the industrial sector.

The World Bank is now developing an SME EE credit line with three Turkish banks—Vakif, Ziraat, and Halk—to support energy efficiency investments among SMEs. Unlike typical credit lines, this project has a specific goal to help lower the transaction costs of such investments while developing alternative business models, such as ESCOs, to support them. In terms of transaction costs, simple, standard product lines would be developed around key technical systems (for example, boilers, kilns and furnaces, motors and drive systems, heating and air conditioning, pumps and fans, lighting, solar water heating, and building envelope measures). The World Bank-administered Energy Sector Management Assistance Program is now supporting the development of a simple EE calculator and project screening tool to help bank loan officers assess such investments. It is also proposed that a portion of the credit line be made available to the banks' subsidiary leasing companies to offer EE equipment leasing as well.

Global Environment Facility funds are being sought to reduce the risks associated with these kinds of new loan products and leasing or ESCO business models. Use of these funds is still under development but may include subordinated debt cofinancing for new product lines and ESCO financing, a loan-loss reserve fund for new EE project types, and initial ESCO subloans to help test these new products without having the private finance initiatives take on excessive risks. Experiences gained under these schemes will help the banks expand their potential markets, improving the prospects for quick use of the International Bank for Reconstruction and Development credit line and continuing with their own capital resources once the project is completed.

instruments in pursuing energy efficiency is still largely untapped, particularly outside the OECD countries, including those in Eastern Europe (IEA 2009). Governments in ECA would benefit from energy efficiency communication strategies to raise awareness and expand information programs through energy efficiency centers, technical demonstrations, and training to promote and assist the adaptation of energy-efficient technologies. Good benchmark data and case studies that show the business case, financial modalities, and performance should be used to target company managers at the very top.

Notes

1. The industry sector covers the manufacturing sector (the manufacture of finished goods and products), mining and quarrying of raw materials, and construction. Power generation; refineries; and the distribution of electricity, gas, and water are excluded from the industry sector in this analysis.

2. Manufacturing consists of all industrial activity outside of mining and construction. Value added presents the net economic output, or gross economic output less the value of purchased inputs. It is derived by subtracting the cost of materials, supplies, fuel, and purchased electricity from the value of final outputs. For example, the value added of the refining sector excludes intermediate inputs, such as crude oil used to produce refined oil products.

3. The rate of return was based on the 72 projects carried out in 2008 (DuPont 2010).

4. Industrial value-added data are not available for many ECA countries before 1996. The percentage decline in industrial value added is calculated based on a subset of countries for which these data are available. The percentage decline in energy consumption is calculated based on data of all ECA countries.

5. Energy intensity is measured by energy use per unit of value added. Following the IEA, energy use is defined as final energy consumption per million tons of oil equivalent (mtoe), excluding energy losses in the generation, transmission, and distribution of purchased electric power.

6. The EU-15 countries include Austria, Belgium, Denmark, Finland, France, Germany, Greece, Ireland, Italy, Luxembourg, the Netherlands, Portugal, Spain, Sweden, and the United Kingdom.

7. The Central Asian countries include Kazakhstan, the Kyrgyz Republic, Tajikistan, Turkmenistan, and Uzbekistan.

8. The iron and steel industry uses most of its energy in two forms: coal is used as a feedstock to produce coke, which is then used to form steel, and coal and other fuels are used to produce heat.

9. Feedstocks—raw materials that fuel machines or industrial processes—account for about half the energy used in the chemical industry. For example, natural gas is a principal feedstock for the production of ammonia, which is used as a fertilizer. The remainder is used for process heat, motor drive, and a variety of other uses.

10. Because steel or cement production both have a lower economic value and a higher energy input, the energy intensity of these basic manufacturing industries is higher than many industries producing finished goods.

11. See detailed results in Zhang (2012). This is done for countries where energy consumption and value-added data are not always available at the subsector level, and we focus on sectors that are energy-intensive.

12. Data from World Steel Association Statistics (http://www.worldsteel.org/statistics/statistics-archive/annual-steel-archive.html).

13. Most of these studies are based on ex ante simulation; few ex post empirical studies exist.

14. This finding is consistent with previous decomposition analysis, which suggests that most of the reduction in energy intensity has occurred

because of improvements in energy efficiency as opposed to shifts from energy-intensive to less-intensive manufacturing activity.

15. The chemical sector encompasses a broad range of manufacturing processes and products. For lack of publicly available data on the product mix in the sector, we do not provide benchmarking based on physical energy intensity. Cross-country comparison of the economic energy intensity of the chemical sector can be found in Zhang (2012).

16. World Steel Association Statistics: http://www.worldsteel.org/statistics /statistics-archive/annual-steel-archive.html.

17. The choice between BOF and EAF also depends on the existence of scrap and iron ore supply in the country. Because scrap and iron ore supply are limited, there is a limit to the proportion of total steel output that can be produced by the EAF route.

18. The share of open-hearth furnaces has declined in Russia from 26.3 percent in 2000 to 9.8 percent in 2010 and in Ukraine from 48.2 percent to 26.2 percent.

19. The low efficiencies of ECA steel production also reflect the region's shrinking production volumes. Many countries in the region have passed through long recession periods of steel production before the commodity boom, which led to low levels of capacity unitization—with the Czech Republic, Poland, Romania, Russia, and Ukraine being the most affected. For example, Russian steel production in 2005 was only 74 percent of the production volume in 1990 (IEA 2007a).

20. Researchers at the U.S. Lawrence Berkeley National Laboratory have identified 47 cost-effective, energy-efficient measures to reduce energy use of steel production (Worrell et al. 2008).

21. CIS countries include Azerbaijan, Armenia, Belarus, Georgia, Kazakhstan, the Kyrgyz Republic, Moldova, the Russian Federation, Tajikistan, Turkmenistan, Ukraine, and Uzbekistan.

22. Although the use of recovered paper is less energy intensive, the impact of the use of recovered paper on CO_2 emissions is less clear because the energy used from the production of recovered paper pulp comes from fossil fuels. The production of raw pulp can be based on biomass.

23. JI is relevant for countries (like Belarus and Turkey) listed in Annex B of the Protocol, which sets binding emissions targets. CDM is designed for developing countries without firm targets.

References

Balezentis, A., T. Balezentis, and D. Streimikiene. 2011. "The Energy Intensity in Lithuania during 1995–2009: A LMDI Approach." *Energy Policy* 39 (11): 7322–34

Bernstein, L., J. Roy, K. C. Delhotal, J. Harnisch, R. Matsuhashi, L. Price, K. Tanaka, E. Worrell, F. Yamba, and Z. Fengqi. 2007. "Industry." In *Climate Change 2007: Mitigation. Contribution of Working Group III to the Fourth Assessment Report of the Intergovernmental Panel on Climate Change*, ed. by B. Metz, O. R. Davidson, P. R. Bosch, R. Dave, and L. A. Meyer. Cambridge and New York: Cambridge University Press.

CSI (Cement Sustainability Initiative). 2009. Global Cement Database, CSI, World Business Council for Sustainable Development, Geneva. http://wbcsdcement.org/GNR-2009/index.html.

Davis, S. J., and K. Caldera. 2010. "Consumption-Based Accounting of CO_2 Emissions." *Proceedings of the National Academy of Sciences* 107 (12): 5687–92.

DuPont. 2010. "Improving Energy Efficiency & Profitability with DuPont." Energy-efficiency case study, DuPont, Wilmington, DE. http://www2.dupont.com/DuPont_Sustainable_Solutions/en_US/assets/downloads/DuPont_Energy_Efficiency_Case_Study.pdf.

EBRD (European Bank for Reconstruction and Development). 2010. "Transition Report 2010: Recovery and Reform." Report of the Office of the Chief Economist, EBRD, London.

European Commission. 2011. "EU Industrial Structure 2011: Trends and Performance." DG Enterprise and Industry report, European Commission, Brussels.

———. n.d. "Mining, Metals, and Minerals—Cement." Enterprise and Industry (online database), European Commission, Brussels. http://ec.europa.eu/enterprise/sectors/metals-minerals/non-metallic-mineral-products/cement/index_en.htm.

Enerdata. 2011. *Global Energy Statistical Yearbook*. Grenoble, France: Enerdata. http://yearbook.enerdata.net/.

ERRA (Energy Regulators Regional Association). n.d. Erranet tariff database. http://www.erranet.org/Products/TariffDatabase.

Financial Times. 2012. "Guest Post by Vladimir Putin: Russia Needs More Technology and Less Corruption." January 30. http://www.ft.com/intl/comment/blogs.

Fischer, Carolyn, and Louis Preonas. 2012. "Feed-In Tariffs for Renewable Energy: Effectiveness and Social Impacts." Background paper, Europe and Central Asia Region, World Bank, Washington, DC.

Grossman, G. M., and E. Helpman. 1991. *Innovation and Growth in the Global Economy*. Cambridge, MA: MIT Press.

Havrylyshyn, O., I. Izvorski, and R. Rooden. 1998. "Recovery and Growth in Transition Economies 1990–97: A Stylised Regression Analysis." Working Paper WP/98/141, International Monetary Fund, Washington, DC.

IEA (International Energy Agency). 2006. *Energy Technology Perspectives: Scenarios & Strategies to 2050*. Paris: Organisation for Economic Co-operation and Development and IEA.

———. 2007a. "Recent Analysis into Indicators for Industrial Energy Efficiency and CO_2 Emissions." Report, IEA, Paris.

———. 2007b. *Tracking Industrial Energy Efficiency and CO_2 Emissions*. Paris: Organisation for Economic Co-operation and Development and IEA.

———. 2009. *Energy Technology Transitions for Industry: Strategies for the Next Industrial Revolution*. Paris: Organisation for Economic Co-operation and Development and IEA.

———. 2011. *World Energy Outlook 2010*. Paris: Organisation for Economic Co-operation and Development and IEA.

IISI (International Iron and Steel Institute). 2011. *Steel Statistical Yearbook 2011*. Brussels: IISI.

Intelligent Energy. 2009. "Energy Efficiency Trends and Policies in the Industrial Sector in the EU-27." Lessons from the ODYSSEE-MURE Project, Intelligent Energy Europe and ADEME Editions, Paris.

Myers, N., and J. Kent. 2001. *Perverse Subsidies: How Tax Dollars Can Undercut the Environment and the Economy*. Washington, DC: Island Press.

Odyssee. 2009. Odyssee Energy Efficiency Database for Europe. http://www.odyssee-indicators.org/database/database.php.

Price, L. 2005. "Voluntary Agreements for Energy Efficiency or GHG Emissions Reduction in Industry: An Assessment of Programs around the World." Proceedings of the 2005 American Council for Energy Efficient Economy (ACEEE) Summer Study on Energy Efficiency in Industry, West Point, NY.

Reinaud, J. 2008. "Climate Policy and Carbon Leakage: Impacts of the European Emissions Trading Scheme on Aluminum." International Energy Agency (IEA) Information Paper, Organisation for Economic Co-operation and Development and IEA, Paris.

RUSEFF (Russian Sustainable Energy Financing Facility). 2011. "40% Reduction in Energy Costs at Metal Production Company." http://www.ruseff.com/Practical-Cases/SVERDLOVSK/41.

Sachs, J. D. 1995. "Reforms in Eastern Europe and the Former Soviet Union in Light of the East Asian Experiences." *Journal of the Japanese and International Economies* 9 (4): 454–85.

Tanaka, K. 2009. "A Review of Energy Efficiency Policy in Industry." International Energy Agency (IEA) Information Paper, Organisation for Economic Co-operation and Development and IEA, Paris.

U.S. Congress, Office of Technology Assessment. 1993. *Energy Efficiency Technologies for Central and Eastern Europe*. OTA-E-562. Washington, DC: U.S. Government Printing Office.

UNEP (United Nations Environment Programme). 2003. *Energy Subsidies: Lessons Learned in Assessing Their Impact and Designing Policy Reforms*. Sheffield, U.K.: Greenleaf Publishing.

UNIDO (United Nations Industrial Development Organization). 2008. "Policies for Promoting Industrial Energy Efficiency in Developing Countries and Transition Economies." Policy paper, UNIDO, Vienna.

———. 2010. "Global Industrial Energy Efficiency Benchmarking: An Energy Policy Tool." Report, UNIDO, Vienna.

———. 2011. Industrial Statistical Databases. UNIDO, Vienna. http://www.unido.org/statistics.

World Bank. 2008. "Energy Efficiency in Russia: Untapped Reserves." Study report, International Finance Corporation, World Bank Group, Washington, DC.

———. 2010a. *Climate Change and the World Bank Group—Phase II: The Challenge of Low-Carbon Growth*. Independent Evaluations Group (IEG) Study Series. Washington, DC: World Bank.

———. 2010b. *Lights Out? The Outlook for Energy in Eastern Europe and the Former Soviet Union*. Washington, DC: World Bank.

Worrell, E., L. Price, M. Neelis, C. Galitsky, and N. Zhou. 2008. "World Best Practice Energy Intensity Values for Selected Industrial Sectors." Report, Ernest Orlando Lawrence Berkeley National Laboratory, Berkeley, CA.

Yulkin, Michael A. 2005. "Russia and the Kyoto Protocol: How to Meet the Challenges and Not to Miss the Chances." Online article, Climate Change Global Services. http://rusbiocenter.ru/eng/doc/Yulkineng17062005.pdf.

———. 2010. "Climate and Energy Efficiency, or a True Story of a Pulp and Paper Mill." Online article, Climate Change Global Services. http://ccgs.ru/en/publications/articles/.

Zhang, Fan. 2012. "Industrial Energy Efficiency in the Europe and Central Asia Region." Background paper, Europe and Central Asia Region, World Bank, Washington, DC.

MAP 8.1

Greenhouse Gas Emissions from Road Transportation

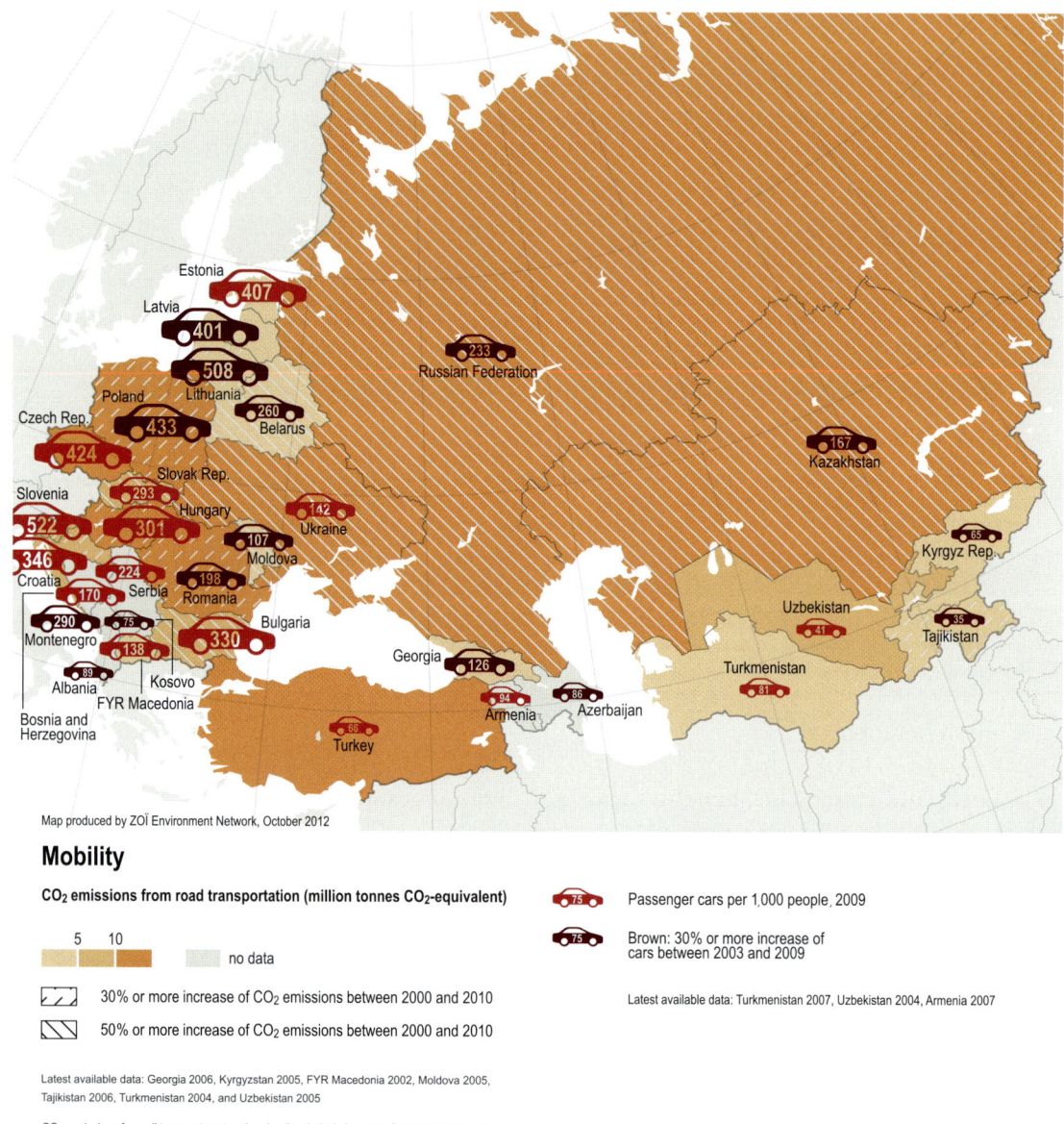

Map produced by ZOÏ Environment Network, October 2012

Mobility

CO₂ emissions from road transportation (million tonnes CO₂-equivalent)

5   10

[ ] no data

[///] 30% or more increase of CO₂ emissions between 2000 and 2010

[\\\] 50% or more increase of CO₂ emissions between 2000 and 2010

Latest available data: Georgia 2006, Kyrgyzstan 2005, FYR Macedonia 2002, Moldova 2005, Tajikistan 2006, Turkmenistan 2004, and Uzbekistan 2005

CO₂ emissions from all transport sectors (road, rail, aviation): Armenia, Georgia, Kyrgyzstan, Moldova, Tajikistan, Turkmenistan, and Uzbekistan

[car 75] Passenger cars per 1,000 people, 2009

[car 75] Brown: 30% or more increase of cars between 2003 and 2009

Latest available data: Turkmenistan 2007, Uzbekistan 2004, Armenia 2007

Source: World Bank, World Development Indicators (http://data.worldbank.org); Eurostat 2011 Energy, transport and environment indicators, Luxembourg (http://ec.europa.eu/eurostat); United Nations Framework Convetion on Climate Change (http://unfccc.int/ol/FlexibleQueries.do).

CHAPTER 8

Mobility[1]

Main Messages

- Local and immediate benefits from reduced congestion, lower health impacts from air pollution, fewer accidents, and less noise pollution alone make the economic case for climate action in road transport. Predicted climate change impacts represent only about 5 percent of these external costs from transport that can add up to almost US$0.50 per kilometer.

- Europe and Central Asia (ECA) countries still have relatively compact cities, a legacy of good public transit, and high rail shares that enable a more carbon-efficient transport sector. But recent policies have favored increasing the use of cars and trucks. A further shift toward more energy-intensive transport options will be hard to reverse.

- A broad range of well-understood policies support sustainable urban and regional transport. These can slow but are unlikely to reverse the increase in transport-related emissions that come with rising wealth. However, they will promote more sustainable and efficient multimodal mobility.

Mobility is essential for economic development. It fosters growth by matching people and firms with economic opportunities and by enabling the geography of production to adapt to market demand (World Bank 2009). Europe and Central Asia (ECA) countries have benefited from greater economic integration made possible by improved transport links. The EU-10 countries are now fully tied into the European Union's (EU) production networks.[2] For instance, the Slovak Republic's foreign direct investment (FDI)-driven vehicle manufacturers more than doubled production between 2004 and 2008. EU candidate countries are upgrading their transport systems to support the economic convergence that membership promises.[3] Better transport infrastructure in Southeastern Europe will promote domestic integration and provide crucial links between Western Europe, the Balkan countries, Turkey, and beyond.[4] On a broader scale, Central Asia is becoming a bridge between Asia and Europe.[5] The entire region will benefit from interregional transport investments connecting China with Europe. On a more local scale, economic development comes with greater concentration of economic activity in cities. Growing cities—whether from population growth or migration—require smart transport policies to ensure efficient and sustainable urban mobility.

But a transport sector that benefits development also has significant costs. Transport accounts for 22 percent of global carbon dioxide (CO_2) emissions from fuel combustion, three-quarters of which come from road transport. With rising motorization, emissions from transport are bound to grow further both in absolute and relative terms. Countries and cities face the challenge of reducing these emissions without jeopardizing mobility. This requires three main changes: (a) fewer or shorter trips, for instance through better logistics and land use planning; (b) lower fuel use per kilometer (km) traveled or shipped through efficiency gains and shifts to greener forms of transportation; and (c) a shrinking share of fossil fuels in powering transport. These improvements will contribute to climate change mitigation. But the same policies that promote climate goals also help achieve three important objectives with large and more immediate benefits:

- *First, transport sector policies aimed at emission reductions have significant welfare benefits, especially in urban areas.* CO_2 emissions are only one of the negative externalities—or unwanted side effects—from transport, and they are not even the most significant one. The largest costs come from the loss of time and cost of extra fuel on congested roads, the local health effects from vehicle noise and air pollution, and the damages from traffic accidents. Together these account for as much as 95 percent of the hidden cost of road transport, climate change impacts accounting for the remaining 5 percent (Proost and Van Dender 2011). Reducing these local impacts of transport raises the quality of life in cities. As economies become more knowledge intensive, cities need to compete for qualified workers and entrepreneurs. They will want to live in cities that provide a clean, safe, and efficient environment for work and life. Sustainable transport becomes ever more important. Cities from New York to Copenhagen are moving away from car-centered to people-centered transport planning to strengthen their economic competitiveness.

- *Second, a more efficient transport sector supports economic growth.* Lowering transport emissions by reducing fuel use and congestion raises the efficiency of the sector. Lower transport costs facilitate inter- and intraindustry trade, labor market pooling, and the exchange of knowledge and ideas. These "agglomeration economies" favor concentration of economic activities and lead to productivity improvements. As the ECA region continues to diversify its economies—away from being resource based in the east and toward closer integration with the EU in the west—steady improvements in transport will be necessary to promote the emergence of productive cities across the urban hierarchy. Growth may well increase the demand for mobility overall, even with good policies, which would make it even more important to shrink the footprint of transport services.

- *Third, climate-smart transport policies that reduce fossil-fuel consumption provide insurance against future economic shocks.* Even if "peak oil" remains in the distant future, fossil-fuel prices will likely increase as production becomes more expensive and global demand increases. And even if a global climate agreement is unlikely in the near future, some form of carbon price will become likely once climate change impacts become more apparent. It is important to act early because transport infrastructure is subject to three types of long-term inertia (World Bank 2012a; Shalizi and Lecocq 2009):

  - *Infrastructure inertia* favors improving existing (carbon-intensive) infrastructure over replacement with new (low-carbon) options. Once a dominant road-oriented transport network is established, it becomes difficult to move to an integrated, multimodal structure that also provides other mass-transit options like light rail or buses.

  - *Technology inertia* implies that major road investments favor additional investments in infrastructure or services that build on them and make it harder for alternatives to compete. For instance, a powerful trucking industry will encourage future investments in roads rather than rail.

o *Built-environment inertia* means that transport investments guide consumers' and firms' investments in housing and commercial real estate. Car-oriented development encourages low urban densities, which make public transport less sustainable.

There is immense potential for ECA countries to pursue CO_2 emission-reducing transport policies, local welfare benefits, and economic productivity gains. Like their Western European counterparts, ECA cities are still relatively compact and have a legacy of extensive and efficient public transportation. If the recent underinvestment in buses, metros, and light rail systems can be reversed, the region's cities will achieve greater and cleaner mobility—focusing on moving people rather than moving cars.

At the interregional level, ECA has an extensive rail network, and many of the railways are relatively well managed and well used. In 2009, railways accounted for an average of 30 percent of total inland freight transport among the region's EU members versus 13 percent in Western Europe.[6] But much of the recent investment has been in interregional roads and highways, leading to an increasing road share in freight transport. Western European countries that favored road investments for too long are now trying to shift more freight and passenger traffic back to railways. ECA can still limit the shift to increased dependence on roads in the first place. Supply-side policies and investments in public transport will be necessary but are not sufficient. They need to be complemented by demand-side policies that reflect the full economic and social costs of car and truck use, encouraging the switch to public transportation and fewer trips in more fuel-efficient vehicles.

Although the principles of sustainable mobility are well known, putting them into practice is challenging. In the absence of technical breakthroughs, such as inexpensive electric cars charged by renewable energy, they require behavioral changes that policies encourage but do not guarantee. Transport volumes tend to increase with rising incomes, and many will pay a premium to be able to travel independently. Economic analysis suggests that with a general carbon price, transport emissions would fall less than those in other sectors such as power generation and industrial production (where emission reductions are often cheaper and fewer actors make coordination easier). However, given the broader benefits of an efficient transport sector, climate-smart transport policies are worth pursuing. Rather than following the car-oriented, energy- and emission-intensive mobility patterns of North American and some Middle Eastern countries, ECA could emulate high-income European and Asian countries that have a much smaller transport footprint (figure 8.1). This chapter discusses the policies that promote this goal.

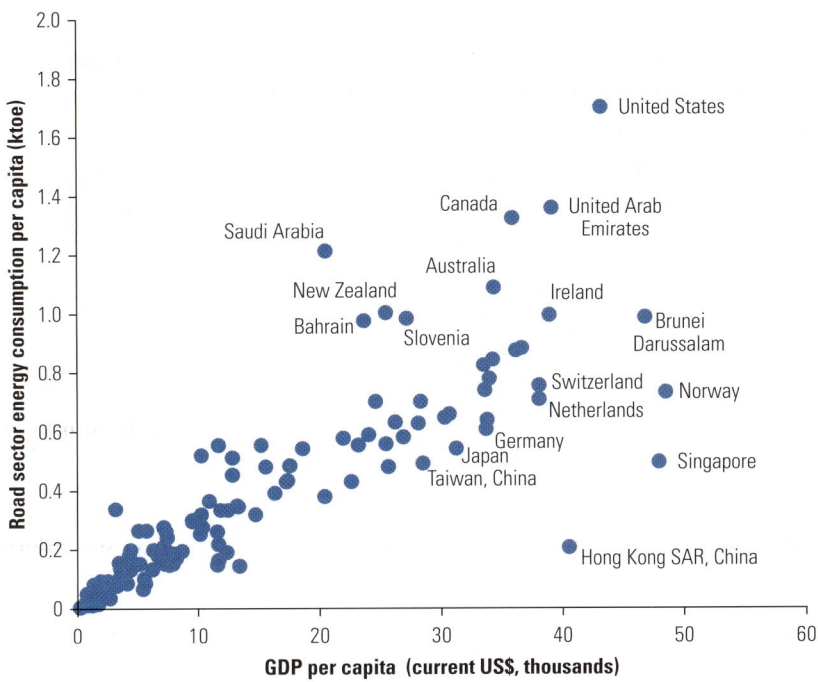

FIGURE 8.1

Road Sector Energy Consumption in Relation to Per Capita Income, Selected Countries, 2008

Source: World Bank *World Development Indicators.*
Note: ktoe = kilotons of oil equivalent.

Emissions from the Transport Sector

Transport-sector CO_2 emissions generated in ECA are a fraction of EU-15 emissions, but they are rising fast (see box 8.1 for a cautionary note on the accuracy of transport emission statistics).[7] In 2000, emissions from the region's transport sector were equivalent to 47 percent of EU-15 emissions. This level rose to 64 percent by 2008, as shown in figure 8.2. This rising trend, as figure 8.3 illustrates, suggests convergence with EU-15 levels within a decade. Emissions are heavily concentrated in just a few countries: the Russian Federation generated 48 percent, followed by Turkey, Poland, and Ukraine, as figure 8.4 shows. Together these four countries account for 72 percent of transport sector emissions. Road transport emissions are the largest contributor, exceeding 70 percent, as seen in figure 8.5. Data from the EU-27[8] suggest that the fastest-growing emissions come from air transport, a trend that is likely to be replicated in the ECA region as a whole.

The large increase in transport emissions in ECA is being driven by growth in road transport demand. Passenger vehicle registrations in

BOX 8.1

Improving Measurement of Transport Sector Emissions

Information on emissions from transport in ECA is limited. The data reported here on CO_2 and GHG emissions from transport are based on fuel consumption—a top-down approach to measuring emissions. This approach can be problematic because there is no direct correspondence between changes in transport activity and changes in fuel use; there is no official breakdown of fuel use by vehicle type; and not all countries in the region publish data on the total kilometers traveled per vehicle or per passenger. The situation is somewhat better for rail and air transport.

Good transport policy analysis requires far better information. For example, to assess policies aimed at improving vehicle fuel economy (kilometers traveled per liter of fuel consumed) requires data on fuel consumed for each type of vehicle-fuel combination (whether diesel, gasoline, or natural gas)—data that are not currently available in most ECA countries. Measuring emissions from road transport properly would require a bottom-up framework to collect information on (a) the number of vehicles by fuel type and vehicle type; (b) the average annual number of kilometers traveled for each vehicle type; and (c) the total number of passengers and the total freight transported by each mode (ADB 2009). Tools such as the World Bank's Transport Activity Measurement Toolkit support collection of such data.[a]

a. The World Bank Latin America and the Caribbean Region Sustainable Development Department Transport Cluster, in conjunction with the World Bank's Environment-Climate Change (ENV-CC) Department, sponsored the development of the Transport Activity Measurement Toolkit (TAMT). TAMT provides a framework that helps local institutions to collect high-quality, low-cost vehicle activity data in a simplified standardized manner. Further information can be found at http://code .google.com/p/tamt/.

Poland, Russia, and Turkey have grown dramatically this decade, exceeding 50 percent, as shown in figure 8.6. The new EU member states are not far behind, with 43 percent growth; this contrasts with the 13 percent growth rate for the EU-15 countries. High growth reflects large differences in motorization rates: 352 vehicles per 1,000 people in EU-10 in 2008, compared with 501 for EU-15, as figure 8.7 indicates. The motorization rates are much lower in the non-EU ECA countries. In the case of Turkey, the rate is only 95 per 1,000 people, although rising rapidly. This rapid motorization trend translates into rising CO_2 emissions, as shown in figure 8.8. Growth rates of road transport CO_2 emissions vary widely across countries—from negative values to triple-digit growth over the 2000–08 period. Total transport sector emissions reveal wide dispersion, seen in figure 8.9—reflecting in part the size of countries, the structure of their economies, and their levels of development.

There is a close correlation between per capita income and the increasing number of cars and trucks on the roads in ECA, as figure 8.10 shows. As incomes rise above about US$5,000, vehicle

FIGURE 8.2

Volume of Transport Sector CO_2 Emissions, Europe and Central Asia, Relative to EU-15, 2000–08

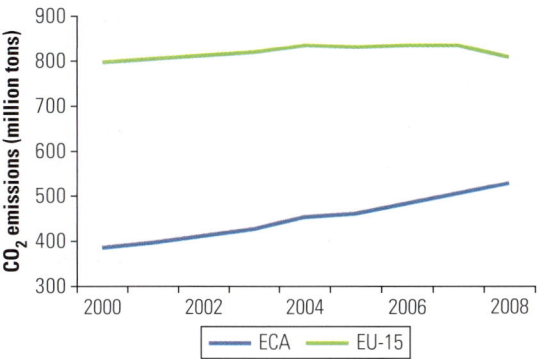

Source: International Energy Agency (IEA) "CO_2 Emissions from Fuel Combustion" statistics from OECD iLibrary, http://stats.oecd.org.
Note: EU-15 countries include Austria, Belgium, Denmark, Finland, France, Germany, Greece, Ireland, Italy, Luxembourg, the Netherlands, Portugal, Spain, Sweden, and the United Kingdom. CO_2 = carbon dioxide; EU = European Union.

FIGURE 8.3

Growth of Transport Sector CO_2 Emissions in Europe and Central Asia, Relative to EU-15, 2000–08

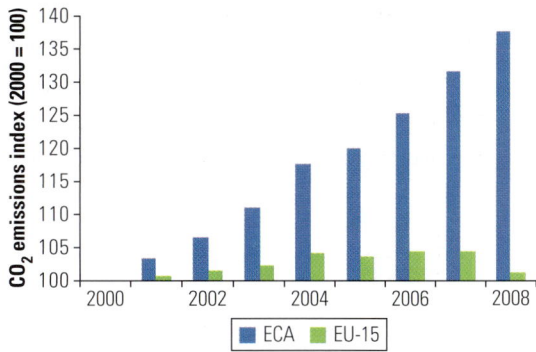

Source: International Energy Agency (IEA) "CO_2 Emissions from Fuel Combustion" statistics from OECD iLibrary, http://stats.oecd.org.
Note: EU-15 countries include Austria, Belgium, Denmark, Finland, France, Germany, Greece, Ireland, Italy, Luxembourg, the Netherlands, Portugal, Spain, Sweden, and the United Kingdom. CO_2 = carbon dioxide; EU = European Union.

FIGURE 8.4

Top Four Transport CO_2 Emitters in Europe and Central Asia, 2008

percentage

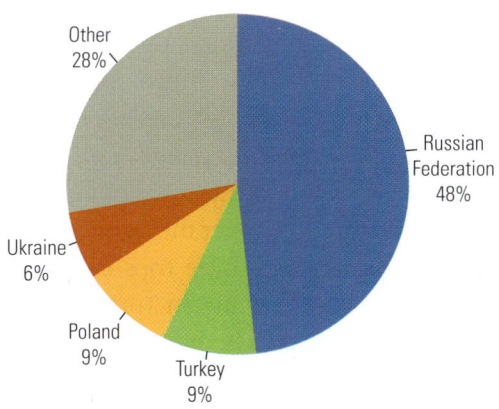

Source: International Energy Agency (IEA) "CO_2 Emissions from Fuel Combustion" statistics from OECD iLibrary, http://stats.oecd.org.

FIGURE 8.5

Road Emissions as Percentage of Transport Sector CO_2 Emissions in Europe and Central Asia, 2000–08

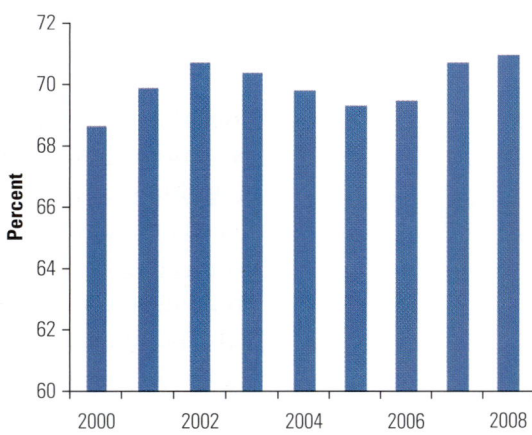

Source: International Energy Agency (IEA) "CO_2 Emissions from Fuel Combustion" statistics from OECD iLibrary, http://stats.oecd.org.

ownership increases rapidly. But there are nuances. For instance, among Group of Seven (G-7) economies, such as Canada, France, Germany, Italy, and the United Kingdom, with broadly comparable gross domestic product (GDP) per capita, motorization rates—the number of cars per 1,000 persons—vary from a low of 399 for Canada to a high of 596 for Italy. Such variation suggests a large

FIGURE 8.6
Increase in Passenger Vehicle Registration, Selected Countries, 2000–08

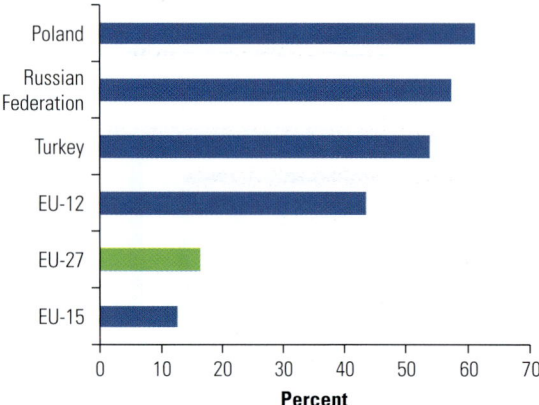

FIGURE 8.7
Motorization Rate in Selected EU Regions, 2000–08

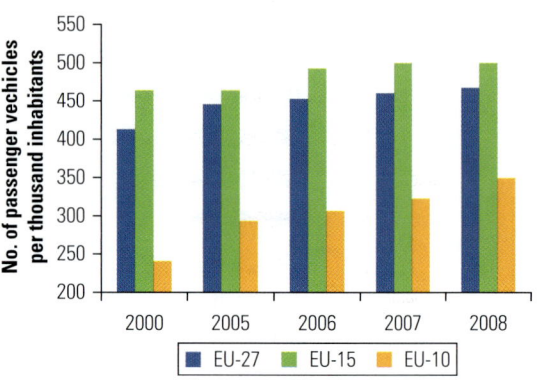

Source: UNECE 2010.
Note: EU = European Union. EU-27 countries include all current EU members. EU-15 countries include Austria, Belgium, Denmark, Finland, France, Germany, Greece, Ireland, Italy, Luxembourg, the Netherlands, Portugal, Spain, Sweden, and the United Kingdom. EU-10 countries include Bulgaria, the Czech Republic, Estonia, Hungary, Latvia, Lithuania, Poland, Romania, the Slovak Republic, and Slovenia.

Source: UNECE 2010.
Note: EU = European Union. EU-27 countries include all current EU members. EU-15 countries include Austria, Belgium, Denmark, Finland, France, Germany, Greece, Ireland, Italy, Luxembourg, the Netherlands, Portugal, Spain, Sweden, and the United Kingdom. EU-10 countries include Bulgaria, the Czech Republic, Estonia, Hungary, Latvia, Lithuania, Poland, Romania, the Slovak Republic, and Slovenia.

role for country-specific factors and a role for transport policies. Policy measures that affect the cost of car ownership and the desirability of alternative transportation options can affect motorization trends and therefore road sector emissions.

Cars and trucks account for an increasing share of all trips in Eastern Europe—what transport experts call the modal split. The share of vehicle passenger trips among the EU-10 has risen from 68 percent in 2000 to 77 percent in 2008, with passenger rail transport dropping from 13 percent to 6 percent and the share of bus and coach rides falling from 22 percent to 14 percent over the same period. In the case of Poland, the share of passenger vehicle trips increased from 65 percent in 2000 to 85 percent in 2008, while for Turkey, vehicle trips jumped from 35 percent to 50 percent over the same period.

Road transport also accounts for a rising share of ECA's freight, replacing rail. Among the EU-10, rail freight transport has seen a sharp decline, with the share of traffic declining from 43 percent in 2000 to 26 percent in 2008, paralleled by a rise in road freight transport from 55 percent to 71 percent over the period, as shown in figure 8.11. Poland's use of rail dropped from 57 percent in 2000 to 26 percent in 2008. In the case of Russia—the largest greenhouse gas (GHG) emitter in the region—the use of rail has remained largely unchanged, from 15 percent in 2000 to 16 percent in 2008, while for Turkey, use of freight is very limited, at 5 percent in 2008, down from 6 percent in 2000. For a number of ECA countries, use of rail freight

FIGURE 8.8

Growth in Road CO$_2$ Emissions, Selected Europe and Central Asia Countries Relative to Region and EU Groups, 2000–08

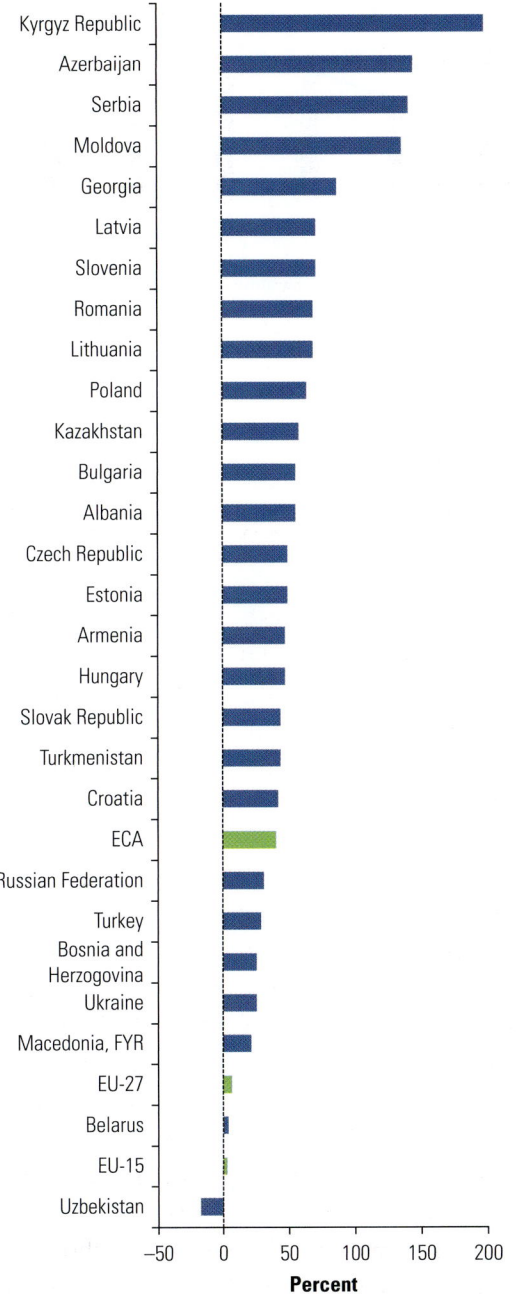

Percent

FIGURE 8.9

Transport Sector CO$_2$ Emissions, Selected Europe and Central Asia Countries, 2008

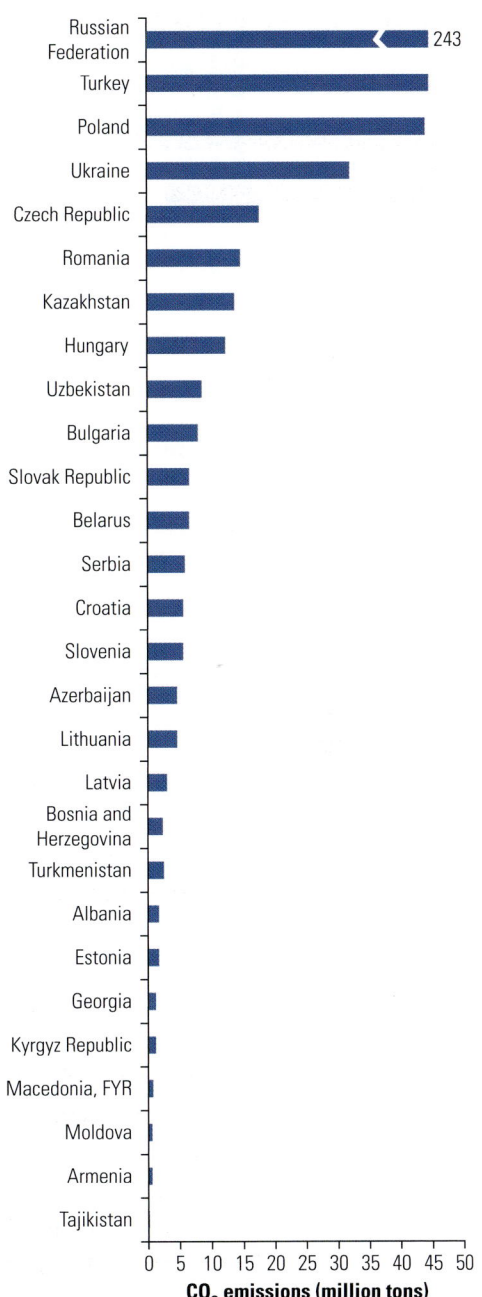

CO$_2$ emissions (million tons)

Source: IEA "CO$_2$ Emissions from Fuel Combustion" statistics from OECD iLibrary, http://stats.oecd.org.
Note: Tajikistan, excluded from the chart, had emissions growth of 600 percent. EU-27 countries include all current EU members. EU-15 countries include Austria, Belgium, Denmark, Finland, France, Germany, Greece, Ireland, Italy, Luxembourg, the Netherlands, Portugal, Spain, Sweden, and the United Kingdom. CO$_2$ = carbon dioxide; EU = European Union.

Source: IEA "CO$_2$ Emissions from Fuel Combustion" statistics from OECD iLibrary, http://stats.oecd.org.

FIGURE 8.10

Motorization Rates in Europe and Central Asia Relative to Nominal GDP per Capita, 2008

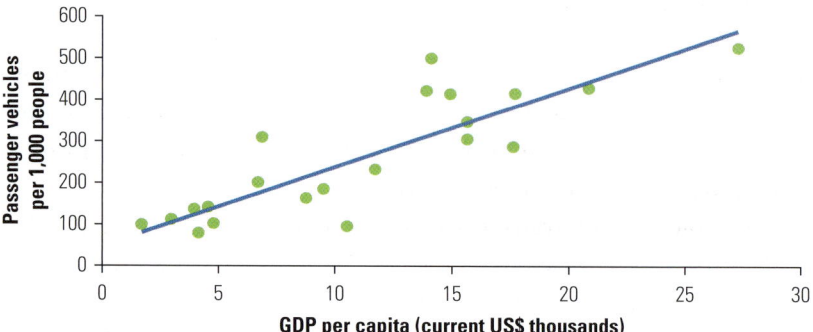

Source: UNECE 2010.

FIGURE 8.11

Land Modal Split for Freight Transport, EU-15 and EU-10 Countries, 2000–08

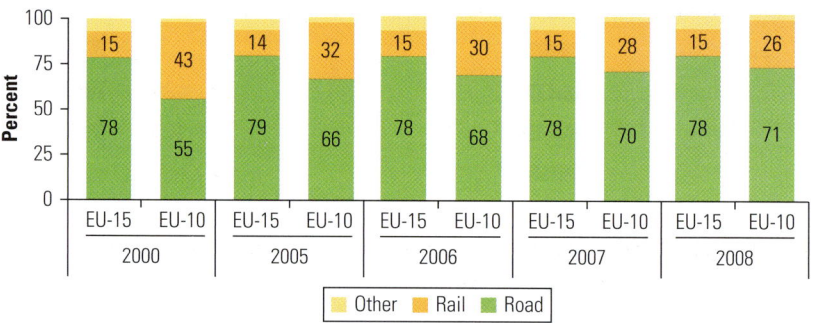

Source: European Commission 2010.
Note: Modal shares based on road, rail, and inland waterway transport services. EU = European Union. EU-15 countries include Austria, Belgium, Denmark, Finland, France, Germany, Greece, Ireland, Italy, Luxembourg, the Netherlands, Portugal, Spain, Sweden, and the United Kingdom. EU-10 countries include Bulgaria, the Czech Republic, Estonia, Hungary, Latvia, Lithuania, Poland, Romania, the Slovak Republic, and Slovenia.

transport is higher than for the EU-15. But this gap is narrowing, and for the largest emitters, rail carries relatively low levels of freight except in Poland.

EU-27 air transport CO_2 emissions increased from 8.8 percent of total transport emissions in 1990 to 12.5 percent by 2007 and are the fastest growing source of emissions in the sector. The situation in ECA varies by country, with air transport emissions accounting for only 3.5 percent of the total in Poland, compared with 11.9 percent in Turkey. Future aircraft emissions are expected to increase more rapidly than transport emissions in general because of projected air passenger growth, in part due to the continued strength of low-cost airlines. To date, the energy efficiency of the aviation system has not kept pace with the growth of the sector, and unlike other transport modes, GHG emissions vary according to flight distance because the largest emissions occur at takeoff. In contrast to land transport modes, the impact of aviation is also compounded by the fact that GHG emissions are released directly into the atmosphere (Ross 2009).[9] Before the international economic crisis that started in 2008, air transport emissions were going up in the EU-15 countries. They were also rising more sharply in the EU-10 countries but from a much lower base, as figure 8.12 shows.

What role will the transport sector play in supporting global GHG reductions? The answer depends on the mitigation approach. The International Energy Agency's (IEA) scenarios assess *what is possible* from a technical-engineering perspective. They investigate the benefits of numerous feasible improvements in fuel efficiency and fuel switching, coupled with demand-side policies that shift passengers and freight to more sustainable transport modes (IEA 2009). Globally, the result is a reduction in transport CO_2 emissions of 40 percent by 2050 compared with 2006 levels—and a massive reduction of 70–80 percent compared with business-as-usual scenarios.

An alternative approach, using a so-called integrated assessment model, estimates *what is likely* given an economywide carbon price that would limit warming to 2 degrees Celsius (Clarke et al. 2007).[10]

FIGURE 8.12

Air Transport GHG Emissions in Selected Europe and Central Asia Countries Relative to EU-15, 2000–09

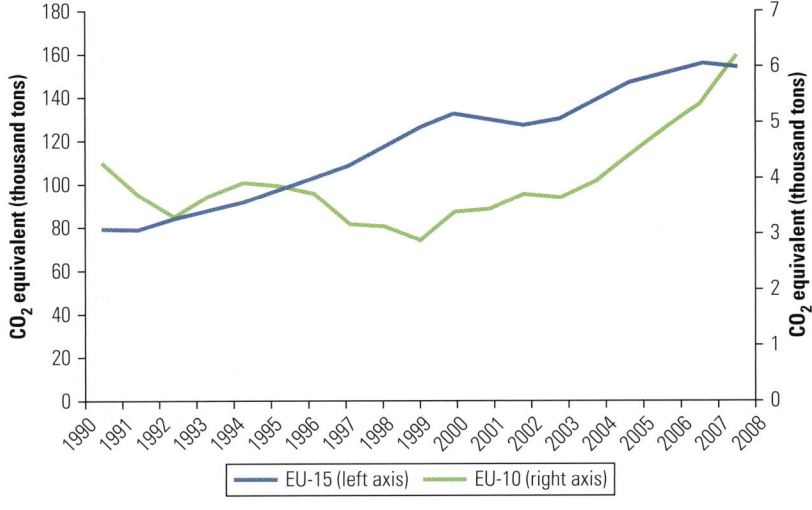

Source: Eurostat (European Commission n.d.).
Note: "Selected Europe and Central Asia countries" include the region's EU-member states plus Turkey. EU-15 countries include Austria, Belgium, Denmark, Finland, France, Germany, Greece, Ireland, Italy, Luxembourg, the Netherlands, Portugal, Spain, Sweden, and the United Kingdom. GHG = greenhouse gas; EU = European Union.

A universal carbon price shifts reductions to sectors where mitigation is easiest and cheapest. In these scenarios, emissions from transport, rather than falling, surprisingly increase by 47 percent over 2005 levels, and transport surpasses the power sector as the largest emitter.

Both of these studies make strong assumptions. The IEA predicts rapid development and adoption of advanced vehicle technologies. The integrated assessment model assumes that carbon capture and storage enables inexpensive mitigation in the power sector that makes cuts in transport emissions less pressing (see chapter 6). It is also unlikely that a global carbon price will be the main, let alone the only, climate mitigation instrument, at least in the short to medium term. In the absence of a carbon tax or a cap-and-trade system that includes transport, countries are best off designing policies that seek to lower all external costs of the transport sector—with emission reductions as one of several benefits.

The Broader Benefits of Sustainable Mobility

Passenger and freight transport by road accounts for the largest share of transport energy use and consequently of emissions. Road transport by car or truck is usually preferred because it is often faster and always more flexible, not dependent on fixed routes or schedules. With rising incomes, households can afford the purchase of a car, which supports a greater range of activities such as commuting to distant work places, shopping, or recreation. Globally, a rapid takeoff in car ownership tends to happen once countries reach an average per capita GDP of about US$5,000, a level that most ECA countries have already reached, as shown in figure 8.13.

This rising use of cars and trucks can have broader economic as well as individual welfare benefits. But as the number of drivers increases, damages from the road transport sector rise as well. GHG emissions are one problem—each liter of gasoline burned by a vehicle emits 2.3 kilograms (kg) of CO_2 (2.6 kg for diesel). But estimated damages from carbon emissions are dwarfed by those from other unwanted side effects from road transport. Costs or damages from congestion, air pollution, traffic accidents, and even noise pollution are larger. At the high end of available estimates, climate change impacts add US$0.037 per mile, but congestion adds US$0.357 and air pollution US$0.148 (see table 8.1). The range of estimates is broad, reflecting their sensitivity to specific conditions and analytical methods. Total external costs might therefore vary between about US$0.07 and US$0.74 per mile (US$0.04 and US$0.46 per km). These costs exceed typical fuel taxes. One study for the United States estimates that

FIGURE 8.13

Car Ownership in Relation to Per Capita Incomes, Selected Europe and Central Asia Countries and Others, 2008

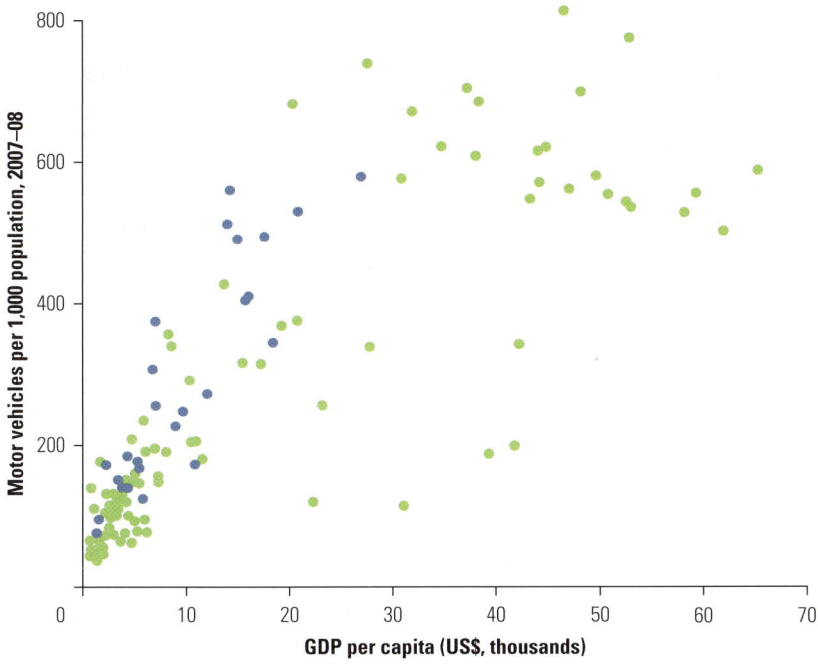

Source: World Bank's *World Development Indicators.*
Note: Blue dots indicate ECA countries; green dots indicate other countries.

accounting for externalities could add US$2.28 to a gallon of gasoline, well above the average fuel tax of US$0.40 per gallon.[11] Many ECA countries have notably higher fuel taxes.

Addressing these additional externalities, especially congestion and health, offers climate change mitigation benefits. For example, Washington, DC—site of the World Bank's headquarters—is considered the most congested metro area in the United States (TTI 2011). Its approximately 2.6 million commuters each spent an average of 73 hours annually in traffic jams in 2010, each burning an extra 37 gallons (140 liters) of gasoline. In total, this wastes 95 million gallons (360 million liters) and adds 840,000 tons of CO_2 to the already significant emissions from the region's road transport. The extra gasoline and wasted time cost the regional economy US$3.8 billion or about US$1,495 per commuter per year. No similar studies have been conducted in ECA cities, and comparable data on average travel speed or other indicators of congestion are scarce, but in some cities, costs will likely be high as well.

For example, Moscow now has 3.9 million cars, four times the number in 1990, and 200,000 are added every year. Its traffic jams

TABLE 8.1
Road Transport Externalities

Externality	Source	Nature of costs	Orders of magnitude of costs (U.S. cents per mile, 2005 prices)	Public abatement and supply-type policies	Policies affecting demand and vehicle characteristics
Climate change	GHG emissions from fossil-fuel use	Wide-ranging and uncertain adverse impacts from climate change	0.3–3.7	n.a.	Fuel efficiency standards, CO_2 or fuel taxes, cap-and-trade
Congestion	Volume of use approaches or exceeds design capacity per unit of time	Mainly time and schedule delay costs	4.2–35.7	Network capacity	Congestion charges, fuel taxes, access restrictions, land-use regulation, quantity controls
Air pollution	Fuel combustion and exhaust	Mainly health, loss of life, and environmental degradation	1.1–14.8	n.a.	Standards (vehicle equipment, fuel quality), access charges
Traffic safety	High traffic density and heterogeneity in vehicle weight and speed, increased average accident risk	Mainly health and loss of life; material damage	1.1–10.5	Adaptation of road infrastructure, emergency services, mandatory insurance	Traffic rules and procedures, risk-dependent insurance premiums
Noise	Engines and movement	Health, discomfort	0.1–9.5	Sound barriers, silent road surfacing, curfews	Standards, curfews, tradable permits

Source: Proost and Van Dender 2011.
Note: GHG = greenhouse gas; CO_2 = carbon dioxide; n.a. = not applicable.

are legendary (Gessen 2010). To reduce them, the city recently announced plans to double its geographic footprint and move many government offices to new territory in the southwest of the city.[12] Meanwhile, as Istanbul's population increased by 3.3 percent per year, average motorized travel times increased from 41 minutes in 1996 to 49 minutes in 2007, and estimated CO_2 emissions jumped from 7 million to 9 million tons (Gerçek and Demir 2008).

Regarding the impacts on health, most of the detailed research has been done in the United States. In addition to CO_2, gasoline vehicles emit carbon monoxide (CO), nitrogen oxides (NO_x), and volatile organic compounds. The latter two combine to form particulate matter (PM_{10} and $PM_{2.5}$) that contributes to heart and lung disease. One comprehensive recent study estimates that the road vehicle sector generated US$56 billion in health damages in America in 2005, with US$36 billion from passenger cars and $20 billion from trucks (NRC 2009). This translates to a per-mile cost (at the low end of the estimates in table 8.1) of US$0.012 to US$0.017.

Information on the health impacts of transport in ECA countries is sparse. With the phasing out of inefficient industries after transition, the share of pollution that is from transport rather than industrial activity has increased. In Moscow, motor vehicles now account for 70 percent of total air pollution, exacerbated by high levels of congestion. The European Environment Agency collects data on air pollution for EU and accession countries (see map 8.2). Transport is

MAP 8.2
NO$_2$ Emissions, Selected European Countries, 2009

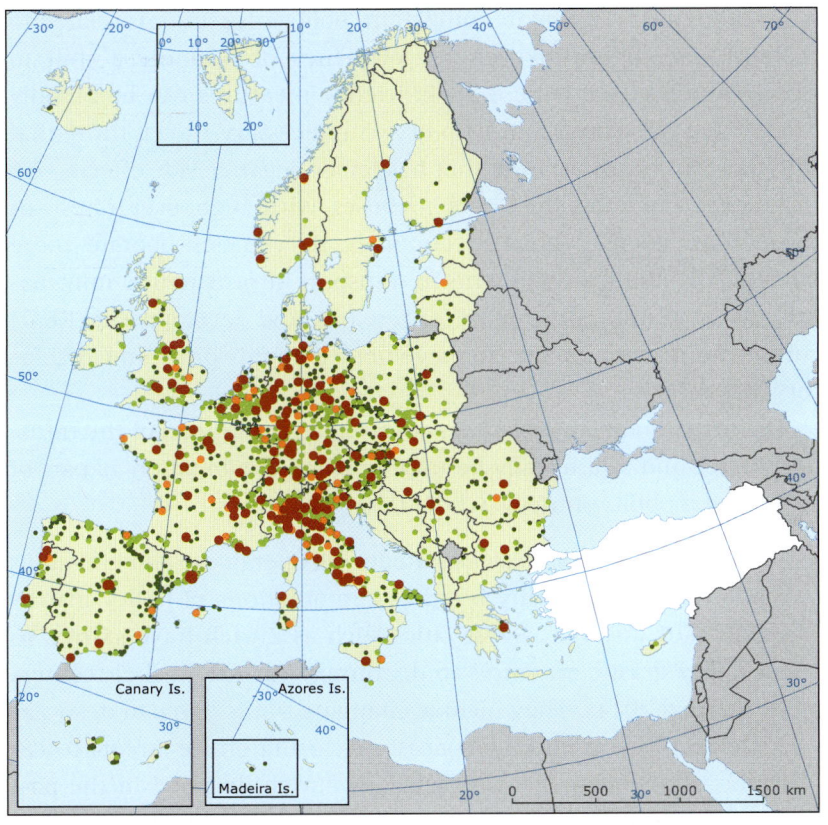

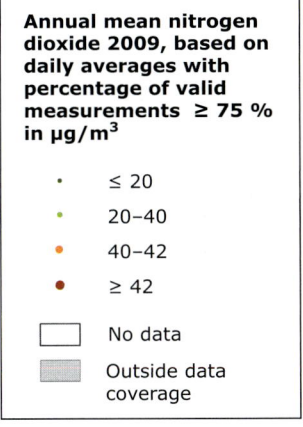

Annual mean nitrogen dioxide 2009, based on daily averages with percentage of valid measurements ≥ 75 % in µg/m^3

- ·   ≤ 20
- ·   20–40
- ●   40–42
- ●   ≥ 42

   No data

   Outside data coverage

Source: EEA 2011a.
Note: NO$_2$ = nitrogen dioxide.

the largest contributor to NO_x emissions (including nitrogen oxide). These still tend to be lower in much of Eastern Europe than in parts of Italy or Germany, but they will rise with more drivers on the road (EEA 2011b).

Climate-Smart Mobility Policies

As in other areas of climate mitigation, policy instruments related to mobility aim to improve energy efficiency or promote a shift to cleaner energy (see table 8.2). Efficiency measures can be broadly categorized into those that help *avoid* unnecessary travel, those that *shift* travel to transportation that has fewer negative side effects—for instance, from individual cars to buses with high user rates—or those that *improve* vehicle technology so vehicles operate more efficiently.[13] Public investments in transport infrastructure, information dissemination, or support for research and development (R&D) complement demand-side management through price and regulatory instruments.

The instruments in table 8.2 address different aspects of the transport sector and will usually be used in some combination as part of an overall mobility strategy, which raises two issues:

• *Policies need to be complementary.* Reducing energy use and emissions while ensuring mobility requires combinations of policy instruments. For instance, Swiss cities such as Zurich have started to use fairly drastic measures to discourage car use in their centers (Rosenthal 2011). They include changing traffic lights to delay car travel, removing parking spaces, or closing entire streets to car traffic. Household car ownership in Zurich dropped in the past 10 years, from 60 percent to 55 percent, as has overall car use. The

TABLE 8.2
Policy Instruments for Sustainable Road Transport

	Energy efficiency			Cleaner energy
	Avoid	Shift	Improve	Fuel switch
Supply-side (investments)	Expand IT access for telework	Improve public transport infrastructure and capital stock, use of IT or smart ticketing	Information campaigns, facilitate improved vehicle technology adoption	Facilitate R&D, tech adoption, support alternative fuel infrastructure
Demand-side (prices and regulations)	Land-use regulations for transport-oriented densification	Fuel taxes, congestion pricing, parking fees; reduce parking availability; traffic management; public transit subsidies	Fuel efficiency standards; registration fees based on emissions	Subsidies or requirements for alternative fuels

Note: IT = information technology. R&D = research and development.

measures to restrict car use were implemented gradually in different parts of the city, but only after significant improvements in public transit were made. If car use is restricted without providing alternative means of transport, mobility is reduced, with adverse economic and welfare consequences. Conversely, if investments in public transit are not complemented by policies reducing private vehicle traffic, then buses or rail systems will be underused and investments will be uneconomical. This has been the experience of rail systems in some U.S. cities such as Los Angeles or Detroit.

- *There is often a trade-off between public acceptability and revenue generation.* To stay with the Swiss example, rather than simply making life more difficult for drivers, some form of congestion pricing and higher parking fees could achieve the same objective and would raise funding that could support public transit investments. However, charging drivers is always unpopular, in part because such charges are regressive—hurting low-income drivers more than wealthier citizens. Therefore, second-best, less-efficient policies often prevail because they are politically more feasible even if economically inefficient.

Policies that Reduce Vehicle Emissions

Fuel Taxes

Fuel taxes are the main instrument to account for the broader costs of vehicle use. By making gasoline or diesel fuel more expensive, taxes discourage driving and encourage more fuel-efficient vehicles. Although the oil price is global and refinery costs do not vary greatly, gasoline and diesel prices at the pump differ greatly among countries. In ECA, they vary by more than an order of magnitude—from US$0.22 per liter in Turkmenistan to US$2.52 in Turkey.[14] The average ECA price has tripled for diesel and more than doubled for gasoline between 2000–10. These prices are below what is considered the unsubsidized fuel price in only four countries for diesel and one for gasoline, as shown in figure 8.14. But 14 of the region's countries still have fuel prices below those of Romania, which has the lowest in the EU.

In principle, fuel taxes should be a combination of (a) a charge per liter to account for carbon emissions and oil dependence, and (b) a charge per kilometer traveled to account for congestion, local pollution, and accidents—representing what economists call a Pigouvian tax reflecting all damages caused by vehicle use. Some countries earmark a share or all of tax revenues for road building and

FIGURE 8.14
Transport Fuel Prices in Selected Europe and Central Asia Countries, 2000–10

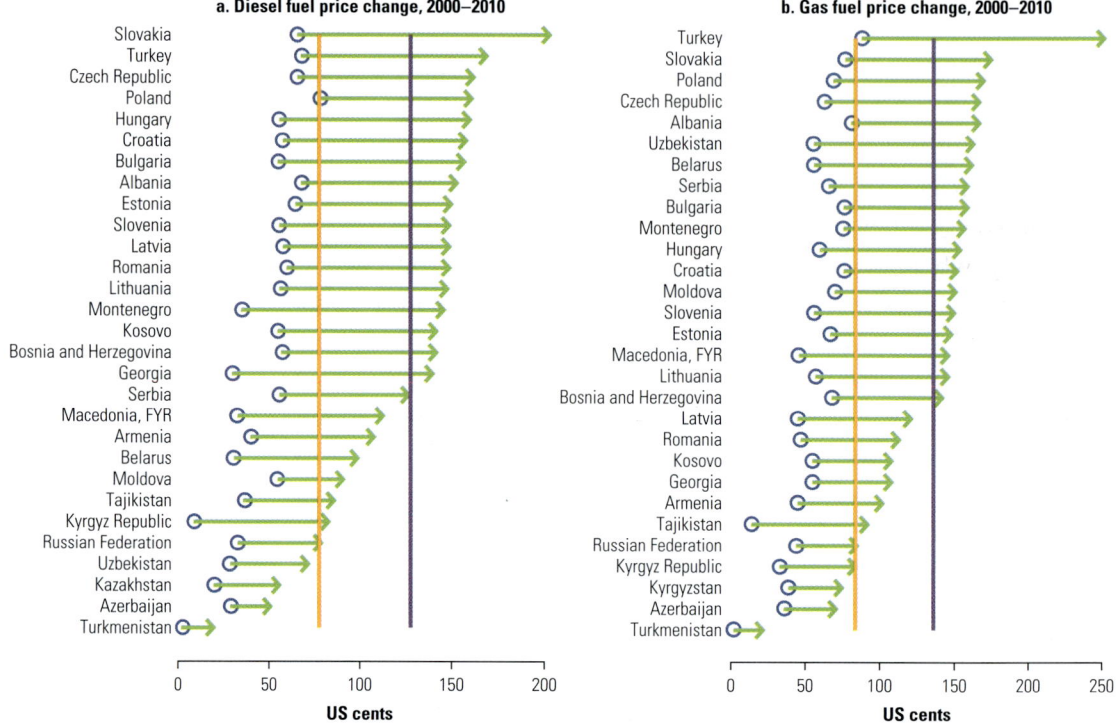

Source: GIZ 2012.

Note: Vertical orange lines indicate 2008 prices in the United States, an international minimum benchmark for a nonsubsidized road transport policy. Vertical purple lines indicate prices in Romania, the lowest in the EU, which could be considered a lower bound for a social price of transport fuel.

maintenance, which increases public acceptance. Others use fuel taxes as a significant source of general revenue that is relatively easy to collect; this is the case in Turkey, where fuel surcharges are higher than environmental objectives would suggest.

Fuel taxes tend to be effective in reducing energy use. Increasing the price of gasoline by 10 percent lowers consumption by about 6–7 percent on average. But fuel taxes can be harmful to the poorest car-owning families because fuel is a larger share of low-income household budgets. Fuel tax revenue can be invested in convenient and affordable public transit, which helps to address this problem. Fuel taxes are also quite unpopular and therefore difficult to increase, in part because drivers are usually more aware of the price at the pump than of electricity prices, for instance.

Fuel tax revenue could grow significantly in some ECA countries, where gasoline and especially diesel prices are still very low. In the extreme case of Turkmenistan, raising gasoline prices from a heavily subsidized US$0.22 per liter to the U.S. level of US$0.56 per liter in 2008 could free up US$375 million per year.[15] Raising fuel prices further to incorporate at least some of their social costs from car use and fossil-fuel burning can generate significant additional revenue. A fuel tax increase corresponding to a carbon charge of US$100 per ton (far higher than current EU Emissions Trading System [ETS] prices) would add about US$0.06 per liter of gasoline. Based on 2008 gasoline consumption, this would raise about US$2.2 billion in revenue in Russia.

But countries should not rely too much on fuel taxes as a revenue income source. If they are successful in achieving their objectives, especially when combined with other policies, this income will decrease. Fuel economy standards that reduce gasoline consumption and demand-side policies that reduce car use more generally further erode the tax base and can create shortfalls in road funding or general revenue. In the United States, fuel tax revenue is earmarked for federal and state road funds. New fuel economy standards will save more than 60 billion gallons (227 billion liters) of fuel between 2012 and 2016 (Van Dender and Crist 2010). With a fixed (and quite low) fuel tax, revenue will fall by US$26 billion or 72 percent of the 2008 U.S. Highway Trust Fund receipts. To maintain revenue that ensures an efficient transport infrastructure and at the same time retain incentives for reducing fuel consumption and emissions, fuel taxes will need to be indexed or taxes levied on distance traveled rather than fuel consumption.

Fuel Economy Standards[16]

New-vehicle buyers are often said to be more concerned about the number of cup holders than the car's fuel efficiency. This presumed market failure, where consumers undervalue vehicle efficiency or have short time horizons in their decisions, justifies fuel economy standards that regulate vehicle sales in most countries. Researchers are divided over whether this market failure really exists. If it did, providing better information to consumers would be a better policy. More likely, it reflects consumer preferences for other vehicle attributes such as styling or horsepower.

Studies on the effectiveness of standards in reducing overall fuel consumption are also inconclusive. Fuel economy has generally improved in most countries, but much of that could be due to fuel costs or changing consumer preferences. Basic economic analysis suggests that higher fuel taxes are more effective than fuel economy standards. Standards also impose the same initial cost on people who drive a lot and those who drive less, so they do not necessarily reduce emissions as much. Fuel economy standards also take time to affect overall emissions because complete vehicle stock turnover takes about 15 years. From a manufacturer's perspective, strict standards are inefficient because they typically face different costs of improvement. Some element of trading of fuel economy credits—or "feebates," as further discussed in box 8.2—would lower these costs overall.

The wide adoption of fuel economy standards shows that they are politically feasible and, in the absence of other measures, likely to have some impact. Most ECA countries have either adopted or plan to introduce fuel economy standards (GFEI 2010). EU members are subject to regionwide standards that target a fuel economy consistent with emissions of 130 grams of CO_2 per km by 2012. In ECA countries for which data are available, average fuel economy has generally improved—by as much as 3 percent in the case of Lithuania in just three years (2005–08), as figure 8.15 shows. Because many of the region's countries import all or most of their cars, including large numbers of used cars, fuel economy regulations typically work through import restrictions on vehicle age, technology, emissions, or import taxes. The scope of these policies varies. Albania has no restrictions, while Russia limits imports to cars that adhere to EU standards (EU 4 norms for domestic and imported cars since 2010).

BOX 8.2

Fuel Economy Standards or Feebates?

Feebates are an alternative to uniform or fleetwide fuel economy standards (Anderson et al. 2011; Parry 2011). The principle is simple: Feebates impose a fee on manufacturers or buyers of new vehicles that have a fuel consumption above some level—the so-called pivot point—and give a rebate to those with consumption below that point. Like fuel economy standards, feebates do not charge in proportion to actual car use (and emissions), so they do not affect the amount of driving. Feebates are revenue-neutral and therefore often more politically acceptable than fuel taxes. The main advantage of feebates over standards is that they allow for differentiated costs of compliance across manufacturers, thus minimizing efficiency losses. Both manufacturers of cars above and below the pivot point have an incentive to innovate to minimize fees or maximize rebates. The greatest improvements will be made where the costs of doing so are lowest.

Canada's feebate system ran from 2007 to 2008 and consisted of two programs:

* *The EcoAuto Rebate* offered rebates from Can$1,000 to Can$2,000 to buyers of fuel-efficient vehicles (better than 6.5 liters per 100 km) (Banerjee 2007).

* *The Green Levy* program imposed a tax on fuel-inefficient vehicles (consuming more than 13 liters per 100 km) from Can$1,000 to Can$4,000.

FIGURE 8.15

Average Fuel Economy and New Vehicle Registrations in Selected Countries, 2005 and 2008

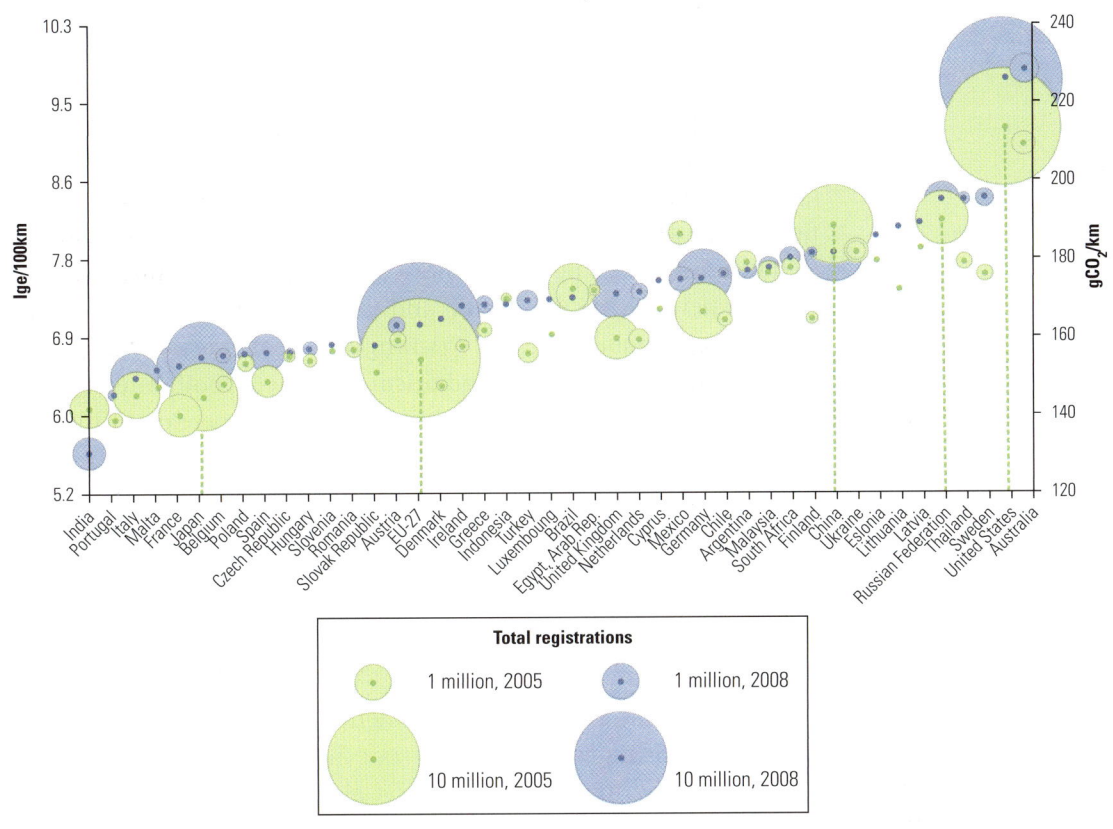

Source: IEA 2011.
Note: lge/100km = liters of gasoline equivalent per 100 km; gCO₂/km = grams of carbon dioxide per kilometer; EU = European Union.

Registration Taxes

Seventeen countries in the EU (among them Latvia and Romania) now base all or part of their registration fees on a vehicle's CO_2 emissions (ACEA 2010). Registration taxes share the disadvantage of fuel economy standards in increasing the cost of a vehicle purchase but not considering its amount of use and therefore emissions. High registration taxes, like high import duties, can also act as a barrier to vehicle fleet turnover. To avoid the high one-time cost, consumers will keep aging and polluting cars on the road longer.

Road Pricing

The typical response to congestion is to build new roads. Post-transition in Moscow, as the number of people owning cars shot up from 60 to more than 300 cars per 1,000 residents, the city government widened the Moscow Ring Road, completed the Third Ring Road, and started construction of a fourth (Gessen 2010). However, congestion continued to worsen, confirming the "fundamental law of road congestion": traffic volumes increase in proportion to additional road capacity (Duranton and Turner 2011). In fact, in U.S. cities, the benefits of a marginal increase in road capacity generally do not justify the costs of construction unless it is accompanied by some form of congestion pricing. This does not imply that road investments are generally bad but that they tend to be ineffective in relieving congestion where demand growth is high.

Rather than adding roads until congestion eases, making driving more expensive reduces traffic and generates revenue that can be used for road maintenance or to strengthen public transit alternatives (Anas and Lindsey 2011). Road pricing is used for both interregional and urban traffic, either as a charge per distance traveled (for example, toll roads) or as a charge to enter or park in a typically high-traffic area (congestion charges or parking fees). London, Singapore, Stockholm, and Milan all introduced a fixed or time-varying charge to enter the busiest zones. They have been generally successful in reducing time lost due to congestion. Traffic volumes in the center of London dropped by 34 percent for cars, and congestion delays fell by 30 percent.

Benefits from reduced environmental impacts are lower than those from easing congestion, but not insignificant. Reduced idling and stop-and-go traffic lowered local air pollution by 8–19 percent and CO_2 emissions by 14–19 percent in London, Stockholm, and Milan.

About 50 countries worldwide use highway tolls, including several in ECA such as Croatia and Hungary, where all highways are toll roads. A disadvantage of toll roads—and, to some extent, also of urban congestion charging—is that they often simply divert traffic, causing higher congestion in other parts of the road network.

Parking Fees

Congestion pricing is effective, but it is complex to implement. An alternative way to reduce traffic in congested areas is to increase the cost of parking by raising fees directly or by reducing the number of parking spaces—for instance, for newly constructed buildings. Some cities in the United Kingdom relate residential parking fees to vehicle CO_2 emissions.

With few exceptions, commercial parking rates in Eastern European cities are lower than in Western Europe (figure 8.16), reflecting lower average incomes. In some places, parking is not

FIGURE 8.16

Daily Commercial Parking Rates in Selected Cities, 2011

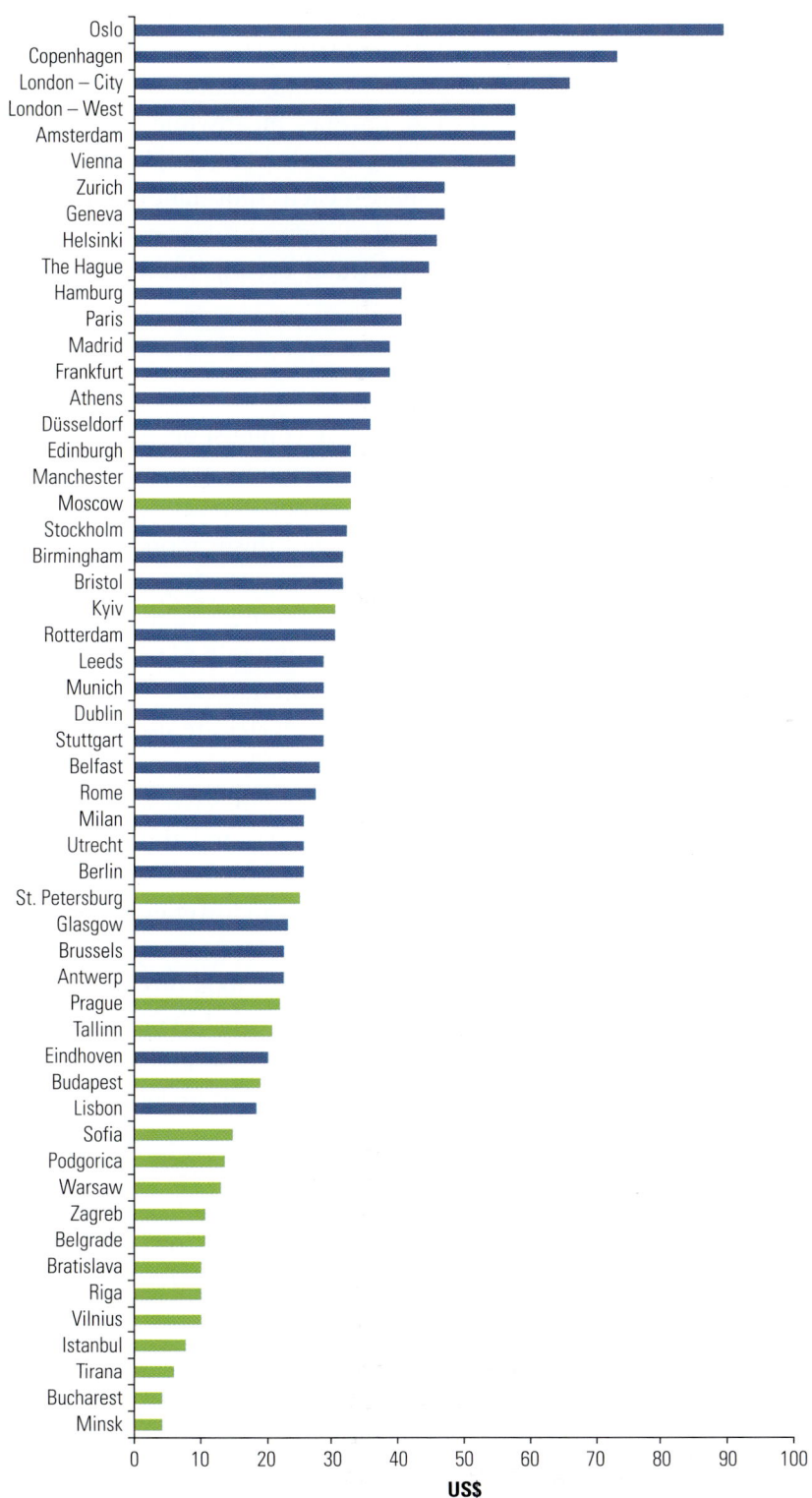

Source: Colliers International 2011.

priced or regulated at all. Municipalities in Kazakhstan, Russia, and some other Commonwealth of Independent States (CIS) countries,[17] lack the legal basis to charge for on-street parking or to set fines for illegal parking (Podolske 2010). The resulting chaotic parking behavior adds to congestion, yet fails to discourage driving into the city center. In Western Europe, meanwhile, cities raise prices and reduce supply of parking as a lower-cost alternative to congestion pricing. Freed-up space—each spot takes up 15–30 square meters—and revenue is used to support lower-carbon transport modes such as public transport and cycling (Kodransky and Hermann 2011).

Lower-Carbon Fuels

Improvements in conventional internal combustion engines will continue to gradually lower fuel consumption and emissions. A higher share of diesel engines yields further decreases. Diesel fuel produces higher CO_2 emissions per liter, but this is more than offset by higher fuel efficiency, so that emissions are 10–20 percent lower. In Western Europe, more than half of all passenger vehicles are now diesel fueled and, through FDI and exports of new and used vehicles, this will also help lower ECA's emissions from transport.

Compressed natural gas (CNG)-fueled vehicles are used in Armenia, Belarus, Moldova, Russia, Tajikistan, and Ukraine, although only in Armenia is their market share over 30 percent. CNG vehicles emit 25 percent less CO_2 than gasoline-powered cars and also emit fewer local pollutants.[18]

Expanded use of biofuels from food crops can lower emissions further. EU countries are subject to a target of 10 percent biofuels in transport energy, although the implications for food production and land use are still debated (Fonseca et al. 2010). In countries with abundant land and large potential to raise agricultural efficiencies, such as Russia and Ukraine, biofuels could play a larger role in domestic fuel markets as well as in exports.

In the future, hybrid electric and fully electric vehicles will capture a larger share of the vehicle market, but they currently need large subsidies to be competitive. They will also need to rely on cleaner energy, such as renewables or coal with carbon capture and storage, to contribute to emission reductions.

Public Transit

Adding road capacity will rarely relieve congestion or reduce fuel consumption and emissions in the long term. Increased road capacity improves speeds and saves fuel temporarily, but it also encourages more driving and a return to congestion. Estimates for São Paulo, for example, suggest that increasing road capacity by 20 percent would increase fuel use by 5 percent and emissions by 3 percent (Anas and Timilsina 2009). Beyond a certain point, the more cars that use the road network, the less efficient it gets as each additional car will reduce average travel speed. The opposite is generally true for public transit, provided it operates on separate tracks or lanes. The more it is used, the more routes can be supported, the more frequently it can run, and the more efficiently it will operate. On average, the GHG intensity (per passenger) of well-used bus transit, for example, is less than a third of a car's, as shown in figure 8.17.

To achieve a sufficiently large shift away from cars, public transit needs to be convenient for residents and financially sustainable for local governments. Convenience means an affordable, accessible network that covers the service area well and frequent service at speeds comparable to cars. Close integration of different transit modes such as rail, bus, and metro attracts riders. Some cities have been introducing "mobility smart cards" for payment of all public transit, as well as car sharing, electric or shared bikes, or discounted taxi rides. Travelers between Szczecin or Warsaw and Berlin can now use a single ticket for the rail portion and public transit at both ends of the trip. Making public transport attractive creates positive feedbacks as increased demand justifies network expansion and service improvements.

ECA countries historically had good public transport, but in many parts of the region, transit systems have had trouble sustaining operations since financial responsibility was transferred from central to

FIGURE 8.17
Greenhouse Gas Intensities, by Transportation Mode, 2005

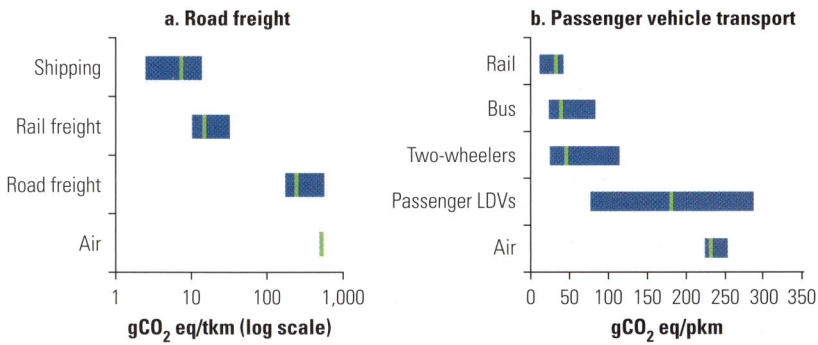

Source: IEA 2009.
Note: Based on global data for 2005. The vertical green line indicates the global average. gCO$_2$ eq/tkm = grams of carbon dioxide equivalent per ton km; gCO$_2$ eq/pkm = grams of carbon dioxide equivalent per passenger km; LDV = light-duty vehicle.

local governments. The share of public transportation in total land passenger transport dropped by 3.4 percent between 2000 and 2008 in the EU-10 countries while it increased slightly (0.1 percent) in the EU-15 (Odyssee n.d.).

To be financially sustainable (and to realize large emission benefits), public transport systems need to be well used. That requires convenient service, complementary demand-side policies discouraging car travel, and land use planning favoring development around transport corridors (see chapter 9). Unconventional financing can make capital investment more feasible, especially in the case of metro construction. In Warsaw, housing within 1 km of a metro station commands a 7 percent value premium (Medda and Modelewska 2009). Various forms of land value capture or property taxation can recover some of these gains from homeowners and businesses that profit from increased traffic around stations to fund the investments.

Even with high ridership and creative financing, public transit will in most cases not be self-financing. Therefore, subsidies may be required, and these can be justified on efficiency grounds given the social benefits of reduced congestion, road safety, reduced local pollution, and emission reductions. In economic analysis, even quite large subsidies (as high as 50 percent of operating budgets) can be welfare improving (Parry and Small 2009). Because these subsidies must come out of the general budget or earmarked taxes, champions for public transit need to build a strong case on the basis of socioeconomic analysis.

Where public transport systems have deteriorated, private operators fill the gap. In cities such as Istanbul or Tbilisi, minibuses or shared taxis ("dolmus" and "marshrutka") account for a high proportion of trips (see box 8.3). They are often uncomfortable and poorly regulated, they tend to be more demand responsive. In Tbilisi, public transit users make but five trips a week on average by minibus, three by bus, and only two by metro (Grdzelishvili and Sathre 2011). Well-regulated minibuses with fixed fares and assigned routes can form an efficient feeder system for bus and rail, as is the case in Hong Kong. But minibuses are a poor substitute for higher-capacity systems for reducing congestion and emissions. Strategies to introduce formal alternatives need to mitigate the employment losses when minibuses and other informal transport services are phased out.

The ECA region has 25 of the world's 140 metro systems, including the second busiest, in Moscow, with 2.3 billion passengers in 2010. Those systems in the former Soviet Union are a legacy of a policy to have a metro in each city above one million inhabitants. However, their share of urban trips is relatively low, in large part because

BOX 8.3

Avoiding Emissions by Reforming Private Transport Services in Tbilisi

Minibuses (locally called "marshrutka") are among the most important transport modes in major cities of Georgia. They are used predominantly by low-income earners, students, and pensioners. For example, in Tbilisi, minibuses carry about 430,000 passengers daily, compared with 260,000 passengers who use metro and 215,000 who use municipal buses. Most of the vehicles are 11–20 years old and highly polluting, and the frequent stops aggravate congestion. The Tbilisi City Hall began implementing its minibus reform program by introducing a zone-based competitive tender. This requires bidders to form limited liability companies (LLCs) and to purchase brand-new vehicles as a requirement for participating in the tender. Some routes have already been tendered under this scheme in 2011, and the operators have subsequently introduced fare increases of more than 50 percent to cover the cost of new vehicles. City Hall plans to extend this scheme to all minibus routes in 2012, but resistance is mounting from citizens.

Reforming this minibus market is challenging and cannot be done by just replacing old vehicles. Better outcomes could be achieved if minibus reforms were geared toward a broader reform effort of the public transport system, not just minibus services. This would require fundamental changes in how public transport services are procured and managed, which in turn requires institutional, legal, and regulatory reforms; initial capital investments; and development and implementation of a long-term fiscal plan.

Source: World Bank 2012b. Contributed by Jen Jung Eun Oh.

the systems tend to be small. Among Russian metros, only Moscow's (301 km) and St. Petersburg's (110 km) have extensive networks. Five others range in length from 9 km to 16 km. On average, ECA metro systems have 40 km of lines with 30 stations, while Western European metros have an average of 73 km and 77 stations. Many of these metro systems are functioning at capacity, and improvement would require more automation. Extending metro lines is far more expensive than light rail or bus rapid transit (BRT) and is therefore justified only where travel volumes are very high.

Compared with other regions, cities in ECA have been less eager to adopt BRT with high-frequency buses on dedicated lanes. Only Turkey has introduced BRT, in Ankara and Istanbul. By comparison, Brazil has 16 systems and Indonesia, 12. Istanbul's BRT system started in 2007, taking only 77 days to set up, and by 2009, transported 800,000 passengers between 31 stations. The system annually avoids emitting an estimated 78.5 tons of carbon monoxide and 283 tons of nitrogen oxide into the air.[19] BRT is usually introduced on existing roads, so it will reduce the space available to other vehicles. BRT and other public transit options need to be an attractive alternative to car use to avoid increased congestion elsewhere, to reach high capacity, and to contribute to overall emission reductions (see box 8.4).

BOX 8.4

How Seoul Effectively Integrated Supply- and Demand-Side Urban Transport Policies

The Republic of Korea moved from low to high income in the second half of the 20th century, averaging about 8 percent annual growth. Car ownership grew in parallel, from 2 cars per 1,000 people in 1970 to almost 350 by 2008. The number of daily trips in Seoul grew fivefold between 1970 and 2002 to almost 30 million. The share of public transport dropped from 75 percent to 60 percent between 1980 and 1996, while the share of private car trips increased from 4 percent to 21 percent, choking the city's arterial roads and contributing to high air pollution levels. Seoul was able to halt, and to some extent reverse, these trends. The city made large investments and operational improvements in public transport and adopted measures that discouraged private car use. By 2002, public transport's share had increased again to 65 percent, while the car share dropped to 18 percent.

The keys to increased public transit use were investments in the metro system and improvements in bus services. The metro network is now 500 km long and serves an increasing share of public transport ridership. Investments in bus rapid transit (BRT); better service coordination; a switch to modern, cleaner vehicles; and a multipurpose smart card for integrated ticket payment made bus use more convenient. After the major bus system reforms in 2004, daily bus ridership increased by 700,000. As public transit improved, private car use was discouraged. Rather than expand parking in the inner city, building policies now discourage setting aside space for vehicles. The direct cost of car use increased with higher fuel taxes, parking fees, vehicle registration fees, and congestion charges.

Seoul's transition from car-oriented to transit-oriented transport has not been easy. The major reform of the bus system initially caused much confusion and debate, showing that such changes need to be accompanied by extensive consultations and information campaigns. But the main challenge is in financing. In the mid-2000s, Seoul's cumulative debt from metro expansion represented 82 percent of its total debt, and the system continues to run deficits. National transfers provide relief. The city makes up for the remaining shortfall in large part with revenue from charging individual car users. By charging the modes of transportation with the highest negative side effects while subsidizing more efficient modes, Seoul is shifting the burden to the sector that is most responsible for the environmental, health, and congestion costs.

Source: Pucher et al. 2005.

The viability of urban public transport depends greatly on urban form and land use. Development along high-density transit corridors supports higher volume and more efficient metro, rail, or bus lines. Sprawling development favors car use, and urban density is negatively correlated with per capita GHG emissions (Kennedy et al. 2009). Because these issues are closely related to urban land use planning, they are discussed further in chapter 9.

Rail and Air Transport

Railways are complex, long-lived systems, so expanding network length or capacity is difficult in the short term but can have large payoffs in the long term. Even without large-scale investments, a shift from road to rail using existing tracks will reduce GHG emissions. Transporting 100 tons of freight by road from Basel in Switzerland to the port of Rotterdam in the Netherlands generates 4.7 tons of CO_2 emissions, but transporting the same load by inland waterways generates only 2.4 tons, and by rail, 0.6 tons (Krohn, Ledbury, and Schwarz 2009).

Rail system electrification—switching from diesel to electric engines—promotes further energy savings and emission reductions of 20–40 percent (Hazeldine et al. 2009). Electric trains are lighter, and electricity generation is typically more efficient than a diesel engine. Rail electrification rates in ECA vary from 7 percent in Lithuania to 75 percent of the network in Bosnia and Herzegovina, which is well above the EU-27 average of 52 percent, as shown in figure 8.18. Increasing electrification of the ECA rail network would be expensive but worth it on high-traffic lines and where integration with the rest of the network is feasible. Energy use and emission reductions in rail operations come from improved aerodynamics, reduction of train weight, regenerative braking and on-board energy storage, lower-carbon electricity generation, and better traffic management.

For rail to play a greater role in freight shipping, it needs to be closely integrated with road and water transport. The share of trucks in inland freight transport is still lower in ECA than in Western Europe. Improving links between different kinds of transport will help rail to retain or expand its share, especially along international freight corridors. Expanding the share of rail use would result in significant emission reduction benefits: shipping a container long haul on rail, with short-haul truck transport at either end, can reduce energy consumption and emissions by half (Krohn, Ledbury, and Schwarz 2009). This change will require not only operational performance improvements and infrastructure rehabilitation but

FIGURE 8.18

Electrification Rate of Rail Networks, Selected Europe and Central Asia Countries and EU-27, 2010

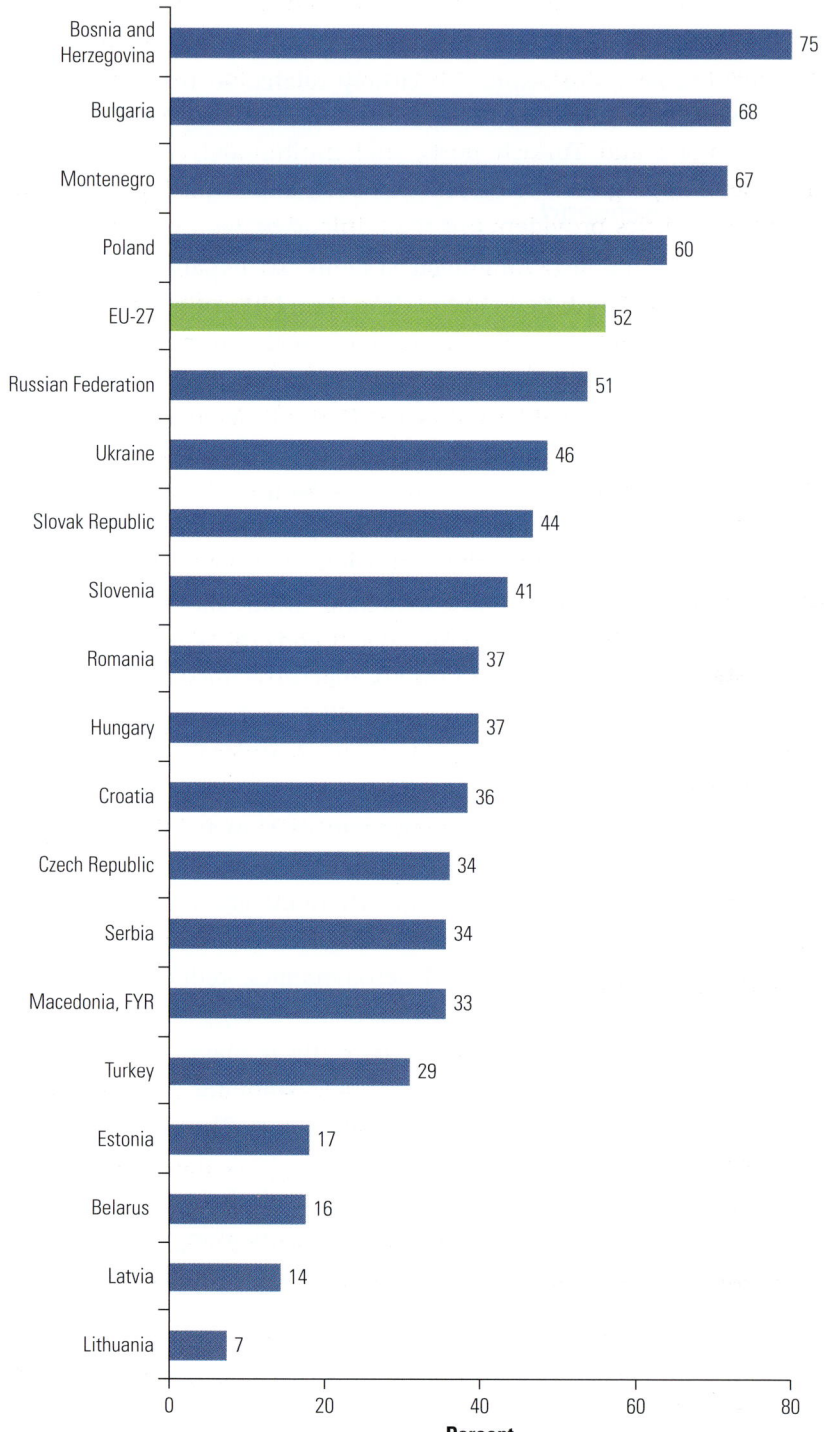

Source: UIC 2012.

also reducing delays at border crossings, for instance in Southeastern Europe and Turkey, as discussed in box 8.5.

Short-sea shipping (moving goods by sea without crossing an ocean) and inland waterway transport up rivers and canals can complement rail improvements but also compete with railways. This includes the Pan-European Corridor VII along the Danube, multimodal corridors with roll-on/roll-off (RoRo) ships between North Adriatic ports and Turkish ports, and multimodal corridors with short-sea shipping between North Sea ports and Turkish ports. Road transport logistics providers use road, inland waterways, and maritime links in Southeastern Europe and Turkey, with prices that are about 15–30 percent lower than rail rates and with significantly shorter transit times.

An important motivation for expanding high-speed passenger rail is to reduce travel time and to compete with air transport. For routes under 600 km, high-speed rail will be faster than air transport, while for distances exceeding 1,000 km, air transport is almost always quicker. Emissions per passenger are lower on high-speed rail—for instance, one-sixth of the carbon emissions of a plane passenger on the Madrid to Barcelona line. But high-speed rail is expensive and only feasible where there is highly concentrated demand within suitable distances and where incomes are high enough that passengers are willing to pay for time savings and convenience.

An economic assessment of potential high-speed rail projects needs to account for benefits from reducing road and air congestion, environmental benefits, and economic impacts in connected cities.

BOX 8.5

Reforming Turkish Railways

The Turkish General Directorate of State Railways Administration (TCDD) has a 5 percent modal share. Currently, only its port operations are profitable. Reform proposals envisage a joint stock company, the Turkish Railway Transportation Corporation (DETAŞ), with separation of passenger and freight operations but retaining state enterprise status. Still, together with other reforms, such as a departure from highway-oriented national transport policies, reorganization could make rail transport more competitive, take traffic off roads, and raise profitability. Reducing the growth of GHG emissions would be a co-benefit of the reform process but not the key driver.

These changes face formidable hurdles. Unprofitable lines would need to be closed and staffing reduced. Turkey's example illustrates the difficulties faced by ECA countries in operating commercially oriented railways.

Source: Monsalve 2011b.

It also needs to assess the relative costs and benefits of 140–160 km per hour versus 200–250 km per hour and 300–350 km per hour services. The amount of fast and high-speed rail should reflect potential traffic along those lines, composition of traffic, and cost-recovery considerations. In the ECA region, both Russia and Turkey—the top two GHG emitters in the transport sector—are currently investing in high-speed train networks.

Multimodal Mobility

The twin objectives of ensuring mobility while reducing environmental and other side effects from the transport sector require a move toward fewer and shorter trips, less fuel consumed per distance traveled, and a lower share of fossil fuels. A national or urban transport strategy needs to combine policy instruments to achieve these goals. Price instruments such as fuel taxes are a priority. As car use gets more expensive, drivers avoid unnecessary trips, residential locations near jobs and amenities are more attractive, relative prices favor public transit, and more fuel-efficient or alternative-energy vehicles become more desirable.

But a fuel tax large enough to trigger significant change will be unpopular, especially in countries that have already raised gasoline and diesel prices. Regulations such as parking restrictions, car-free zones in city centers, or fuel efficiency standards reinforce fuel price signals. All of these raise the cost of driving or make it less convenient. So alternative transportation, public transit, and non-motorized transport (biking and walking) need to be made more attractive while remaining affordable to low-income groups. This change requires investments that user fees will only partially cover. Instead, these investments need to be cofinanced through earmarking fuel tax revenue or other dedicated income sources such as a proportion of local sales taxes. International finance, including climate or carbon finance, can provide further resources, as box 8.6 discusses.

These policies are tried and tested, and many ECA countries already pursue many elements of a smart, multimodal transport strategy. Many cities in Western Europe have significantly shifted their transport policies along these lines. Although there may have been fears that restricting car use obstructs economic development, transit-oriented "green" cities are some of the most attractive and successful in the world. Likewise, by ensuring that rail networks are well managed and that road freight does not have an unfair advantage, a large share

BOX 8.6

Climate Finance Can Catalyze Smart Mobility Strategies

The transport sector has so far made less use of international climate and carbon finance than other sectors. But it can play a catalytic role. There has been only limited funding from the Global Environment Facility (GEF) and the Kyoto Protocol's Clean Development Mechanism (CDM) for transport-related projects. GEF funding leverages about $120 million in transport-related funding per year globally, while the CDM has so far supported only three registered transport projects.

The Clean Technology Fund (CTF), established in 2008 and implemented through the multilateral development banks, has allocated about 16 percent of its funding to transport projects thus far. The CTF supports low-carbon infrastructure investments while encouraging countries to establish broader, low-carbon mobility strategies. The Colombia Strategic Public Transportation Systems Program, for instance, supports the government's national urban transport policy. The program aims to strengthen transport-planning institutions, develop better regulatory frameworks, and promote transit-oriented urban development that is adapted to user needs. The US$20 million CTF grant complements more than US$300 million in Inter-American Development Bank and World Bank lending and will focus on seven medium-size Colombian cities.

Source: http://climatechange.worldbank.org.

of freight can be transported by low-emission modes like rail that reduce congestion, noise, and the risk of accidents on streets and highways.

Unless there are major technological breakthroughs, even quite ambitious policies will not lead to a massive reduction of emissions in the transport sector because overall demand will continue to increase with incomes. In Poland, transport accounts for about 10 percent of overall GHG emissions, but these grew by 74 percent between 1988 and 2006 (World Bank 2011). More than 90 percent of these emissions are from road transport, partially due to a large share of imported used cars with poor fuel consumption and emission controls: 30 percent of cars in Poland are more than 20 years old, 60 percent 10–20 years old. A business-as-usual scenario to 2030 anticipates road sector emissions to almost double. But even with a package of transport sector policy measures, emissions will likely grow by 35 percent. The transport sector is thus unlikely to make a large contribution to badly needed emission reductions. But with climate-smart policies, ECA countries and cities can limit emission growth while also reducing the hidden costs of mobility and increasing the quality-of-life benefits from a sustainable transport sector.

Notes

1. This chapter builds on Monsalve (2011a) and World Bank (2012c).
2. EU-10 countries include Bulgaria, the Czech Republic, Estonia, Hungary, Latvia, Lithuania, Poland, Romania, the Slovak Republic, and Slovenia.
3. EU candidate countries include Albania, Bosnia and Herzegovina, Croatia, Kosovo, the former Yugoslav Republic of Macedonia, Montenegro, Serbia, and Turkey.
4. Balkan countries include Albania, Bosnia and Herzegovina, Croatia, Kosovo, FYR Macedonia, Montenegro, Serbia, and Slovenia.
5. Central Asian countries include Kazakhstan, the Kyrgyz Republic, Tajikistan, Turkmenistan, and Uzbekistan.
6. Eurostat data: http://epp.eurostat.ec.europa.eu/portal/page/portal /transport/data/main_tables.
7. EU-15 countries include Austria, Belgium, Denmark, Finland, France, Germany, Greece, Ireland, Italy, Luxembourg, the Netherlands, Portugal, Spain, Sweden, and the United Kingdom.
8. The EU-27 countries include all current members of the European Union.
9. At the high altitudes flown by large jet airliners, emissions of nitrogen oxides (NO_x) are particularly effective in forming ozone. High-altitude NO_x emissions result in greater concentrations of ozone (O_3) than surface NO_x emissions, and these in turn have a greater global warming effect.
10. An integrated assessment model combines scientific and socioeconomic aspects of environmental problems and embodies numerous empirical relationships between these factors to enable policy evaluation.
11. Parry, Walls, and Harrington (2007, table 2) breaks down these externalities into greenhouse warming (US$0.06), oil dependency (US$0.12), local pollution (US$0.42), congestion (US$1.05), and accidents (US$0.63).
12. For more information, see http://capitalcitiesplanninggroup.com.
13. The avoid-shift-improve paradigm was developed by Holger Dalkmann (see UNEP 2011).
14. See the Deutsche Gesellschaft für Internationale Zusammenarbeit (GIZ) GmbH 2010/2011 fuel price data preview: http://www.giz.de/Themen /en/29957.htm.
15. Based on GIZ data on fuel prices and the World Bank's *World Development Indicators* data on gasoline consumption.
16. Research on fuel economy standards has recently been reviewed by Anderson et al. 2011 and Parry, Walls, and Harrington 2007. Note that "fuel economy" refers to the amount of fuel used to travel a given distance, while "fuel efficiency" refers to the amount of energy yielded by combustion of a given amount of fuel.
17. CIS countries include Azerbaijan, Armenia, Belarus, Georgia, Kazakhstan, the Kyrgyz Republic, Moldova, the Russian Federation, Tajikistan, Turkmenistan, Ukraine, and Uzbekistan.
18. See http://www.afdc.energy.gov/afdc/vehicles/natural_gas_emissions .html.

19. Data from the World Resources Institute's (WRI) EMBARQ (WRI Center for Sustainable Transport): http://www.embarq.org/en/project/istanbul-metrobus.

References

ADB (Asian Development Bank). 2009. "Transportation and CO_2 Emissions: Folding Them into a Unified View for Forecasts, Options Analysis, and Evaluation." Technical note, ADB, Manila.

ACEA, (European Automobile Manufactures' Association). 2010. "CO_2 tax overview" (pdf). http://www.acea.be/images/uploads/files/20100420_CO_2_tax_overview. pdf.

Anas, Alex, and Robin Lindsey. 2011. "Reducing Urban Road Transportation Externalities: Road Pricing in Theory and in Practice." *Review of Environmental Economics and Policy* 5 (1): 66–88.

Anas, Alex, and Govinda Timilsina. 2009. "Sao Paulo Impacts of Policy Instruments to Reduce Congestion and Emissions from Urban Transportation: The Case of Sao Paulo, Brazil." Policy Research Working Paper 5099, World Bank, Washington, DC.

Anderson, Soren T., Ian W. H. Parry, James M. Sallee, and Carolyn Fischer. 2011. "Automobile Fuel Economy Standards: Impacts, Efficiency, and Alternatives." *Review of Environmental Economics and Policy* 5 (1): 89–108.

Banerjee, Robin. 2007. "Deals on Wheels: An Analysis of the New Federal Auto Feebate." Report 108, C. D. Howe Institute, Toronto.

Clarke, Leon E., James A. Edmonds, Henry D. Jacoby, Hugh M. Pitcher, John M. Reilly, and Richard G. Richels. 2007. "Scenarios of Greenhouse Gas Emissions and Atmospheric Concentrations." Synthesis and Assessment Product 2.1a, U.S. Climate Change Science Program, Washington, DC.

Colliers International. 2011. "Global Central Business District Parking Rate Survey." Colliers International, Seattle.

Duranton, G., and M. A. Turner. 2011. "The Fundamental Law of Road Congestion: Evidence from U.S. Cities." *American Economic Review* 101 (6): 2616–52.

EEA (European Environment Agency). 2011a. "Air Quality in Europe—2011 Report." Technical Report 12/2011, EEA, Copenhagen.

———. 2011b. "Exceedances of Air Quality Objectives due to Traffic." EEA, Copenhagen. http://www.eea.europa.eu/data-and-maps/indicators/exceedances-of-air-quality-objectives/.

European Commission. 2010. *EU Transport in Figures: Statistical Pocketbook 2010*. Brussels: European Commission.

———. n.d. Eurostat (online database), European Commission, Brussels. http://epp.eurostat.ec.europa.eu.

Fonseca, María Blanco, Alison Burrell, Hubertus Gay, Martin Henseler, Aikaterini Kavallari, Robert M'Barek, Ignácio Pérez Domínguez, and Axel Tonini. 2010. "Impacts of the EU Biofuel Target on Agricultural Markets and Land Use: A Comparative Modelling Assessment." JRC Scientific and

Technical Report EUR 24449 EN, Joint Research Centre of the European Commission, Seville.

Gerçek, Haluk, and Orhan Demir. 2008. "Urban Mobility in Istanbul." Presentation, "Workshop on Urban Mobility in Istanbul," UNEP/MAP Blue Plan at Istanbul Technical University, June 27.

Gessen, Keith. 2010. "Letter from Moscow: Stuck." *The New Yorker*, August 2.

GFEI (Global Fuel Economy Initiative). 2010. "Cleaner, More Efficient Vehicles: Reducing Emissions in Central and Eastern Europe." Working Paper 3/10, GFEI, London.

G1Z (Deutsche Gesellschaft für Internationale Zusammenarbeit). 2012. GmbH 2010/2011 fuel price data preview: http://www.giz.de/Themen/en/29957 .htm.

Grdzelishvili, Inga, and Roger Sathre. 2011. "Understanding the Urban Travel Attitudes and Behavior of Tbilisi Residents." *Transport Policy* 18 (1): 38–45.

Hazeldine, Tom, Alison Pridmore, Dagmar Nelissen, and Jan Hulskotte. 2009. "Technical Options to Reduce GHG for Non-Road Transport Modes." Paper 3, European Commission Directorate-General Environment and AEA Technology plc. http://www.eutransportghg2050.eu.

IEA (International Energy Agency). 2009. *Transport, Energy, and CO_2: Moving toward Sustainability*. Paris: Organisation for Economic Co-operation and Development and IEA.

———. 2011. "International Comparison of Light-Duty Vehicle Fuel Economy and Related Characteristics." Draft for comment, Working Paper Series IEA/ETP, IEA, Paris.

Kennedy, Christopher, Julia Steinberger, Barrie Gasson, and Yvonne Hansen. 2009. "Greenhouse Gas Emissions from Global Cities." *Environmental Science and Technology* 43 (19): 7297–302.

Kodransky, Michael, and Gabrielle Hermann. 2011. "Europe's Parking U-Turn: From Accommodation to Regulation." Policy paper, Institute for Transportation and Development Policy, New York.

Krohn, Olaf, Matthew Ledbury, and Henning Schwarz. 2009. "Railways and the Environment: Building of the Railways' Environmental Strengths." Report, Community of European Railway and Infrastructure Companies (CER), Brussels. http://www.cer.be/media/090120_railways%20and%20 the%20environment.pdf.

Medda, Francesca Romana, and Marta Modelewska. 2009. "Land Value Capture as a Funding Source for Urban Investment: The Warsaw Metro System." Report, Ernst & Young Better Government Program 2009–10, University College London.

Monsalve, Carolina. 2011a. "Controlling Greenhouse Gas Emissions Generated by the Transport Sector in ECA: Policy Options." Background paper, Europe and Central Asia Region, World Bank, Washington, DC.

———. 2011b. "Railway Reform in South East Europe and Turkey: On the Right Track?" Report 60223-ECA, Transport Unit, Sustainable Development, Europe and Central Asia Region, World Bank, Washington DC.

NRC (National Research Council). 2009. "Driving and the Built Environment: The Effects of Compact Development on Motorized Travel, Energy Use, and CO_2 Emissions." Special Report 298, Transportation Research Board, NRC, Washington, DC.

Odyssee. n.d. "Energy Efficiency Indicators in Europe" (database). http://www.odyssee-indicators.org.

Parry, I. W. 2011. "Designing Fiscal Policy to Mitigate Global Climate Change." Staff Discussion Note, International Monetary Fund, Washington, DC.

Parry, I. W., and Kenneth A. Small. 2009. "Should Urban Transit Subsidies Be Reduced?" *American Economic Review* 99 (3): 700–24.

Parry, I. W., M. Walls, and W. Harrington. 2007. "Automobile Externalities and Policies." *Journal of Economic Literature* 45 (2): 373–99.

Podolske, Richard. 2010. "Urban Transport in ECA." Background note for *Eurasian Cities: New Realities along the Silk Road*. Washington, DC: World Bank.

Proost, Stef, and Kurt Van Dender. 2011. "What Long-Term Road Transport Future? Trends and Policy Options." *Review of Environmental Economics and Policy* 5 (1): 44–65.

Pucher, John, Hyungyong Park, Mook Han Kim, and Jumin Song. 2005. "Transport Reforms in Seoul: Innovations Motivated by Funding Crisis." *Journal of Public Transportation* 8 (5): 41–62.

Rosenthal, Elisabeth. 2011. "Across Europe, Irking Drivers Is Urban Policy." *New York Times*, June 26.

Ross, David. 2009. "GHG Emissions Resulting from Aircraft Travel." Report, Carbon Planet Limited, Adelaide, South Australia. http://www.carbonplanet.com/downloads/Flight_Calculator_Information_v9.2.pdf.

Shalizi, Zmarak, and Franck Lecocq. 2009. "Climate Change and the Economics of Targeted Mitigation in Sectors with Long-Lived Capital Stock." Policy Research Working Paper 5063, World Bank, Washington, DC.

TTI (Texas Transportation Institute). 2011. "2011 Urban Mobility Report." TTI at Texas A&M University, College Station, TX. http://mobility.tamu.edu/ums/report.

UIC (International Union of Railways). 2012. "International Railway Statistics 2010." Statistical tables, UIC, Paris.

UNECE (United Nations Economic Commission for Europe). 2010. "Main Transport Indicators in the UNECE Region." Statistical brochure, UNECE, Geneva.

UNEP (United Nations Environment Programme). 2011. "Towards a Green Economy: Pathways to Sustainable Development and Poverty Reduction—A Synthesis for Policy Makers." UNEP, Paris. http://www.unep.org/greeneconomy.

Van Dender, Kurt, and Philippe Crist. 2010. "What Does Improved Fuel Economy Cost Consumers and Taxpayers? Some Illustrations." International Transport Forum (ITF) discussion paper, ITF-KOTI Joint Seminar on Green Growth in Transport, Paris.

World Bank. 2009. *World Development Report 2009: Reshaping Economic Geography*. Washington, DC: World Bank.

———. 2011. "Transition to a Low-Emissions Economy in Poland." Low-carbon growth study series, Europe and Central Asia Region, World Bank, Washington, DC.

———. 2012a. "Inclusive Green Growth: The Pathways to Sustainable Development." Flagship report, Sustainable Development Network, World Bank, Washington, DC.

———. 2012b. "Turning the Right Corner: Ensuring Development through a Low-Carbon Transport Sector." Report, World Bank, Washington, DC.

———. 2012c. "A Policy Framework for Green Transportation in Georgia." SDN Report 70290-GE, Europe and Central Asia Region, World Bank, Washington, DC.

———. Various years. *World Development Indicators*. Washington, DC: World Bank.

MAP 9.1

Population Trends in Major Cities

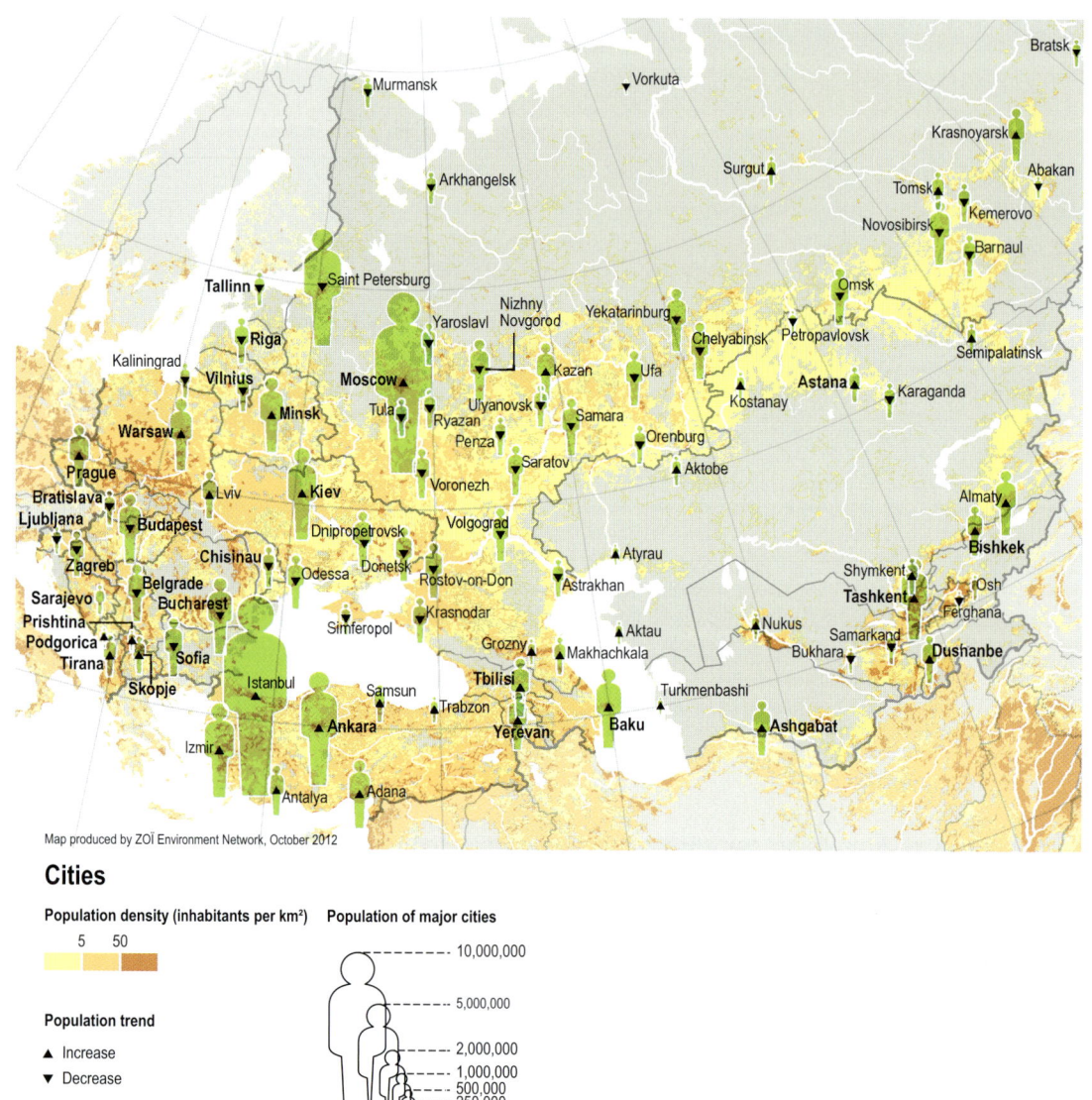

Map produced by ZOÏ Environment Network, October 2012

Cities

Population density (inhabitants per km²)

5   50

Population of major cities

- - - - - - - - 10,000,000

- - - - - - - - 5,000,000

- - - - - - - - 2,000,000

- - - - - - - - 1,000,000
- - - - - - - - 500,000
- - - - - - - - 250,000

Population trend

▲ Increase

▼ Decrease

Source: LandScan Global Population Database 2007, Oak Ridge, TN, Oak Ridge National Laboratory (http://www.ornl.gov/sci/landscan); World Gazetteer 2012 (http://www.world-gazetteer.com)
Note: km² = square kilometer.

CHAPTER 9

Cities[1]

Main Messages

- Cities will be a focus for climate action. They account for the largest share of emissions but also provide concentrated opportunities for large emission reductions. Globally, city leaders have been effective champions for climate action.

- Greening urban areas is not just an environmental policy but also promotes vibrant and innovative cities that attract the best talent for a productive urban economy. That is why Western European cities such as Copenhagen pursue climate action with ambitious goals for emission reductions.

- Energy use in Europe and Central Asia's (ECA's) buildings could be halved, saving up to 315 million tons of carbon dioxide per year, and options for large efficiency gains exist in public services, such as in modernizing district heating systems.

- Even with stable populations, city size will grow with rising wealth. Compact urban areas make public transit and public service provision more efficient. Land use planning and zoning must be closely coordinated with urban transport policies.

A healthy competition is unfolding among European cities to become the "greenest" place to live and work. Hamburg, Europe's green capital in 2011, is developing an entire district—the "HafenCity"—along ecological principles. Its predecessor, Stockholm, despite its cold climate, managed to reduce per capita carbon dioxide (CO_2) emissions to 3.4 tons, with a goal below 3 tons by 2015; this is half the European and a third of the Organisation for Economic Co-operation and Development (OECD) average. Amsterdam has an ambitious goal for 100 percent emission-free urban transport by 2040. Tallinn introduced an energy efficiency action plan in 2009 with the goal of reducing greenhouse gas (GHG) emissions by at least 20 percent by 2020. Copenhagen has set the goal of being carbon neutral by 2025. Moscow, Warsaw, and Istanbul are all members of the C40 Cities Climate Leadership Group that promotes carbon emission reductions and energy efficiency in large cities around the world.[2] More than 3,000 cities, large and small, including 250 from 23 Europe and Central Asia (ECA) countries, have signed the Covenant of Mayors, committing to implementation of sustainable energy policies.[3] Although progress toward global agreements among nations has been slow, climate action at the city level—where 70–80 percent of all carbon emissions originate—is strong (Hoornweg, Sugar, and Lorena 2011).

Reducing the likelihood of dangerous climate change is a major motivation for cities' efforts to reduce carbon emissions. Urban areas face adaptation challenges as climate change impacts intensify. There is also evidence that local CO_2 emissions do not just contribute to global GHG concentrations and global warming but also form "urban CO_2 domes" that aggravate local ozone and particulate matter concentrations (Jacobson 2010).

But environmental concerns are only one of the reasons for greening cities. Another is that green investments save money in the long term. Inefficient use of energy in buildings and in the provision of public services is expensive. Greener transport reduces costs from time lost and health impacts.

A third reason for greening is equally important. Climate-friendly policies help cities improve quality of life for their residents. Energy-efficient buildings increase comfort, cleaner energy improves air quality, and efficient transport avoids nerve-racking time in traffic jams. As urban economies transition from industrial to knowledge-intensive activities, they compete for investments by innovative companies and for highly skilled workers—those most attracted to a city with impeccable green credentials. Amenities matter. A study in the Russian Federation found a clear correlation between migration rates and an index summarizing urban amenities including air and water quality (Berger, Blomquist, and Peter 2008). Environmental quality is more and more a competitiveness issue.

Greening a city requires action on many fronts. This chapter focuses on three:

- First, residential and commercial buildings account for 40 percent of all primary energy consumption but also offer the greatest opportunities for energy savings and thus emission reductions while increasing the comfort of their occupants.

- Second, the public sector is a large energy consumer. The energy performance of public buildings can set an example for others, and more efficient public service delivery raises quality and reduces emissions and costs. Among urban services, a priority in ECA countries is to preserve and improve widespread district heating systems.

- Finally, rising wealth tends to drive urban sprawl. City size usually grows faster than population, and even cities with stable or slightly declining populations often expand. How this growth is managed will determine a city's carbon footprint for generations.

One barrier to greener policy making is the dearth of high-quality information. The chapter concludes with a discussion of efforts to develop comparable urban GHG inventories and an approach for urban energy audits that let cities quickly determine priority areas for action.

Buildings

At the time of transition, 80 percent of the building stock in ECA had been constructed after the Second World War, mostly consisting of large apartment blocks housing a total of 170 million people—more than 3 billion square meters in the former Soviet Union alone.[4] Most of these buildings were constructed with prefabricated concrete panels, creating identical buildings from hot Central Asian deserts to Siberia. Concrete has a poor thermal performance, conducting cold or heat six to seven times more efficiently than wood or brick. Flat roofs, rather than insulated attics, and poor-quality doors and windows reduced heating and cooling efficiency even further. Much has improved in the two decades since transition, but the legacy of low building standards continues to contribute to the ECA region's high energy intensity—the amount of energy consumed per unit of gross domestic product (GDP).

In total, buildings in the ECA region consume an estimated 6.2 million terajoules (TJ) of energy, as shown in table 9.1. Energy consumption per dwelling has come down slightly in the European

TABLE 9.1

Potential Energy Savings from Buildings in Europe and Central Asia, with 2008 as a Base Year

Country	Energy consumption in buildings (TJ)[a]	Total CO_2 emissions (kt)	Potential energy savings (TJ)[b]	CO_2 emissions abatement potential (kt)		Percent of total CO_2 emissions	
				High (energy from brown coal)	Low (energy from gas)	High (brown coal)	Low (gas)
Albania	10,081	4,239	5,040	509	283	12.01	6.67
Armenia	8,161	5,053	4,080	412	229	8.16	4.53
Azerbaijan	43,938	31,749	21,969	2,219	1,232	6.99	3.88
Belarus	170,095	66,747	85,047	8,590	4,771	12.87	7.15
Bosnia and Herzegovina	20,012	29,001	10,006	1,011	561	3.48	1.94
Bulgaria	80,996	51,739	40,498	4,090	2,272	7.91	4.39
Croatia	50,089	24,820	25,045	2,530	1,405	10.19	5.66
Czech Republic	161,830	124,862	80,915	8,172	4,539	6.55	3.64
Estonia	34,793	20,456	17,396	1,757	976	8.59	4.77
Georgia	18,392	6,027	9,196	929	516	15.41	8.56
Hungary	113,434	56,426	56,717	5,728	3,182	10.15	5.64
Kazakhstan	110,855	227,208	55,428	5,598	3,109	2.46	1.37
Kyrgyz Republic	8,258	6,075	4,129	417	232	6.87	3.81
Latvia	38,589	7,819	19,294	1,949	1,082	24.92	13.84
Lithuania	50,190	15,268	25,095	2,535	1,408	16.60	9.22
Macedonia, FYR	18,201	11,267	9,100	919	511	8.16	4.53
Moldova	15,228	4,701	7,614	769	427	16.36	9.09
Poland	446,666	317,119	223,333	22,557	12,529	7.11	3.95
Romania	120,936	94,106	60,468	6,107	3,392	6.49	3.60
Russian Federation	3,898,061	1,536,099	1,949,030	196,852	109,341	12.82	7.12
Slovak Republic	67,552	36,955	33,776	3,411	1,895	9.23	5.13
Slovenia	27,529	15,096	13,765	1,390	772	9.21	5.12
Tajikistan	12,276	7,222	6,138	620	344	8.58	4.77
Turkey	289,195	288,445	144,598	14,604	8,112	5.06	2.81
Turkmenistan	6,613	45,771	3,307	334	186	0.73	0.41
Ukraine	378,200	317,277	189,100	19,099	10,609	6.02	3.34
Uzbekistan	37,969	115,995	18,985	1,917	1,065	1.65	0.92
Total	**6,238,139**	**3,467,540**	**3,119,070**[b]	**315,026**	**174,980**	**10.10**	**5.61**

Sources: IEA 2011a; World Bank *World Development Indicators.*
Note: TJ = terajoules; kt = kilotons; CO_2 = carbon dioxide.
a. Refers to electricity and heat delivered during the operational phase of a building.
b. Potential energy savings are estimated at 50 percent of the 2008 energy consumption levels.

Union (EU)-10 countries for which data are available—and where energy consumption per dwelling is, in fact, lower than in many EU-15 countries (see figure 9.1).[5] This seemingly favorable comparison is likely because of smaller dwelling sizes, as energy use per unit of area is generally higher. Polish buildings, for instance, consume 240 kilowatt-hours (kWh) per square meter per year, twice the Danish average. Energy consumption in Russian buildings is double that of Canadian ones.

Assuming that energy use in buildings can be reduced by half, CO_2 emission reductions of 175 million to 315 million tons per year are possible depending on the energy source being replaced. Most potential energy savings are in space heating in both residential and commercial buildings, followed by appliances, water heating, and air conditioning, as shown in table 9.2. Although energy consumption

FIGURE 9.1

Energy Consumption (Electricity and Heating) per Dwelling, Selected Countries, 2010, and Annual Change, 1990–2010

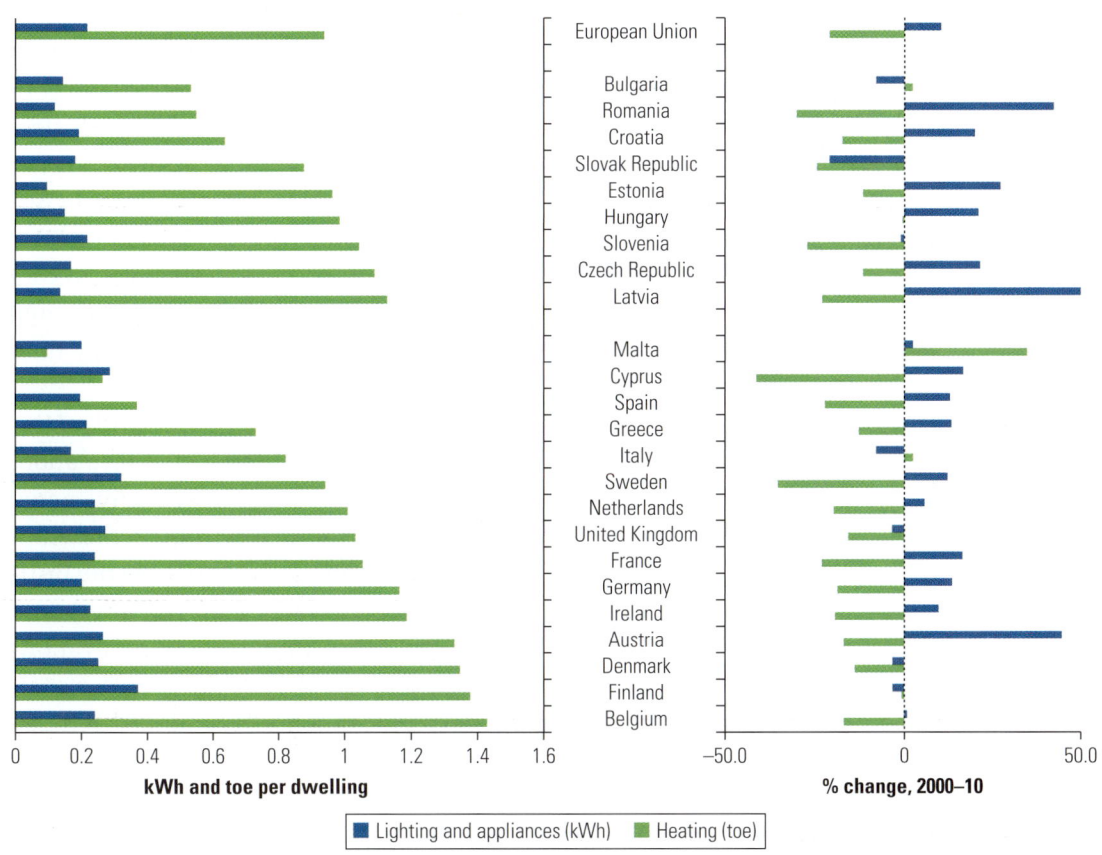

Source: Odyssee n.d.
Note: kWh = kilowatt hours; toe = tons of oil equivalent.

TABLE 9.2

Prospective Energy Savings in Buildings

percent

	Residential	Commercial
Space heating	55	40
Appliances	21	—
Water heating	15	16
Air conditioning	6	13
Lighting	3	—
Lighting and miscellaneous electricity end uses	—	32

Source: IEA 2011a.
Note: — = not available.

for heating has fallen in both EU-10 and EU-15 countries, it has increased for lighting and electric appliances, as figure 9.1 illustrates.

Significantly lower building energy consumption is possible. This goal is easier to achieve in new buildings where comprehensive energy saving adds 5–10 percent to building costs. The EU's revised Directive on Energy Performance in Buildings requires all public buildings to follow nearly zero-energy standards by 2018, followed by all new buildings by 2020.[6] Low-energy buildings consume 40–60 kWh per square meter per year. The next level of efficiency is achieved by so-called passive buildings that use less than 15 kWh per square meter per year.[7] The most efficient ones do not require a heater even in cool climates. In the long term, positive energy buildings with integrated solar panels or geothermal heating could generate more energy than the buildings consume.

Upgrading existing buildings is far more difficult and costly. Limited, "shallow" upgrades—such as caulking, adding attic insulation, or replacing windows—are easy to implement but leave many energy saving opportunities untapped. "Deep" thermal retrofits represent a far greater investment. They involve significantly increasing the insulation of the building envelope (roof, walls, and windows) and upgrading heating systems and appliances. Both types of renovations will deliver energy savings—in the range of 30–50 percent for shallow versus 70–90 percent for deep renovations. Depending on energy price levels, they will often cost less than the value of energy they save even without financial incentives. However, only "deep" renovations can deliver the scale of energy-use reductions that ambitious mitigation scenarios envisage—such as the German goal of reducing primary energy demand by 80 percent by 2050 (Hermelink and

Müller 2010).[8] Likely future energy price rises will make the economics of efficiency upgrading even more compelling. Significantly higher comfort levels further boost the case for building improvements, as box 9.1 discusses further.

Obstacles to Greater Energy Efficiency

Despite seemingly attractive benefits for homeowners, it turns out to be quite difficult to get a large number of them to invest in energy-efficient upgrades. For several reasons, many homeowners appear to discount the payoffs. A large share of the ECA building stock has been privatized, so ownership rates are quite high by international standards—above 90 percent in Albania, Armenia, Bulgaria, Georgia, Hungary, Lithuania, and Romania, for instance. But renting remains common in other countries. More than half of all dwellings are rented in the Czech Republic and the Slovak Republic. In cases where tenants pay energy bills directly or the cost of energy is added to the rent, there is little incentive for the landlord to invest in energy efficiency because the cost is simply passed on. This landlord-tenant

---

BOX 9.1

Thermal Efficiency Upgrading in Romania: Demonstration of Benefits Fosters Widespread Adoption

After transition, 82 percent of Romanian households lived in large housing estates, many constructed with prefabricated concrete blocks. This inexpensive and quick building technique helped relieve housing shortages but at the expense of quality—in particular, thermal insulation. Apartments are cold and drafty in the winter and hot in the summer. It took some time after transition before serious efforts were made to make improvements, in part because of the reluctance of homeowners who didn't have the funds for upgrades and were unsure about the benefits.

When the Romanian city of Cluj-Napoca introduced a government-sponsored energy-efficient renovation program that required only a 20 percent contribution, uptake at first was low. Only once several building associations implemented the upgrades and residents saw the results did the program take off. Besides the energy savings of 40–55 percent, improvements in building aesthetics encouraged participation. Within two years, more than 20,000 apartments were renovated—three-quarters with private funding after the initial grant program ran out.

Source: Contributed by Marcel Ionescu-Heroiu.

problem (also called the principal-agent or split incentive problem) can be a significant barrier to energy efficiency improvements. One study in the Netherlands estimates that a quarter of all residential energy used for space heating is potentially caused by the landlord-tenant problem (IEA 2007a). A study in the United States found that landlords provide significantly less-energy-efficient appliances such as refrigerators or washing machines when renters pay the utility bills (Davis 2010). Defining rules for landlords to transfer investment costs to tenants as controlled rent increases helps address these problems, which are also significant in the commercial building sector and for household appliances.

Where apartments have been privatized, owners face a coordination problem because many upgrades will affect the entire building and everyone needs to participate. This is a particular problem where households of different means live in the same building. Some may simply not be able to afford the investments. A further problem is that homeowners tend to focus more on the short-term costs of energy-efficient upgrading than on the long-term savings. Achieving high investment rates requires extremely short payback periods. Economists call these incentive problems *behavioral failures*—similar to market failures—that often require some form of public sector involvement to resolve. Getting more homeowners to invest in energy efficiency will require a combination of financial incentives, regulations, and information dissemination.

Financial Incentives for Energy Efficiency

Financial incentives lower the cost of large up-front investments that pay dividends over time in the form of lower energy bills. One common instrument is a tax incentive in the form of accelerated depreciation for commercial investments or simple tax write-offs or credits. They are available in many countries for energy efficiency investments, mostly for single-measure or shallow renovations. Good design can prevent people from taking advantage of the system, where incentives go to households that would have upgraded appliances or replaced windows anyway. For instance, only the most energy-efficient, and likely more expensive, products should qualify.

Comprehensive renovation will almost always require credit financing. In principle, banks should provide medium- or long-term financing to homeowners. However, they are often unfamiliar with energy efficiency economics and are risk averse as a result. Loan guarantees reduce the risk for local lenders, but implementation of these programs has been more difficult than expected, as discussed in box 9.2. Some countries use more direct financial support such as grants or subsidized loans—often at considerable expense to the public purse. These can be justified, for example, where energy efficiency upgrades for low-income homeowners reduce long-term energy consumption subsidies or where ambitious energy efficiency and climate goals will otherwise not be achieved.

As an example, Germany's publicly owned Kreditanstalt für Wiederaufbau (KfW), a development bank with roots in the postwar Marshall Plan, plays an important role in the national energy

---

BOX 9.2

Mixed Success of Partial Credit Guarantees for Home Energy Efficiency Investments

Upgrading building energy efficiency is capital intensive, with high up-front costs and long-term savings. Financing will almost always be required. Private lenders are often reluctant to fund energy efficiency upgrades in buildings because they lack the expertise to evaluate the credit risk. The World Bank and its International Finance Corporation, often in collaboration with the Global Environment Facility, have used partial credit guarantees to reduce banks' risk and leverage private funding. The outcome has been mixed, with generally poor uptake in the ECA region, and there is little clarity as to why this is the case. One possibility is that there is asymmetric information: lenders may think that only less desirable borrowers purchase guarantees, while borrowers may think their use of guarantees reflects uncertainty about energy efficiency project quality. A second possibility is that the guarantees applied only to a narrow niche where fundamental risk dominates—for instance, where there was incomplete information about the expected benefits of the investments rather than just a perceived risk. A third possible reason is that, given the small scale of many energy efficiency projects, transaction costs become too high with the added expense for a guarantee. Finally, there may be competing financing instruments such as grants or concessional loans available to homeowners.

The World Bank's Independent Evaluation Group found that loan guarantees are most effective when targeting less-creditworthy borrowers or when lending in underdeveloped markets (World Bank 2010b). For example, a guarantee program in Hungary successfully supported the retrofit of apartment blocks by homeowner associations where the market and borrowers were new and unconventional. Guarantee programs were less effective in other ECA countries where inadequate lending was due to wider credit market failures rather than banks' unfamiliarity with energy efficiency projects. Banks were not looking at energy efficiency investments as project finance that generates a return in the form of energy efficiency savings. They were more concerned with the overall creditworthiness of the borrowers.

Source: World Bank 2010b.

efficiency policy. Complementing broader regulation and information provision, KfW has, since 2001, financed building efficiency upgrades and new construction of low-energy buildings at very low interest rates.[9] Between 2006 and 2009, these programs supported the construction or modernization of more than 1.4 million houses and apartments, avoiding almost 4 million tons of CO_2 emissions per year and yielding estimated lifetime reductions of about 72 million tons (Mt) of CO_2 (KfW 2010; Schröder et al. 2011). These programs contributed to a halving of building energy use to 60 kWh per square meter per year in new homes and reductions to 80 kWh per square meter per year in renovated buildings. Residents in KfW-supported dwellings saved €1 billion in energy costs per year. Furthermore, the program created or secured numerous jobs, mostly in small and medium-size firms.

KfW had been long established before it engaged in residential sector energy efficiency financing. In countries where no suitable institution exists, special entities or funds can fill that role. The World Bank has supported such funds in a number of countries, including Bulgaria, Lithuania, Romania, and Turkey (Sarkar and Singh 2010). However, most of these funds have distributed relatively limited resources. One reason is that they were not embedded in a more comprehensive strategy to nudge building owners toward investments.

Codes, Regulations, and Ratings for Energy Efficiency

Building codes and similar regulations that set efficiency standards for buildings helped achieve significant energy savings in many countries. In the EU, the Directive on Energy Performance of Buildings (last updated in 2010) regulates energy standards in member countries. By themselves, however, building codes will not achieve scale in energy savings quickly because, in any given year, at most 3 percent of the building stock gets added or upgraded, and major renovations are made only every 25–40 years (Ries, Jenkins, and Wise 2009). Because of this low turnover, low efficiency standards at the time of construction also have a long-term impact on average building efficiencies (Costa and Kahn 2011).

Building codes will set minimum standards that can be tightened over time as new techniques become available and costs drop (Jakob 2006; Galvin 2009). Most building codes set out clear, firm guidelines and regulations. This makes them easier to implement and encourages materials suppliers to standardize production. But prescriptive codes can inhibit innovation. Performance-based standards are preferable in principle but are much harder to implement and monitor. Complementary initiatives will ensure that building standards are effective. Enforcement requires well-trained inspectors and measures to reduce or eliminate informal arrangements. Governments can make it easier for building material suppliers to deliver more energy-efficient products—for example, by reducing import duties or providing technical assistance. They can also help ensure that low-income homeowners can afford compliance with efficiency codes.

Information programs can promote energy efficiency investments and ensure access to reliable information about technical options, costs, and benefits. Media campaigns, advisory services, and energy audits lower information barriers. Larger, publicly supported investments can be packaged with an independent professional review to ensure deployment of the best possible options.

It is also important to ensure that energy efficiency performance becomes embedded in home and building values. Building certification programs provide official ratings. They also reduce the landlord-tenant problem because homeowners can more easily capitalize on the improvements through higher rents or resale values.[10] An asset-based rating is determined by a building's design characteristics. Operational ratings reflect a building's measured performance, which can be monitored relatively easily. Of course, ratings will always be an incomplete estimate done in advance of actual energy use, since actual consumption depends on behavior of the residents. One study of 42 identical low-energy houses in Germany showed actual energy use varying by a factor of seven (Ries, Jenkins, and Wise 2009). More and larger studies that include both predictions and actual measurement of energy use would help evaluate energy efficiency program design.

This discussion raises a broader question about performance monitoring. Although the benefits of energy efficiency investments are rarely questioned, there is surprisingly little empirical evidence of the actual costs and benefits of specific policies. Project preparation requires estimates of benefits, but few projects have a monitoring and evaluation component that follows up to confirm that the savings were realized and cost effective. An emerging literature has started to tackle these questions, often using indirect measurement through utility bills, for instance. But much of this work has been in the United States or other wealthy countries (for example, Costa and Kahn 2010; Jacobsen and Kotchen 2010). Clearly there is much that governments and international financial institutions can do to improve project and program design by investing relatively modest funds to strengthen impact evaluation of building energy efficiency.

Public Services

The energy efficiency of buildings, as discussed in the previous section, is a priority for private residences and commercial structures. It also represents a large opportunity in the public sector. Energy efficiency investments in public buildings, such as office space, schools, or hospitals, generate emission reductions and long-term savings. Given the large number of public buildings, these investments will also generate local demand for green building products and services, making them more easily available and cheaper for all. Public sector energy efficiency measures will also have a large demonstration effect, encouraging similar efforts by private building owners.

Public buildings should be subject to at least the same energy efficiency requirements as private buildings but ideally should adhere to more ambitious efficiency goals. The EU requires that public buildings in member countries display an energy certificate indicating the building's energy rating. This is meant to encourage public institutions to invest in improvements but also to increase citizen awareness about energy efficiency issues. Innovative project design and financing terms can also generate large energy savings in low- and middle-income countries, as further discussed in boxes 9.3 and 9.4.

Beyond public buildings, large energy-use and emission reductions are possible across public services including water supply and sanitation, street lighting, and solid waste (see also box 9.5). Improvements of district heating systems, most of which are owned and operated by municipalities, represent perhaps the largest opportunity for efficiency gains.

Water Supply and Sanitation

Water supply is one of the main energy consumers in the public sector, consuming an estimated 2–3 percent of all energy. The energy savings potential is often high because systems that have grown over decades are frequently undermaintained, with high nonmetered water losses and aging, inefficient equipment. Energy use may not be the highest concern in water supply operations, especially where the utility does not pay the full electricity market price, but energy is an important cost factor. In the lifetime cost of a pump, only 3 percent accounts for the purchase price, while almost three-quarters are for energy. Therefore, even small efficiency gains have large payoffs. Table 9.3 shows some measures that utilities can implement that offer attractive payback rates. Additional opportunities exist—for

BOX 9.3

Energy-Efficient Services in the Public Sector

Given the large scale and common characteristics of public buildings and services, there are opportunities for bundling energy efficiency measures in government-operated facilities. Achieving this can act as a catalyst by showing the benefits of investing in energy efficiency and thus help to develop a local market for such services. But the public sector is rarely at the forefront of energy efficiency. Not being commercially oriented, the sector's energy price signals are less effective, and financial management and budgeting can make even profitable investments cumbersome. One approach to overcoming these difficulties is an energy savings performance contract. It consists of a bundle of goods and services that improve energy efficiency in a set of public facilities and can be contracted to suitable energy services providers—energy suppliers, engineering firms, equipment manufacturers, or building management firms—in addition to more traditional energy services companies (ESCOs).

There is now a large body of experience with different variations of energy services contracting—from purely private sector provided to publicly owned energy service providers, and from more traditional fee-for-services contracting to performance-based agreements. Projects have been implemented in some ECA countries including Croatia, the Czech Republic, Hungary, Poland, and Ukraine. The World Bank Group has supported such projects for about 10 years through the International Bank of Reconstruction and Development, the International Development Association, the Global Environment Facility, and the Clean Technology Fund, using instruments such as credit lines, credit guarantees, and direct lending to public energy service providers.

These new energy services procurement models can help overcome inertia in the public sector and deal with problems like the large number of relatively small projects that nevertheless add up to big energy savings. Efficiency gains in the public sector should then also have spillovers to the private sector that can benefit from an emerging market for energy services.

Source: ESMAP 2010.

example, in wastewater treatment—for heat recovery or methane capture for power generation.

Cities with declining populations face particular challenges because infrastructure may be built to support a larger population than it needs to. A large share of service provision is determined by long-term fixed costs (as is the case for up to 75 percent of wastewater treatment costs in Germany, for instance). With a shrinking population, this fixed cost is imposed on a smaller number of ratepayers, so per capita costs go up—in a place where incomes are typically low. Removing parts of a networked infrastructure system will not be easy

BOX 9.4

Innovative Financing of Energy Efficiency Investments in Public Buildings in Armenia

Armenia's reliance on imported fuels (accounting for 90 percent of the country's energy needs) and undermaintained transmission and distribution assets put the country at risk of price fluctuations, supply interruptions, and power outages. Over 50 percent of energy is consumed in buildings, and energy bills, at up to 20 percent, are often the second-highest operating cost. Public and social buildings have suffered from chronic underinvestment because of budget shortages and are particularly inefficient. The price shock from recent energy price reforms—gas prices have increased by almost 60 percent since 2008 and electricity by about 25 percent—triggered chronic underheating, with the result that only 40 percent of the population felt their homes were heated to a comfortable level.

A 2008 World Bank study found that energy efficiency investments could save Armenia 132 billion drams (more than US$360 million) annually, equivalent to 4.3 percent of its 2009 GDP. The Renewable Resources and Energy Efficiency Fund (R2E2 Fund) was established in 2005 as an independent entity to promote renewable energy and energy efficiency. Under a recently approved project, the R2E2 Fund will offer energy service agreements (ESAs), rather than loans or grants, to public facilities such as schools. Under this scheme, a school agrees to pay its baseline energy costs into an escrow account for an agreed period of time, typically 7–10 years. The Fund conducts a preliminary diagnosis, prepares tender documents, oversees construction of the retrofit, and monitors savings. The Fund will use the escrow account balance to pay the school's energy bills and recover its investment and fees. The ESA will be flexible, so the Fund can reduce the contract length if it recovers its investment earlier, which provides an incentive for the school to save energy. The Fund expects to sign its first ESA in 2012 and sign US$2 million to US$3 million in ESAs each year.

Another innovative element relates to procurement. The Fund wanted to pass on some of the project performance risks to private contractors. So, rather than have the usual design contract followed by a construction works contract, the project has combined them in a small design-build contract. Rather than select the cheapest option, selection will be made based on the contractor offering the best long-term value to the customer. The benefit of this approach is that a net present value calculation combines the up-front investment cost with the energy, maintenance, and replacement costs in one simple number and provides a transparent way for selection of the bid that offers the overall best value to the client. Further, contract payments will be partially based on demonstrated energy savings. That is, a portion of the payments will be made only if the contractor can verify that the completed project lowers the client's energy costs to the levels promised in its original bid.

Source: Contributed by Jas Singh.

BOX 9.5

TRACEing Energy Efficiency Opportunities in the Urban Public Sector

With the large number of potential entry points for energy efficiency improvements in public services, it can be a daunting prospect to decide where to begin. A tool developed by the World Bank-administered Energy Sector Management Assistance Program (ESMAP) assists city managers in identifying priorities. The Tool for Rapid Assessment of City Energy (TRACE) provides a structured way to collect basic data on energy consumption in six sectors: urban transport, municipal buildings, street lighting, water and wastewater, power and heating, and solid waste.[a] Information collected from other cities allows for benchmarking of performance to identify areas of greatest potential efficiency gains (see figure B9.5.1). As more cities participate, this function will become more and more valuable. Once information has been collected and analyzed, TRACE also provides initial pointers and recommendations for implementing improvements. The tool is intended as a way to structure an interactive process that results in a comprehensive report and action plan in a short time period.

As one of its first activities, the ECA Region's Sustainable Cities Initiative (further discussed in box 9.7) applied TRACE in the city of Gaziantep, Turkey. The Gaziantep TRACE report provides local authorities with recommendations on improving energy efficiency, ranking actions according to the highest savings potential, as shown in table B9.5.1. The pilot demonstrates the value of this approach as a tool for learning and investment prioritization at the city level across ECA

FIGURE B9.5.1
Benchmarking Energy Use for Street Lighting in Gaziantep, Turkey, and Other Selected Cities, 2011

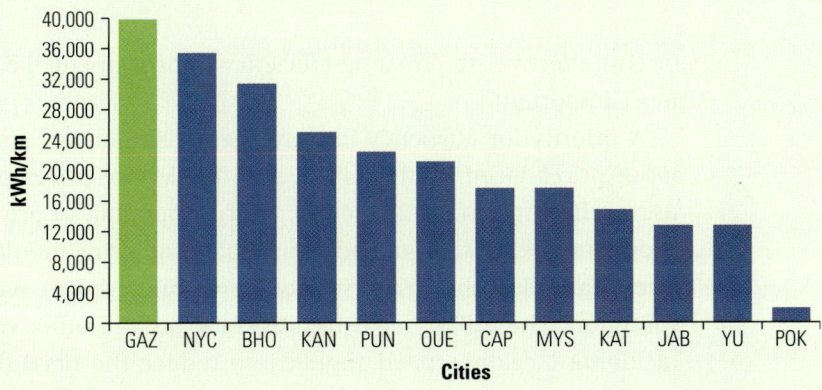

Source: World Bank 2012.
Note: km = kilometer; kWh = kilowatt-hours.

continued

BOX 9.5 *continued*

TABLE B9.5.1
Ranking of Energy Efficiency Savings Potential in Gaziantep, Turkey, 2011

Sector	Energy consumption (US$)	Relative energy intensity (%)	Level of local control (0 = no control; 1 = full control)	Savings potential (US$) [Priority]
Potable water	20,046,760	86.1	0.80	13,819,468 [Priority 1]
Public transit[a]	53,775,872	40.6	0.55	12,015,546 [Priority 2]
Private vehicles[a]	199,442,747	36.5	0.15	10,930,996 [Priority 3]
Municipal buildings	13,836,029	54.8	1.00	7,586,851 [Priority 4]
Street lighting	12,999,355	51.2	0.90	5,998,875 [Priority 5]
Wastewater	1,194,840	5.0	0.90	53,767
Solid waste[a]	500,000	48.2	0.75	180,803
Electricity	538,517,487	31.5	0.01	1,701,657

Source: World Bank 2012.
a. Sectors for which energy consumption figures have been estimated.

cities. Other opportunities emerged for collaboration with sectoral departments and financial intermediaries who were seeking an effective analytical tool to improve energy efficiency diagnostics and municipal investment prioritization. This collaboration set the stage for the formulation of a strategic pillar and a new investment lending program in Turkey under the current Country Partnership Strategy.

a. For more about ESMAP's Tool for Rapid Assessment of City Energy (TRACE), see http://www.esmap.org/esmap/node/235.

or cost effective, so pursuing efficiency improvements becomes even more important.

A priority for efficiency improvements is to ensure that both the input prices faced by the utility and the service prices paid by end users reflect true costs. Therefore, energy for public or private utilities should not be subsidized, and households and firms should pay water rates that reflect the cost of service provision to reduce wasteful consumption. The utility needs to prevent leaks and other water losses, including illegally tapped supplies, to reduce the need to treat and pump water. Water losses also mean energy losses because the energy used to pump and treat the water is wasted if it does not reach the end user.

Updating or upgrading equipment is the next step in realizing energy savings. For example, the Armenian capital Yerevan used an innovative performance-based contracting arrangement for a private

TABLE 9.3
Efficiency Measures in Water Supply and Wastewater Treatment

Area	Function	Typical payback period (years)
Electricity rates	Reduce demand during periods of peak electricity rates	0–2 depending on storage capacity
Electric installations	Power factor optimization with capacitors	0.8–1.5
	Reduction in voltage imbalance	1–1.5
Operations and maintenance	Routine pump maintenance	2
	Deep-well maintenance and rehabilitation	1–2
Production and distribution	Use automation, for example to control pressure and output in the networks and to optimize the operation of pumping equipment	0–5
	New efficient pump	1–2
	New efficient motor	2–3
	Replace impeller	0.5
	Optimize distribution network (for example, removing unnecessary valves, sectoring, installing variable speed drives, and regulating valves)	0.5–3
End use	Incentive program for the use of efficiency technologies	1–3
	Effective metering of consumption	1–2

Source: Barry 2007.

operator to upgrade water and sewerage services. Besides general service improvements, pump upgrades and replacements, more efficient network management, and greater use of gravity-fed water led to a decline in energy use by 30 percent in just four years, from 240 million kWh in 1999/2000 to 169 million kWh in 2003/04 (ESMAP 2011b). Similar performance-based contracts can be structured to overcome capital constraints by financing investments in part with future savings. Big savings are also possible under a public utility model. A US$15 million investment in Mostar, Bosnia and Herzegovina, rehabilitated infrastructure and introduced an antileakage program that contributed to 40 percent energy savings (ESMAP 2011a).

Street Lighting

Street lighting is a straightforward energy efficiency task because the savings of higher-performance lights can be fairly easily predicted. Still, too often, cities base investment decisions on initial acquisition costs rather than operational life-cycle costs.

Upgrading mercury vapor lamps can reduce energy costs by 30–40 percent and has additional benefits such as higher-quality light, less light pollution outside public spaces, and fewer light poles required. A study for the former Yugoslav Republic of Macedonia suggests that an investment of about €6.5 million in better street lighting would generate €1.16 million in annual savings for a payback period of 5–6 years and emission reductions of almost 5,000 tons of CO_2.[11]

Solid Waste

Approximately 3 percent of global GHG emissions are from solid waste, mostly methane (CH_4) from landfills and wastewater, nitrous oxide (N_2O), and CO_2 emissions from incineration of high-carbon waste such as plastics (Bernstein et al. 2007). A wide range of techniques for mitigation in the waste sector include material management that avoids waste generation in the first place, recycling, composting of organic waste, landfill gas recovery, and modern incineration with electricity or heat generation. Some of these techniques are commercially viable, such as gas recovery in large landfills.

Western European cities incinerate an increasing share of waste in advanced facilities that filter harmful emissions. The city of Copenhagen supplies electricity to 60,000 households and heat to 120,000 households from waste incinerators located within city limits, reducing the use of fossil fuels such as coal and fuel oil. The city's share of waste going to landfills has dropped to 2 percent from more than 40 percent 20 years ago.

EU-wide, waste reduction goals have not been achieved because per capita waste generation has increased in line with materials use in almost all countries, as shown in figure 9.2. On average, each person in the EU uses 16 tons of materials per person per year, of which 6 tons end up as waste. But the management of waste has improved. A smaller amount of waste is going into landfills and a higher share is recycled or composted (including for the production of biogas), with the remainder burned in modern incinerators that generate electricity and heat. ECA countries still produce less waste than Western European countries, but they also saw greater increases between 2003 and 2010.

Most ECA countries lag in the modernization of solid waste management. Waste often ends up in dump sites that are either illegal or are not up to standard; there are not enough modern incineration facilities; and far more waste could be recycled. In Romania, for instance, only 2 percent of municipal waste gets recycled (World Bank 2011). Improvements would significantly contribute to EU-wide climate goals (EEA 2010). By 2020, better municipal waste management could result in additional reductions of 44 million tons of carbon dioxide equivalent (CO_2e) compared with 2008 in Europe. If all countries complied with the EU Landfill Directive's targets, savings would rise to 62 million tons.[12] Adding a ban on landfilling

FIGURE 9.2

Municipal Waste Generation in EU, EFTA, Turkey, and Western Balkan Countries, 2003 and 2010

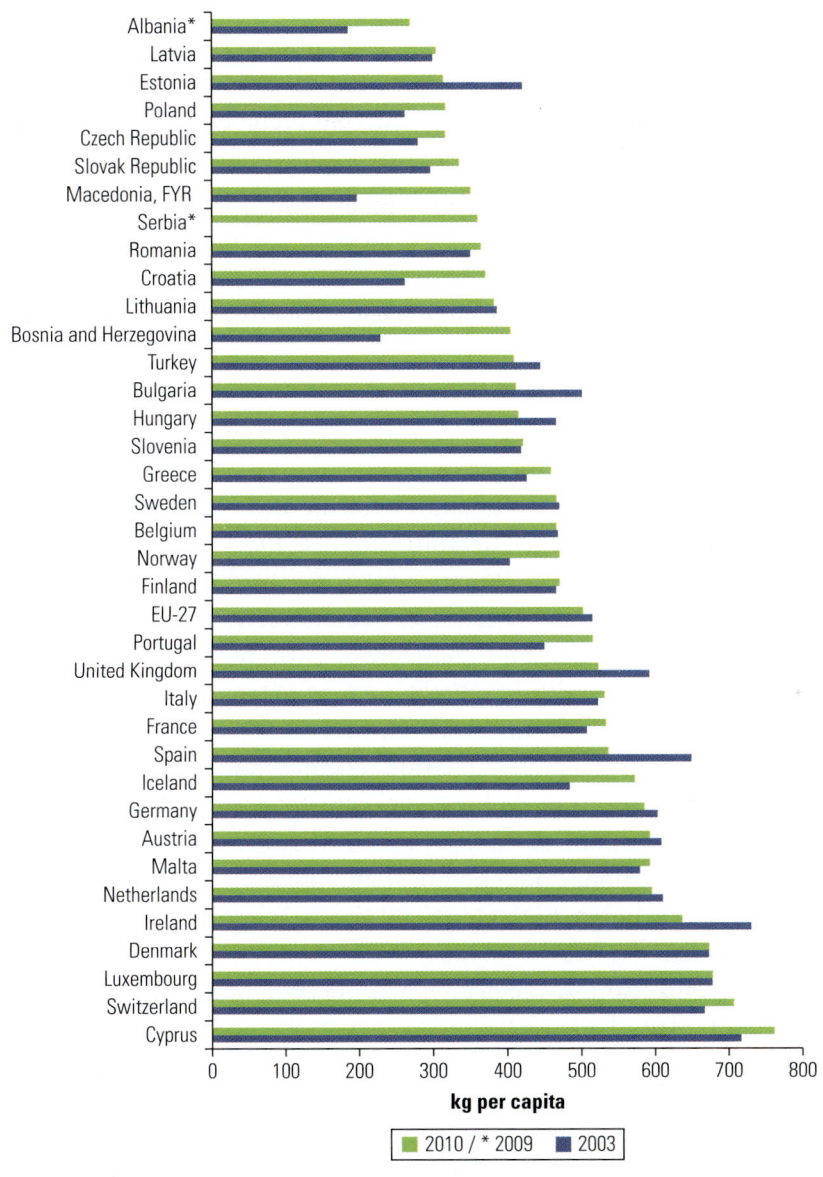

Source: EEA 2010.

Note: The star refers to countries for which only 2009 data are available. EU = European Union; EFTA = European Free Trade Association (Iceland, Liechtenstein, Norway, and Switzerland); kg = kilogram.

biodegradable waste would increase emission reductions to 77 million tons—comparable to the savings from full implementation of the EU's building energy efficiency directive.

Climate change concerns are generally only a minor motivation for improvements in solid waste disposal, including for the EU's landfill directive. More immediate benefits dominate, such as preventing pollution and odors; maintaining air, soil, and water quality; and addressing a shortage of space for landfills. Emission reductions come as an additional benefit. Still, the mitigation aspects are important enough that carbon finance has been used to cofund solid-waste management improvements, in particular for capturing landfill gases such as in Russia and Ukraine. For ECA EU members, carbon finance does not play a major role because the EU landfill directive considers methane capture to be a baseline case rather than an additional emission reduction. EU funding for new member states is available, but implementation is lagging behind in areas such as project preparation, site selection for new landfills, procurement, and monitoring of project implementation.

District Heating Systems

A positive legacy of central planning in the ECA region is the large number of urban district heating systems. Because of their large potential contribution to climate action, they are discussed here in more detail. These systems generate heat in a central facility and distribute it to residential and commercial customers through a network of pipes. Some systems distribute heat for space heating as well as hot water or steam for industrial processes requiring very high temperatures. Others also provide cooling during hot summer months, often by using lower-temperature lake or ocean water.[13] For heat generation, so-called heat-only systems solely generate heat for distribution. Combined heat and power (CHP) or cogeneration systems generate both electricity and heat either by design or as a by-product.

The main advantages of district heating systems are efficiency and flexibility. CHP plants typically convert 75–85 percent of the fuel's embedded energy into useful energy—the best of them as much as 90 percent—compared with 20–35 percent for conventional thermal condensing power stations. These efficiency gains translate into significant CO_2 emission reductions. Because the operational characteristics of CHP and district heating systems differ greatly, as do the systems they replace, CO_2 emission reductions will also vary widely, but they can be as high as 42–52 percent compared with realistic alternatives.[14] In the United States, CHP represents around 13 percent of all profitable CO_2 emission reductions for buildings by 2030 and 53 percent for industry (IEA 2007b).

District heating and cogeneration systems are flexible in terms of heat source or fuel use. Some Western European cities now use waste incineration in CHP plants equipped with advanced filters to prevent toxic emissions. These plants are clean enough to be located within city limits in Copenhagen and Hamburg. Plants can also burn biomass, as is common in Scandinavia, or use other renewable energy sources such as geothermal or solar collectors. Waste heat can come from large power stations, including nuclear reactors, or large-scale industrial plants.

The share of district heating in residential heat supply is highest in Eastern European and Scandinavian countries, as figure 9.3 illustrates. Geothermal district heating in Iceland has the highest coverage; Norway has among the lowest because most houses are heated with hydro-generated electricity. District heating is common in the former Soviet Union and other Eastern European countries, where its share is around 40 percent and 60 percent, respectively; Russia alone accounts for 72 percent of the ECA region's district heating

FIGURE 9.3

Share of Population Served by District Heating, Selected Countries, 2009

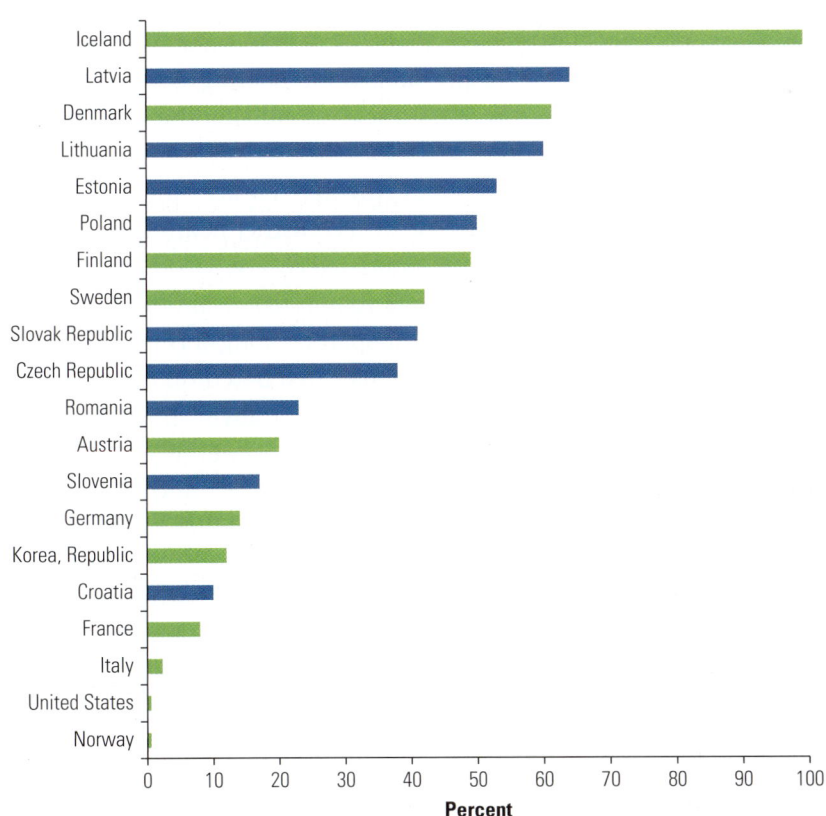

Source: Euroheat & Power (http://www.euroheat.org).
Note: The blue bars indicate ECA countries; the green bars indicate other countries.

capacity (World Bank 2010c). The region is therefore in a good position to achieve high levels of efficiency in the heating sector if utilities can reduce waste and recover costs. The three policy priorities will be

- Pricing reform to enhance cost recovery and encourage efficient energy use;

- Regulatory reform to improve the operational efficiency on the supply side; and

- Investments to upgrade power and heat generation equipment.

Average cost recovery in 2009 among a sample of 35 Ukrainian district heating utilities was 79 percent (for heat only) and as low as 55 percent for individual plants (Semikolenova, Pierce, and Hankinson 2012; Brown 2010). Part of the reason is wasteful consumption because customers do not face realistic prices. Many district heating systems set rates based on the amount of space heated, not actual heat consumption. There is therefore no incentive for households to conserve energy, and suppliers lack the information to set appropriate tariffs. Fewer than half of the households in Ukraine have meters for heat, and even fewer have meters for hot water. In privatized apartment buildings, the responsibility for heating common areas and maintenance of pipes is often unclear, reducing the incentive to heat those spaces efficiently.

Explicit and implicit subsidies further reduce incentives for demand-side efficiency and cost recovery. District heating systems in Ukraine pay only a quarter of the estimated cost of natural gas, which is the most common fuel used. Undermaintenance and underinvestment further keep costs down but jeopardize long-term sustainability. Demand-side subsidies are often allocated based on belonging to a certain group (for example, war veterans or government employees) rather than need. Instead of subsidizing energy use directly, households will more likely conserve if they face the cost-recovery price of supply. As discussed in chapter 5, affordability issues among poor and vulnerable groups are better addressed through the general social safety system or by providing assistance for household energy efficiency investments. Individual metering and subsidy reform help reduce inefficient heat consumption, but those measures cannot be seen in isolation from the more general building energy efficiency measures discussed before. The best-run district heating system will remain inefficient if windows are leaky and walls lack insulation.

District heating systems in ECA tend to operate as regulated monopolies. The design of the overall regulatory framework and particularly the choice of a rate- and tariff-setting model will greatly influence incentives for efficiency improvements. District heating services are provided by numerous municipal or commercial operators. Ukraine has about 900 operators, with about 7,000 heat-only boilers and 250 CHP plants, most of them using natural gas. However, only about 200 of those systems are large-scale systems. In the past, Ukrainian municipalities both operated and regulated their district heating systems.

Although a local government may feel more accountable to local customers, there can be a conflict of interest where, for instance, general municipal funds are used to avoid unpopular rate hikes. An independent regulator with high technical competence and detached from political processes is preferable. Centralized regulation has advantages over decentralized regulation in many circumstances. It is more sheltered from noneconomic considerations, concentrates resources and expertise, and is more likely to provide a uniform regulatory environment that is attractive to investors. Ukraine therefore recently set up a National Commission on the Regulation of the Utilities Market.

There are three dominant regulatory approaches for rate setting.[15] In rate of return (ROR) pricing, the utility is allowed to recover a predetermined rate of return after adjusting for normal capital investment, depreciation, and operational expenses. A price cap (PC) model addresses the problem of potential overinvestment in the ROR model, since the utilities' profits depend in large part on the level of capital investment. PC rates are determined by the utilities' cost basis adjusted for inflation and a factor reflecting productivity increases. PC is the model most commonly used because it is simple and less subject to manipulation. However, it requires the regulator to be correct about the cost structure, and it is relatively inflexible. The third approach is to introduce a revenue cap (RC) instead. Rather than capping the price the utility can charge, an RC sets the revenue level it can recover. The price is determined by the cost basis adjusted for efficiency gains. So in contrast to the other two approaches, the utility's profit does not depend exclusively on the quantity of heat and hot water it sells. In fact, the utility has an incentive to encourage energy efficiency among its customers because efficiency increases its income and profit.

Fixing the district heating business model will increase the ability of utilities to make necessary capital investments. Network losses in aging systems in Central Europe and the former Soviet Union are more than three times those of best-practice utilities in Scandinavia. Investments should be sequenced to achieve the highest gains first.

For instance, upgrading 10–20 percent of the distribution network can often halve the total thermal losses (World Bank 2010c).

Urban Form

Urban density—the number of people living and working per unit of a city's area—affects its GHG emissions. A smaller area means destinations are closer, so the average trip distance in more-compact cities is shorter. This shorter distance also means that people make fewer trips in cars because it is just as easy to walk or bike, and public transport is typically more efficient. Moreover, the high fixed cost of public transit is spread over more people because any given route is accessible by more users, which means that trams or buses are better used.

More-compact cities also have smaller dwelling sizes, with many residents living in apartments or row houses rather than single-family houses. These use less energy for cooling and heating. Consistent GHG inventories are not available for many cities, as box 9.6 discusses, but for those where estimates exist, CO_2 emissions per capita are indeed generally lower in denser European and Asian cities than in more sprawling North American cities, as figure 9.4 shows.[16]

The concentration of both production and consumption in city centers encourages higher population densities. Traditionally most industrial jobs were located in city centers. As economies shifted toward higher shares of value added from services, jobs in retail, finance, and other services replaced factory work. But despite the often-forecast "death of distance" due to advances in telecommunications, service sector firms also benefit from clustering in defined urban areas. Face-to-face communication becomes even more important, and the centers of many leading world cities are dominated by highly productive financial, legal, and other business services firms (Storper and Venables 2004).

Where most commuters travel to the same or only few centers of work, public transit becomes more efficient than driving. The best-known example is New York City, where the subway system has supported a high concentration of service sector jobs for more than 100 years. As one transport expert observed, if rush-hour subway commuters heading to Manhattan shifted to cars, the city would need 84 Queens Midtown Tunnels, 76 Brooklyn Bridges, or 200 Fifth Avenues—in addition to new parking spaces three times the size of Central Park.[17] Instead, during morning rush hour, a train carrying

BOX 9.6

Monitoring Urban Emissions: You Can't Manage What You Can't Measure

Good estimates of GHG emissions at the city level, such as those used to illustrate the emissions-density relationship, exist for relatively few cities. To better understand such relationships, data for many more cities would be desirable. More important, for cities to play a central role in climate action, city managers need better information about the sources of local emissions.

The Intergovernmental Panel on Climate Change has developed protocols for measuring GHG emissions at the national level. The same principles also apply at the local level. Inventories need to be *transparent* so results can be easily understood and replicated; *complete*, including all relevant sources of emissions; *consistent* between years and with national GHG inventories; *comparable* across cities; and *accurate* so that they do not over- or underestimate actual emissions.

For an inventory to be complete, it is not enough to count the emissions originating within the city limits. Urban activities also cause emissions outside city limits, from electricity generation in remote power plants or from waste disposal in landfills outside the city. For an inventory to be comparable and consistent, cities need to follow standard conventions. The United Nations Environment Program and UN-HABITAT (the United Nations Human Settlements Program) are preparing an international standard that builds on previous efforts by academic researchers such as ICLEI-Local Governments for Sustainability and the C40 Clinton Climate Initiative, for instance (Kennedy et al. 2009). The methodology will be relatively simple, building on data generally available at the city level. Acceptance of such inventories would also facilitate citywide approaches to carbon finance (World Bank 2010a).

Source: ICLEI-Local Governments for Sustainability (http://www.iclei.org/).

more than 1,000 people crosses into the central business district every six seconds. A similar thought experiment for Moscow, where the metro system handles almost 9 million trips per day, suggests that if a large proportion of travelers switched to cars, the city would need additional parking spaces the size of 1,000 Red Squares.[18]

Although local urban development strategies focus mostly on attracting producers, cities increasingly also cast themselves as "consumer cities" that provide high-quality amenities. Many cities have greatly reduced the negative aspects of urban life (pollution, crime, and congestion) and promoted the positive ones (entertainment, retail, and urban green space). This attracts skilled workers, affluent residents, and tourists. "Greening," in particular, attracts new residents who are willing to forgo extra space and businesses that require creative and innovative employees. Countering these trends toward dense cities is

FIGURE 9.4

Greenhouse Gas Emissions Relative to Population Density, Selected Cities, 2010

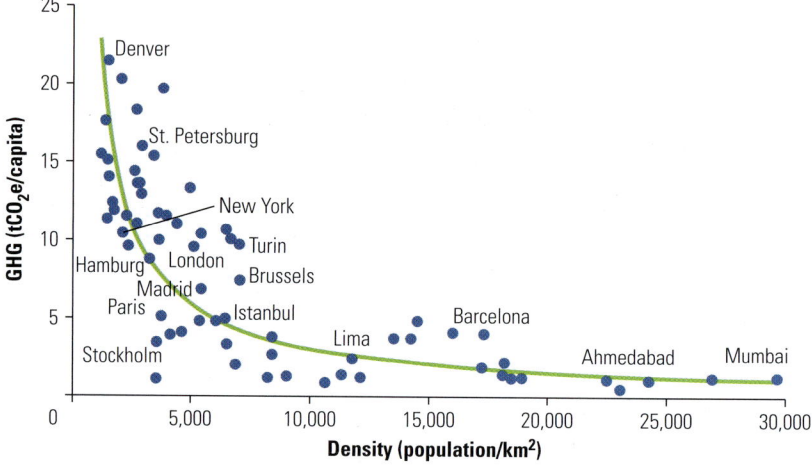

Sources: World Bank *World Development Indicators* and Citymayors.com.
Note: GHG = greenhouse gases; tCO_2e = tons of carbon dioxide equivalent; km^2 = square kilometer.

the desire for larger living spaces, especially among families, and for more convenient travel by private car. Where public policies did not discourage this trend toward sprawl, rising incomes and motorization led to far greater suburbanization, as in most North American cities.

Growth of the ECA Urban Footprint

ECA cities have generally maintained the relatively compact urban form that is a legacy of low car ownership, relatively good public transport, and state-directed housing policies favoring multiunit structures close to work places. But urban footprints have grown since 1990, not just in cities experiencing fast population growth, such as Istanbul, but even in some cities that have stable or declining populations, such as Octyabrsky in Russia and Shimkent in Kazakhstan (see map 9.2). The reasons are economic, social, and institutional. As incomes rise, per capita housing area tends to go up because people desire more living space. Average household sizes have been decreasing, as in Western Europe, and the strict institutional restrictions to urban expansion have greatly diminished, allowing new developments in the areas around cities.

Changes in urban form—if they are in the form of unplanned sprawl rather than deliberate expansion—make it harder to sustain public transit as a popular choice among commuters. The urban extent of Skopje, the capital of FYR Macedonia, almost doubled in

MAP 9.2

Population Change in Selected Urban Areas of Kazakhstan, Russia, and Turkey, 1980s–2001

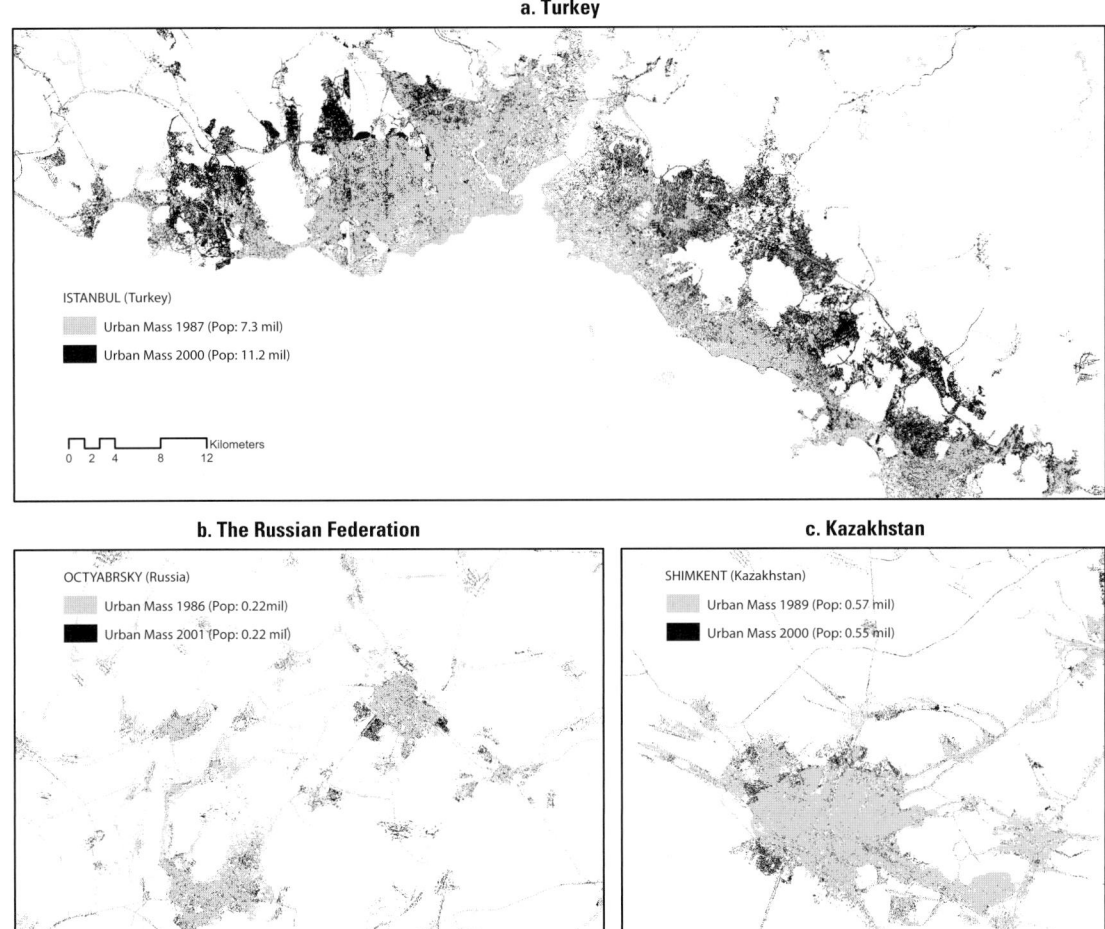

a. Turkey

ISTANBUL (Turkey)

Urban Mass 1987 (Pop: 7.3 mil)

Urban Mass 2000 (Pop: 11.2 mil)

Kilometers
0 2 4   8   12

b. The Russian Federation

OCTYABRSKY (Russia)

Urban Mass 1986 (Pop: 0.22mil)

Urban Mass 2001 (Pop: 0.22 mil)

Kilometers
0 2 4   8   12

c. Kazakhstan

SHIMKENT (Kazakhstan)

Urban Mass 1989 (Pop: 0.57 mil)

Urban Mass 2000 (Pop: 0.55 mil)

Kilometers
0 1.5 3   6   9

Source: Angel, Sheppard, and Civco 2005.
Note: pop = population; mil = millions.

the transition years. Between 2006 and 2010, growth in the number of new dwellings outstripped the rise in the number of households by a factor of almost four (Republic of Macedonia 2011). Public transit ridership dropped 60 percent, from a peak of 164 million in 1990 to 64 million in 2010, as transit routes reached fewer people and more people could afford cars. The decrease in ridership numbers led to a continuous reduction in the size of the existing urban transport network. Between 2006 and 2010, the number of transit lines in Skopje fell by 25 percent, while their total length fell by 42 percent. Over the same time period, the number of public transport vehicles and overall seating capacity dropped by 16 percent.

Policies to Manage Growth and Curtail Emissions

Policies can influence urban densities and consequently carbon emissions. It is difficult to retrofit a city once it has grown in a car-oriented, sprawling pattern. In fact, even aggressive policies would not significantly reduce per capita emissions in low-density U.S. cities anytime soon (NRC 2009).[19] But policies can reduce the risk of excessive suburbanization in the first place in cities that are still relatively compact, as is the case in many ECA cities. The following principles promote greener, carbon-efficient urban development:

- Effective zoning and other land management instruments help encourage a high quality of life in cities, making the center attractive for dense development. Such instruments include separating land uses with negative spillovers (such as polluting or noisy manufacturing) from residential and commercial spaces; encouraging mixed-use development that allows easy access to nearby shopping, entertainment, education; and sensibly preserving historical assets while also making space for new development.

- Maintaining a high modal share of public transit requires that zoning, land development, and transport policies are closely coordinated. Good planning enables people to use public transit. Complementary transport-demand management provides the incentives to do so (as discussed in chapter 8).

- A city must provide room to grow within its existing borders, on brownfields or by facilitating redevelopment (World Bank 2010d). Well-intentioned restrictions on dense development such as building-height limits lead to rising housing prices, encouraging residents to seek cheaper options outside the city. Instead, cities should foster high-density development in areas within easy access to public transit.

- Tax preferences such as mortgage interest deductions encourage home ownership that allows people to build assets and can promote stable communities. But if they are too generous, they encourage overconsumption of housing, which often means large dwellings away from city centers. Similarly, zoning and development fees will promote sprawl if they do not reflect negative side effects of spread-out development.

- Where growth cannot be accommodated within cities, good planning can guide urban expansion along a more sustainable path. Copenhagen and Stockholm are examples where transit corridors set the parameters for new developments.

These general principles make sense even if emission reductions are not the main goal because compact, efficient, and attractive cities are also competitive cities in emerging service economies. Most ECA cities still have an opportunity to preserve compact, historically grown urban form, in part because, with low population growth, they face less pressure for expansive development. But policy making needs to preserve a fine balance. It should nudge residents into embracing a low-carbon lifestyle, but if it becomes overly restrictive and prescriptive, it will drive residential and commercial uses into suburban or exurban areas. Drastic policies are justified only if the co-benefits—in the form of much-reduced congestion or air pollution, for instance—are very large or if the benefits from carbon emission reductions are valued very highly. Otherwise, other policies that reduce emissions from transport and buildings are less intrusive and more cost effective (Parry, Walls, and Harrington 2007; Proost and Van Dender 2011).

BOX 9.7

The ECA Sustainable Cities Initiative

The ECA Sustainable Cities Initiative (SCI) was formally launched in May 2010 with a knowledge exchange that attracted participants from nine ECA countries.[a] The Initiative started from the premise that city-level sustainable development actions can set the stage for worldwide sustainable development. Home to over 50 percent of the world's population, and accounting for an even larger share of global GDP, employment, and innovation, cities are engines of economic growth. They are also responsible for 60–80 percent of global energy demand and for more than 70 percent of GHG emissions.

The SCI is particularly relevant for ECA cities. Cities in ECA face challenges as they continue to transition from economies mainly fueled by manufacturing and heavy industry to more service-driven market economies. Dilapidated factories and underused urban land in the form of brownfields make many cities unattractive as places to live and work. But those areas also present an opportunity to redevelop cities around more efficient transit corridors that reduce commute times and expand desirable commercial and residential space. The housing stock is often poorly constructed and maintained, and it is aging fast. Old concrete-panel apartment buildings score badly on energy efficiency, but thermal insulation and service delivery upgrading can make these dwellings more comfortable while also reducing utility bills. Overall, to contribute to economic development, cities need to be attractive and efficient places to live and do business. Efforts to raise their sustainability are therefore also part of an urban economic growth strategy.

continued

BOX 9.7 *continued*

The economic and social realities in ECA cities therefore call for appropriate sustainable development solutions. To address some of the issues ECA cities are facing, the SCI applies the following approach:

- *Knowledge and awareness raising*—for example, general orientation workshops, learning materials and case studies, knowledge exchange and learning tours, profiling global best practice, peer learning, and innovative applications

- *Diagnostic assessment*—for example, baseline surveys and benchmarking, urban planning audits, carbon footprint calculation, energy efficiency diagnostics, shadow credit ratings, life-cycle costing, and traffic system management studies

- *Policy reform and investment strategies*—for example, updating master plans, updating urban planning regulations, setting emissions targets, city energy efficiency targets, and sustainable city investment strategies

- *Financing*—for example, specific investment financing, results-based financing, private sector finance (ESCOs), carbon financing, output-based aid, and donor cofinancing

a. SCI participants included cities from Armenia, Azerbaijan, Bosnia and Herzegovina, Georgia, FYR Macedonia, Romania, Russia, Turkey, and Ukraine.

Notes

1. This chapter builds on Ionescu-Heroiu (2011) and Kahn (2011).
2. For more about the C40 Cities Climate Leadership Group, see http://www.c40cities.org/.
3. For more on the Covenant of Mayors and its policies, see http://www.eumayors.eu.
4. Estimates from the United Nations Economic Commission for Europe, the United Nations Environment Programme, and the European Environment Agency (see Ionescu-Heroiu 2011).
5. The EU-10 countries include Bulgaria, the Czech Republic, Estonia, Hungary, Latvia, Lithuania, Poland, Romania, the Slovak Republic, and Slovenia. The EU-15 countries include Austria, Belgium, Denmark, Finland, France, Germany, Greece, Ireland, Italy, Luxembourg, the Netherlands, Portugal, Spain, Sweden, and the United Kingdom.
6. For more information about the Energy Performance of Buildings Directive—the main EU legislative instrument to reduce the energy consumption of buildings—see http://ec.europa.eu/energy/efficiency/buildings/buildings_en.htm.
7. Passive houses exploit passive solar gain, triple-pane windows, airtight building envelopes, and thermal bridge-free construction with heat recovery in the air exchanger. Passive houses require minimal or no heating or cooling.

8. This raises the question of whether policies should promote fewer deep renovations or many shallow ones to achieve the highest energy and emissions savings.

9. See the Climate Policy Initiative's policy map for a sketch of the German building efficiency program: http://climatepolicyinitiative.org/wp-content/uploads/2011/12/Policy-Map.pdf.

10. Eichholtz, Kok, and Quigley (2010) show that this is indeed the case with LEED (Leadership in Energy and Environmental Design)-certified buildings in the United States. For more information, see http://new.usgbc.org/leed.

11. Task Force for Central and Eastern Europe (http://www.taskforcecee.com) and World Bank staff calculations.

12. For more on the Landfill Directive, see http://ec.europa.eu/environment/waste/landfill_index.htm.

13. Demand for cooling is still low in most of ECA, but it may be increasing, especially with more frequent hot summers. Warsaw's electric power consumption for cooling quadrupled in just four years (http://c40citieslive.squarespace.com/storage/summit-presentations/Warsaw_Dist%20Heating%20%20Cooling.pdf).

14. "CHP Emission Reductions," UK Department of Energy & Climate Change (http://chp.decc.gov.uk/cms/chp-emission-reductions/).

15. See, for instance, http://www.regulationbodyofknowledge.org for a general discussion.

16. See, for instance, the influential paper by Newman and Kenworthy (1989). City dwellers tend to be richer, therefore they consume more products made outside the city, take more trips by airplane, and so on. But the comparison here is between dense versus (equally wealthy) sprawling cities, where these caveats apply equally.

17. Michael Frumin (http://frumin.net/ation/2009/08/whats_capacity_go_to_do_with_m.html), based on data from the annual "Hub Report" summary of New York City downtown travel (http://www.nymtc.org/data_services/HBT.html).

18. Assuming 1.5 million vehicle trips replace up to 9 million daily metro trips (since there are multiple metro rides per person and more than one person per vehicle). The size of Red Square is about 22,500 square meters, and a parking space requires about 15 square meters.

19. An ambitious scenario in which 75 percent of new and replacement housing units in the United States are located in more-compact developments and where residents of compact communities will drive 25 percent less would reduce vehicle miles traveled and associated fuel use and CO_2 emissions of new and existing households by about 7–8 percent relative to base case conditions by 2030, rising to 8–11 percent in reductions by 2050.

References

Angel, Shlomo, Stephen C. Sheppard, and Daniel L. Civco. 2005. "The Dynamics of Global Urban Expansion." Study report, Transport and Urban Development Department, World Bank, Washington, DC.

Barry, Judith. 2007. "Watergy: Energy and Water Efficiency in Municipal Water Supply and Wastewater Treatment—Cost-Effective Savings of Water and Energy." Paper, Alliance to Save Energy, Washington, DC.

Berger, Mark C., Glenn C. Blomquist, and Klara Sabirianova Peter. 2008. "Compensating Differentials in Emerging Labor and Housing Markets: Estimates of Quality of Life in Russian Cities." *Journal of Urban Economics* 63 (1): 25–55.

Bernstein, L., J. Roy, K. C. Delhotal, J. Harnisch, R. Matsuhashi, L. Price, K. Tanaka, E. Worrell, F. Yamba, and Z. Fengqi. 2007. "Industry." In *Climate Change 2007: Mitigation. Contribution of Working Group III to the Fourth Assessment Report of the Intergovernmental Panel on Climate Change*, ed. by B. Metz, O. R. Davidson, P. R. Bosch, R. Dave, and L. A. Meyer. Cambridge and New York: Cambridge University Press.

Brown, Ashley C. 2010. "Regulation of Communal Services in Ukraine." Policy paper, International Resources Group for the United States Agency for International Development, Washington, DC.

Costa, Dora L., and Matthew E. Kahn. 2010 "Energy Conservation 'Nudges' and Environmentalist Ideology: Evidence from a Randomized Residential Electricity Field Experiment." Working Paper 15939, National Bureau of Economic Research, Cambridge, MA.

———. 2011. "Electricity Consumption and Durable Housing: Understanding Cohort Effects." *American Economic Review: Papers & Proceedings* 101 (3): 88–92.

Davis, Lucas W. 2010. "Evaluating the Slow Adoption of Energy Efficient Investments: Are Renters Less Likely to Have Energy Efficient Appliances?" Working Paper 16114, National Bureau of Economic Research, Cambridge, MA.

EEA (European Environment Agency). 2010. "Natural Resources and Waste." In *The European Environment—State and Outlook 2010*. Copenhagen: EEA.

Eichholtz, Piet M. A., Nils Kok, and John M. Quigley. 2010. "Doing Well by Doing Good? Green Office Buildings." *American Economic Review* 100 (5): 2492–509.

ESMAP (Energy Sector Management Assistance Program). 2010. "Public Procurement of Energy Efficiency Services." Briefing Note 09/10, ESMAP, World Bank, Washington, DC.

———. 2011a. "Good Practices in City Energy Efficiency: Mostar, Bosnia & Herzegovina—Post-Conflict Water and Sewerage Rehabilitation Project." ESMAP report, World Bank, Washington, DC. http://www.esmap.org /sites/esmap.org/files/DocumentLibrary/EECI_Mostar_Water_Case _Study_Final.pdf.

———. 2011b. "Good Practices in City Energy Efficiency: Yerevan, Armenia— Water and Sewerage Management Contract." ESMAP report, World Bank, Washington, DC. http://www.esmap.org/node/1172.

Galvin, Ray. 2009. "Thermal Upgrades of Existing Homes in Germany: The Building Code, Subsidies, and Economic Efficiency." Working Paper EDM 09-1, Centre for Social and Economic Research on the Global Environment, University of East Anglia, Norwich, UK.

Hermelink, Andreas H., and Astrid Müller. 2010. "Economics of Deep Renovation: Implications of a Set of Case Studies." Report by Ecofys, Berlin, for the European Insulation Manufacturers Association (Eurima), Brussels.

Hoornweg, Daniel, Lorraine Sugar, and Claudia Lorena. 2011 "Cities and Greenhouse Gas Emissions: Moving Forward." *Environment and Urbanization* 23 (1): 207–27.

IEA (International Energy Agency). 2007a. *Mind the Gap: Quantifying Principal-Agent Problems in Energy Efficiency*. Paris: Organisation for Economic Co-operation and Development and IEA.

———. 2007b. *Recent Analysis into Indicators for Industrial Energy Efficiency and CO₂ Emissions*. Paris: Organisation for Economic Co-operation and Development and IEA.

———. 2011a. "Technology Roadmap: Energy-Efficient Buildings." Report, IEA, Paris. http://www.iea.org/publications/freepublications/publication/name,3983,en.html.

———. 2011b. *World Energy Outlook 2010*. Paris: Organisation for Economic Co-operation and Development and IEA.

Ionescu-Heroiu, Marcel. 2011. "Addressing Climate Challenges in ECA Cities." Background paper, Sustainable Cities Program, Europe and Central Asia Region, World Bank, Washington, DC.

Jacobsen, Grant D., and Matthew J. Kotchen. 2010. "Are Building Codes Effective at Saving Energy? Evidence from Residential Billing Data in Florida." Working Paper 16194, National Bureau of Economic Research, Cambridge, MA.

Jacobson, Mark Z. 2010. "Enhancement of Local Air Pollution by Urban CO₂ Domes." *Environmental Science and Technology* 44 (7): 2497–502.

Jakob, Martin. 2006. "Marginal Costs and Co-Benefits of Energy Efficiency Investments: The Case of the Swiss Residential Sector." *Energy Policy* 34 (2): 172–87.

Kahn, Matthew E. 2011. "Reducing the Carbon Footprint from Urban Transportation in European and Central Asian Cities." Background paper, Europe and Central Asia Region, World Bank, Washington, DC.

Kennedy, Christopher, Julia Steinberger, Barrie Gasson, Yvonne Hansen, Timothy Hillman, Miroslav Havránek, Diane Pataki, Aumnad Phdungsilp, Anu Ramaswami, and Gara Villalba Mendez. 2009. "Greenhouse Gas Emissions from Global Cities." *Environmental Science and Technology* 43 (19): 7297–302.

KfW (Kreditanstalt für Wiederaufbau). 2010. "Nachhaltigkeitsbericht 2009." (Sustainability Report 2009.) Kreditanstalt für Wiederaufbau, Frankfurt.

Newman, Peter W. G., and Jeffrey R. Kenworthy. 1989. "Gasoline Consumption and Cities." *Journal of the American Planning Association* 55 (1): 24–37.

NRC (National Research Council). 2009. "Driving and the Built Environment: The Effects of Compact Development on Motorized Travel, Energy Use, and CO₂ Emissions." Special Report 298, Transportation Research Board, NRC, Washington, DC.

Odyssee. n.d. Energy Efficiency Indicators in Europe (database). http://www.odyssee-indicators.org.

Parry, I. W., M. Walls, and W. Harrington. 2007. "Automobile Externalities and Policies." *Journal of Economic Literature* 45 (2): 373–99.

Proost, Stef, and Kurt Van Dender. 2011. "What Long-Term Road Transport Future? Trends and Policy Options." *Review of Environmental Economics and Policy* 5 (1): 44–65.

Republic of Macedonia. 2011. *Statistical Yearbook of the Republic of Macedonia.* Skopje: State Statistical Office, former Yugoslav Republic of Macedonia.

Ries, Charles P., Joseph Jenkins, and Oliver Wise. 2009. "Improving the Energy Performance of Buildings: Learning from the European Union and Australia." Technical report, Rand Corporation, Santa Monica, CA.

Sarkar, Ashok, and Jas Singh. 2010. "Financing Energy Efficiency in Developing Countries: Lessons Learned and Remaining Challenges." *Energy Policy* 38 (10): 5560–71.

Schröder, Mark, Paul Ekins, Anne Power, Monika Zulauf, and Robert Lowe. 2011. "The KfW Experience in the Reduction of Energy Use in and CO_2 Emissions from Buildings: Operation, Impacts, and Lessons for the UK." Report, UCL Energy Institute, University College London, London.

Semikolenova, Yadviga, Lauren Pierce, and Denzel Hankinson. 2012. "Modernization of the District Heating Systems in Ukraine: Heat Metering and Consumption-Based Billing." Energy Sector Management Assistance Program report for Europe and Central Asia Region, World Bank, Washington, DC.

Storper, Michael, and Anthony J. Venables. 2004. "Buzz: Face-to-Face Contact and the Urban Economy." *Journal of Economic Geography* 4 (4): 351–70.

World Bank. 2010a. "A City-Wide Approach to Carbon Finance." Policy document, Carbon Finance Unit and Urban Development Unit, World Bank, Washington, DC.

———. 2010b. *Climate Change and the World Bank Group—Phase II: The Challenge of Low-Carbon Growth.* Independent Evaluations Group (IEG) Study Series. Washington, DC: World Bank.

———. 2010c. *Lights Out? The Outlook for Energy in Eastern Europe and the Former Soviet Union.* Washington, DC: World Bank.

———. 2010d. "The Management of Brownfields Redevelopment: A Guidance Note." Europe and Central Asia Region, World Bank, Washington, DC.

———. 2011. "Solid Waste Management in Bulgaria, Croatia, Poland, and Romania: A Cross-Country Analysis of Sector Challenges towards EU Harmonization." Report 6078-ECA, Europe and Central Asia Region, World Bank, Washington, DC.

———. 2012. "Improving Energy Efficiency in Gaziantep." TRACE pilot full report, ECA Sustainable Cities Program, World Bank, Washington, DC.

MAP 10.1
Distribution of Natural Resources

Map produced by ZOÏ Environment Network, October 2012

Farms and forests

▨	Forest	1	Slovenia	5 Montenegro
▨	Agriculture	2	Croatia	6 Albania
⸜	Rangeland	3	Bosnia and Herzegovina	7 Kosovo
		4	Serbia	8 FYR Macedonia

Source: Food and Agriculture Organization of the United Nations (http://www.fao.org/geonetwork/srv/en/main.search?themekey=Managing%20Systems%20at%20Risk).

CHAPTER 10

Farms and Forests

Main Messages

- A more productive agriculture and forestry sector in Europe and Central Asia (ECA) contributes to climate change mitigation and benefits rural economies. Better-managed farms retain more carbon in soils. Additional production on ECA's vast abandoned and degraded lands increases global food supplies and avoids land conversion and associated emissions in other world regions. Reclaiming just the land abandoned since 2001 in the western portion of the Russian Federation could yield 11 million tons of grain at relatively low economic and ecological cost.

- Increasing the productivity of agriculture is a priority. It could be doubled in Kazakhstan or Ukraine and raised by two-thirds in Russia, for instance. Done right, increased productivity will also reduce emissions from land use. More directly, climate action in agriculture will involve restoring degraded land to increase carbon sequestration; reducing direct emissions from unsustainable land use, livestock production, and use of inefficient farm equipment; and exploring the potential for sustainable bioenergy production.

- Forest areas in ECA, while expanding, are threatened by increased fire frequency and locally unsustainable deforestation. Better management could increase forest carbon sequestration while increasing economic use of forest resources and significantly raising employment in forestry and wood processing.

Growing population and rising wealth will not just increase the demand for energy-consuming goods and services. It will also raise demand for products and ecological services from natural resources such as food from agriculture and timber from forests. Meeting this demand will require continued increases in the productivity of the rural sectors. This will involve trade-offs. Some modern agricultural practices contribute to global greenhouse gas (GHG) emissions through mechanized farming practices that deplete soils; excessive use of fertilizers; and large-scale, concentrated livestock operations. And deforestation, mostly in tropical forests, is the second-largest source of global carbon emissions. Yet, without intensification, agricultural areas would have already expanded much farther into grasslands and forests, which would have added vast amounts of carbon to the atmosphere. Instead, yield increases have avoided (net) emissions of up to 590 billion tons of carbon dioxide equivalent (CO_2e) between 1961 and 2005 at a cost of US$4 invested in yield improvement per ton of avoided CO_2e (Burney, Davis, and Lobell 2010). With rapidly rising demand for food, feed, and fuel, continued productivity improvements in agriculture and forestry will have to be an essential feature of climate action.

Topsoils in Europe and Central Asia (ECA) hold carbon equivalent to about 5 to 10 times global annual GHG emissions. Forest biomass holds another 3 times annual emissions.[1] Although tropical forests are a hot spot for climate mitigation because they contain enormous carbon stocks and are under far greater pressure, ECA's vast land mass has an important role in the global carbon balance as well. Unlike most of the rest of the world, ECA's agriculture and natural resources are, overall, likely a net global carbon sink. There is significant potential to maintain and often increase these carbon stocks while contributing positively to meeting the global food challenge. The region's forests, the largest in the world, are growing along with carbon sequestration and are now the world's second-largest sink. Between 1990 and 2010, carbon stocks in forests rose by about 4 billion tons or 10 percent (Shvidenko et al. 2011). But there is also evidence of greater human-caused fire frequency and unsustainable deforestation practices in parts of the region. This evidence suggests the need for better forest management to maintain economically productive forests and prevent forests from turning from carbon sinks to carbon sources.

ECA's land mass is vast, but unlike other regions, its agricultural land use has been declining as abandoned land reverts back to natural vegetation. The composition of the region's livestock herd has shifted away from cattle, which are emission intensive. ECA's population is also declining and aging, and the legacy of the land distribution and restitution process has put an unusually high share of agricultural land under the ownership or management of older generations. Finally, despite significant progress throughout the region since the onset of the economic transition, ECA's performance in agriculture and natural resources such as forestry remains well below potential.

These trends suggest that ECA's natural resources could further contribute to global carbon sequestration even as productivity and total outputs in the agricultural and forest sectors increase. This chapter first discusses climate action in agriculture, describing the link between productivity, mitigation, and adaptation, followed by a discussion of the main mitigation options on farms. These options include increasing the ability of soils to store carbon (sequestration); reducing direct emissions from land use change, livestock production, and energy use in the sector; and exploring the potential for sustainable bioenergy production. The chapter then turns to forest ecosystems and surveys options for increasing their economic potential while maintaining or increasing carbon sequestration.

Agriculture[2]

ECA's agriculture accounts for 6–8 percent of the region's total GHG emissions or almost 400 million tons (t) of CO_2e in 2005.[3] Although CO_2 is the main GHG from fossil-fuel burning, agricultural activities also emit nitrous oxide (N_2O) and methane (CH_4). CO_2 emissions from agricultural activities are less significant. Globally, methane emissions from agriculture total about 3.3 gigatons (Gt) of CO_2e per year, and nitrous oxide about 2.8 $GtCO_2$e. Comparing the relative global warming contribution of these gases is complicated because they trap different amounts of solar radiation per ton emitted, but they also break down in the atmosphere at different speeds. Methane, for instance, traps heat far more effectively as CO_2 and therefore makes a strong short-term contribution to global warming. However, it breaks down after only 25 years, while CO_2 is much longer lived and influences climate for hundreds of years.

The largest shares of agricultural emissions are from soils, livestock keeping, and energy use, as follows:

- *Agricultural soil management*. Nitrous oxide emissions are produced in soils naturally, but they are greatly increased by the often-excessive application of nitrogen fertilizers for crop cultivation. Globally this accounts for about 38 percent of total non-CO_2 emissions from agriculture. The percentages for ECA countries are likely somewhat higher than the global shares because other sources of GHGs such as burning of biomass—organic matter that can be a source of energy—and rice production are less important in ECA.

- *Livestock*. Methane is produced in the digestive processes of farm animals. Beef and dairy cattle are the largest source. Low-quality feed increases methane emissions. Globally, manure accounts for about 32 percent of emissions from agriculture.

- *Manure management*. If manure from livestock is stored as a liquid or slurry, it decomposes without oxygen (anaerobically) and generates methane emissions. Solid manure or manure spread on fields decomposes aerobically with little methane emission. Globally, this contributes about 7 percent.

- *CO_2 from energy use*. Fossil fuel used for operating vehicles and machinery, for heating buildings, or for keeping livestock accounts for about 7 percent of global agricultural emissions.

- *CO_2 from burning of plant materials and soil organic matter*. These emissions occur, for instance, after land clearing.

Productivity, Mitigation, and Adaptation in ECA's Agriculture

Climate action in agriculture links closely to productivity and adaptation. Good land management practices produce high but sustainable yields while also raising the carbon content of soils, and improved fertilizer use reduces nitrous oxide emissions. More sustainable agriculture should rely on more drought- and pest-resistant crops. Particularly in areas where productivity has not yet recovered from decades of mismanagement in the rural sector, agriculture offers the prospect of "triple wins": raising productivity and incomes, reducing emissions, and increasing resilience.

During transition, agricultural productivity in the ECA region initially dropped. Both farm output and efficiency of production—already low—decreased significantly. Macroeconomic instability, deteriorating agricultural terms of trade, and the breakdown of supply chains, combined with disruptions from land reform and farm restructuring, contributed to this decline (Swinnen and Vranken 2010). Productivity started to recover first in the Balkans and Baltic countries and later in the Commonwealth of Independent States countries.[4]

Since the late 1990s, many countries have achieved significant productivity gains. With the adoption of new seeds and farming practices, yields improved considerably. Productivity in some ECA countries now matches that of some of the most advanced agricultural producers, although in other countries the yield gap is still large, as shown in figure 10.1. Improvements have varied across countries largely depending on prereform distortions and the effectiveness and comprehensiveness of reform implementation, with generally greater improvements in Central Europe and the Balkan countries (see figure 10.2). Only in Azerbaijan and the Central Asian countries of Tajikistan, Turkmenistan, and Uzbekistan were production increases greater than yield improvements, indicating that expansion of agricultural area largely drove output growth.

The large and persistent productivity gaps, especially among some of the largest agricultural producers (Kazakhstan, the Russian Federation, and Ukraine), suggest highly inefficient use of land and potentially avoidable emissions. One study estimates it would be possible to more than double productivity in Kazakhstan and Ukraine and to increase output by 64 percent in Russia (Swinnen and Van Herck 2011).

The implications for carbon sequestration—the trapping of CO_2 emissions in soils—are unclear. In principle, the same harvests could be produced on a far smaller land area, with fields returning to natural vegetation cover, which captures more CO_2. In practice, the goal will more likely be to grow more crops for export. The implications for emissions would depend on production practices, so a trade-off between emission goals and food security is possible but not inevitable. This is clearly an area where more analysis would be beneficial.

Besides intensifying production, the ECA region could also put back into production a large stock of abandoned agricultural lands. Estimates vary, but one detailed study for Russia estimates that 26 million hectares (ha) of cropland have been abandoned since 1991—more than the total area of Romania (Schierhorn et al.

FIGURE 10.1

Cereal Yields in Europe and Central Asia Compared with Argentina, Canada, China, and Colombia, 1992–2009

FIGURE 10.2

Change in Production and Cereal Yields in Europe and Central Asia Compared with Argentina, Canada, China, and Colombia, 1993–2007

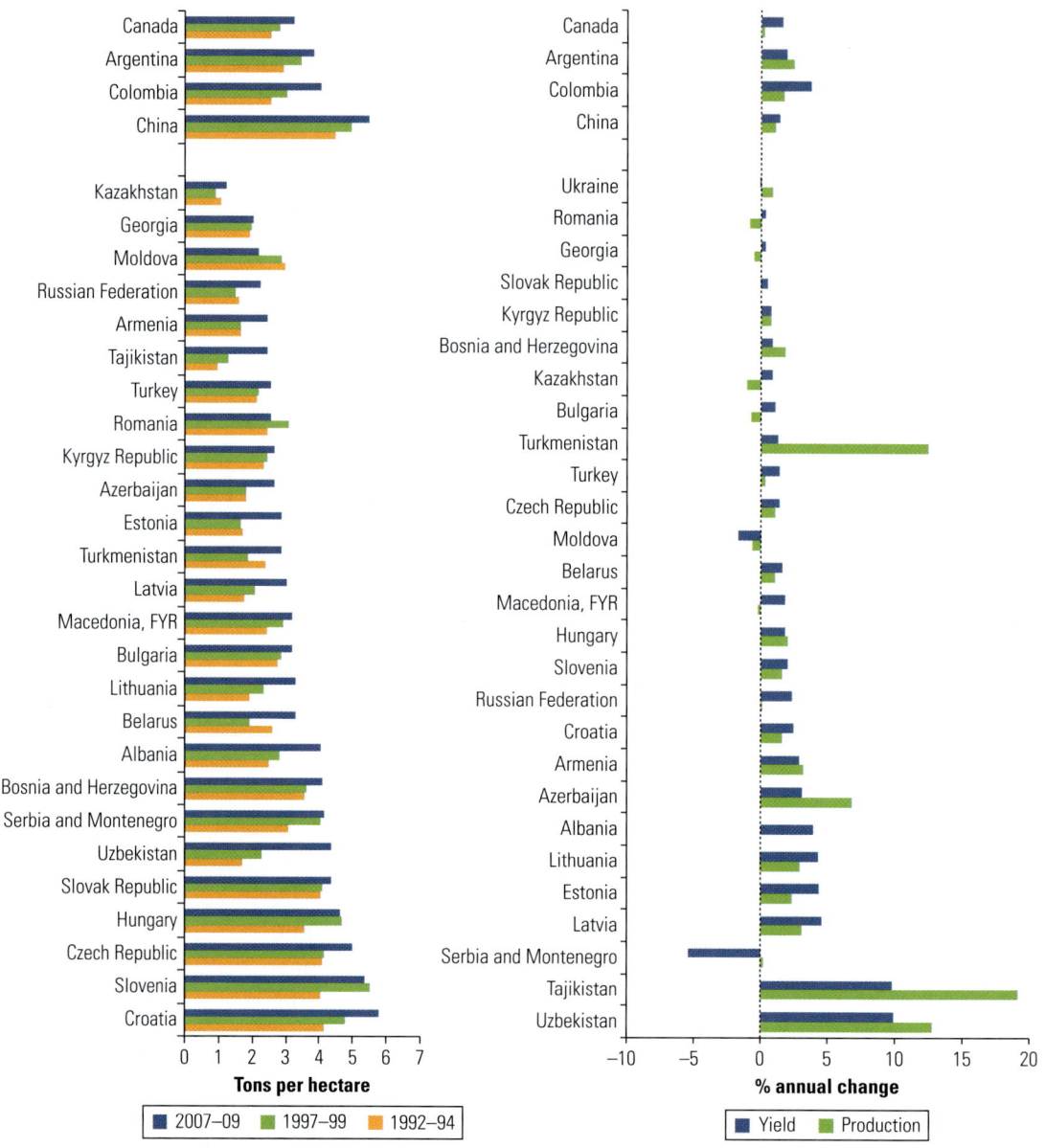

Source: FAO n.d.

Source: Larson, Dinar, and Blankespoor 2012.

2012). Reclamation should favor areas most recently abandoned rather than those where more carbon has already been built up by natural vegetation. Recultivation of the 8 million ha abandoned since 2001 could produce 11.4 million tons of grain with a relatively minor net increase in GHG emissions. Russia—and other ECA countries with similar endowments—could thus play an important role in meeting the rising global demand for food with moderate impacts on the climate compared with production expansion in other world regions.

As with farming, livestock production underwent significant changes after the transition. Figure 10.3 shows how land has shifted out of crop production and into pasture and grazing lands as well as into abandoned land. This trend is consistent with improved production technologies that are land preserving and that accompanied a decline in livestock herd size—as measured in "animal units" (a standard measure of the environmental impact of animals on land). This decline was a correction from a livestock system that was large relative to income levels and dietary needs and heavily reliant on traded feed crops.

Overall, there has been a shift away from cattle and other larger animals to poultry production in the region, a trend that is expected to continue, as shown in figure 10.4. Poultry production is expected to grow by 50 percent in Russia between 2008 and 2015, and pork production is expected to grow by 70 percent during the same period in the eight European Union (EU) members from Eastern Europe (Sutton, Block, and Srivastava 2010).

In general, the shift places less pressure on grazing lands and creates greater opportunities for rotating crops and fallowing. At the same time, commercial poultry and pig operations tend to be confined to smaller areas. This creates its own set of problems, which is why poultry and pig operations are regulated in parts of the United States even when GHG emissions are not. However, the concentrated

FIGURE 10.3

Livestock and Changing Land Use in Europe and Central Asia, 1961–2007

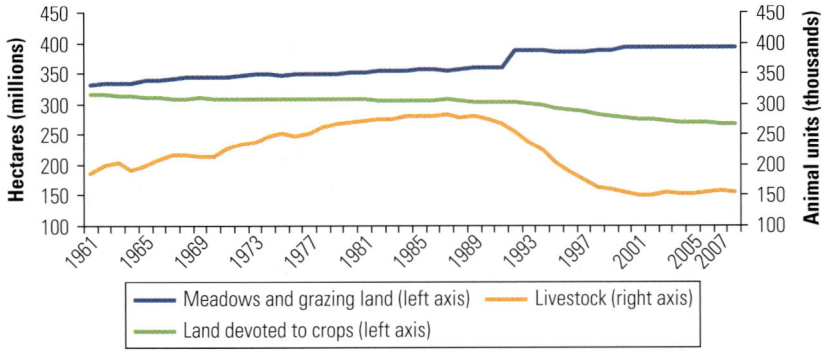

Source: FAO n.d.

Note: An "animal unit" or "livestock unit" is a standard measure that allows aggregation of the numbers of animals of different species into a single summary figure—for instance, to assess the environmental impact of animals on land.

FIGURE 10.4

Changing Composition of Livestock in Europe and Central Asia, 1961–2006

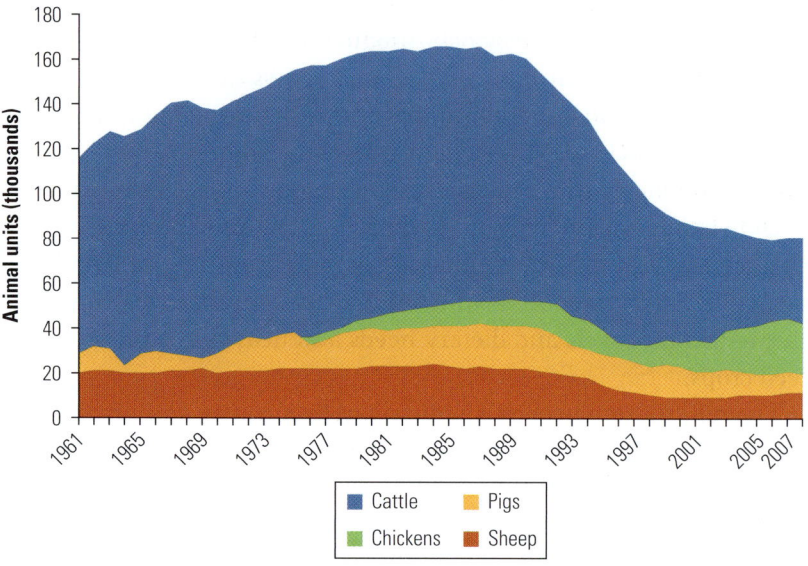

Source: FAO n.d.

Note: An "animal unit" or "livestock unit" is a standard measure that allows aggregation of the numbers of animals of different species into a single summary figure—for instance, to assess the environmental impact of animals on land.

nature of livestock operations facilitates the management of manure and other wastes, a solution that has often been financed through the Clean Development Mechanism of the Kyoto Protocol.[5]

It is still difficult to attribute specific impacts on agriculture to climate change because of the natural variability of temperature and rainfall and because climatic effects are often aggravated by poor management. But continued warming is expected even with more effective future mitigation both globally and locally, with faster warming in higher latitudes. This warming very likely already affects hydrological patterns, causing reduced winter snowfall, increased melting of glaciers, and more frequent and severe winter floods and summer droughts, especially in Southeast Europe, the South Caucasus, and Central Asia (Sutton, Block, and Srivastava 2010).[6] Moderate warming and slightly elevated CO_2 concentrations might benefit agriculture in some areas, although where farm operations are inefficient, they may not be able to take advantage of favorable conditions. With rapid warming, agricultural systems (soils and plants) may also not be able to adapt quickly enough.

Climate change adaptation strategies overlap in large measure with those that raise agricultural productivity. One priority is to close the gap between actual and potential sustainable yields, which would involve reducing the stress on land resources through better management of soil and the nutrients in it as well as livestock. The second priority is to increase potential yields through continued advances in agricultural research to develop higher-yielding crops and livestock species that will thrive in warmer or drier conditions—and by ensuring that these innovations reach farmers. These adaptive and productivity-enhancing practices can also contribute to mitigation (World Bank 2012). More efficient fertilizer use, farming practices such as conservation tillage, and improved irrigation all reduce emissions while improving productivity and resilience.

Mitigation Options in Agriculture

The 2007 Intergovernmental Panel on Climate Change (IPCC) assessment estimated the global technical mitigation potential in agriculture by 2030 at 5,500 million tons (Mt) of CO_2e per year and economic potential at about 1,500 $MtCO_2$e per year at carbon prices of US$20 per ton of CO_2, and at more than 4,000 $MtCO_2$e at prices of US$100 (Smith, Martino et al. 2008). ECA accounts for about 15 percent of global technical mitigation potential, at about 400 $MtCO_2$e in Russia and about 200 $MtCO_2$e each in Central Asia (which in this grouping includes the South Caucasus) and Eastern Europe (which here does not include the Baltic and Western Balkan countries) (see figure 10.5).

There are essentially three ways in which agriculture can contribute to mitigating global climate change, with a large number of individual options available to reduce emissions, as summarized in table 10.1. First, improved farming methods and restoration of degraded lands can increase carbon sequestration—the amount of carbon that remains locked up in soils rather than being released into the atmosphere. Second, farmers can reduce direct GHG emissions from livestock (methane), inefficient or excessive fertilizer use (nitrous oxide), and farm equipment including inefficient irrigation systems (carbon dioxide). Finally, agriculture can produce energy crops, and many agricultural waste products are also suitable for generation of electricity, heat, or transport fuels.

Carbon Sequestration

Global soils contain 3.3 times as much carbon as the atmosphere and 4.5 times as much as all aboveground biological resources (Lal 2004). Soil degradation, mostly through unsustainable land use practices, depletes the soil carbon pool and adds to carbon concentrations in

FIGURE 10.5

Total Technical Climate Change Mitigation Potentials in Agriculture (All Practices, All GHGs), by Global Region, by 2030

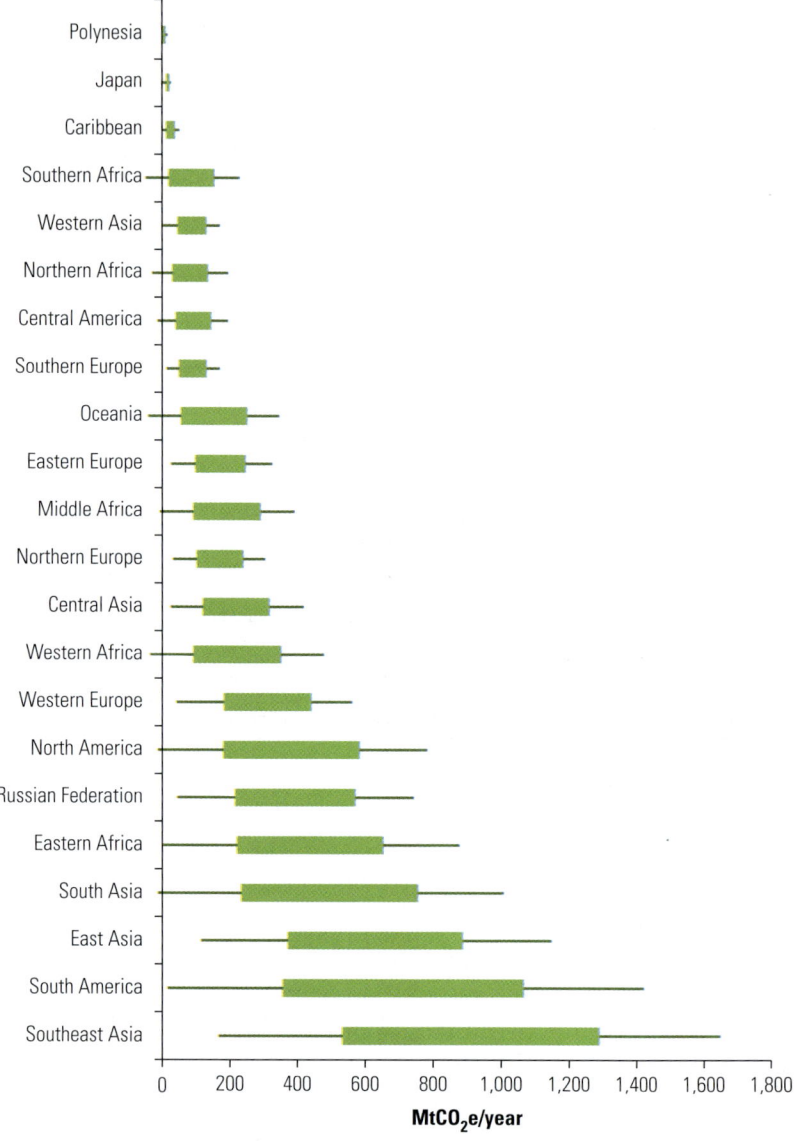

Source: IPCC 2007.
Note: Boxes show 1 standard deviation above and below the mean estimate for per-area mitigation potential, and the horizontal lines show the 95 percent confidence interval around the mean. Calculations are based on the B2 scenario, which assumes increasing population and intermediate economic development, although the pattern is similar for all of the Intergovernmental Panel on Climate Change's (IPCC) Special Report on Emissions Scenarios (SRES) scenarios. GHG = greenhouse gases; $MtCO_2e$ = million tons of carbon dioxide equivalent.

the atmosphere. Soil restoration and sustainable agriculture encourages carbon sequestration—fixing atmospheric CO_2 into long-lived soil carbon pools—which could offset about 5–15 percent of global fossil-fuel emissions. This will also increase soil productivity. An

TABLE 10.1

Opportunities for Carbon Mitigation from Agriculture

Increase carbon sequestration in soils and above-ground biomass
Replace inversion plowing (fully inverting the top soil layer) with conservation- and zero-tillage systems that do not disturb the soil
Adopt mixed crop rotations with cover crops that add nutrients to the soil to increase biomass build-up in the soil
Adopt agroforestry in cropping systems (combining trees and shrubs with crops or livestock) to increase above-ground standing biomass
Minimize summer fallows and periods with no groundcover to keep nutrients and carbon in the soil
Use soil conservation measures such as planting barriers or terracing to avoid soil erosion and loss of organic matter in the soil
Apply compost and manure to increase nutrients in the soil
Improve pastures, and rangelands through grazing, vegetation, and fire management both to reduce degradation and increase organic matter in the soil
Cultivate perennial grasses (60–80 percent of biomass below ground) rather than annuals (20 percent below ground)
Restore and protect agricultural wetlands
Convert low-quality agricultural land to woodlands to increase standing biomass of carbon
Reduce direct and indirect energy use to avoid GHG emissions (CO_2, CH_4, and N_2O)
Conserve fuel and reduce machinery use to avoid fossil-fuel consumption
Use conservation- or zero-tillage to reduce CO_2 emissions from soils
Adopt grass-based grazing systems to reduce methane emissions from ruminant livestock
Use composting to reduce manure methane emissions
Substitute biofuels for fossil-fuel consumption
Reduce the use of inorganic N fertilizers (as manufacturing is highly energy-intensive), and adopt targeted and slow-release fertilizers
Use integrated pest management to reduce pesticide use (avoid indirect energy consumption)
Increase biomass-based renewable energy production to avoid carbon emissions
Cultivate annual and perennial crops, such as grasses and coppiced trees (tree regrowth that can be harvested frequently), for combustion and electricity generation, with crops replanted each cycle for continued energy production
Use biogas digesters to produce methane, thus substituting for fossil-fuel sources

Sources: Pretty 2008; DB Climate Change Advisors 2009.
Note: CO_2 = carbon dioxide; CH_4 = methane; N_2O = nitrous oxide.

additional ton of soil carbon in degraded cropland soils can increase wheat yields by 20–40 kilograms (kg) per hectare or 10–20 kg per ha for maize. Degraded pasture lands can similarly become more productive for raising livestock.

ECA's soils store an immense amount of carbon, somewhere between 55 and 120 Gt, particularly in Russia's boreal forests and nearby areas and in some "hot spots" in Central Asia (see table 10.2 and map 10.2) (Larson, Dinar, and Blankespoor 2012). About 47 percent of ECA's land area is used for agricultural purposes, although only about 24 percent is used for moderate to high-intensity crop or livestock activities. Estimates of topsoil carbon in agricultural areas range from 21 to 47 Gt of carbon (GtC), or 37–39 percent of the total. More than half the soil carbon is contained in soils that are unmanaged. Moreover, of the 28–60 GtC in unmanaged soils, roughly half is

TABLE 10.2

Carbon Stored in Topsoils, by Land Use Type, in Europe and Central Asia, 2010

Land use type	Area (km^2, thousands)	Minimum carbon (MtCO$_2$e)	Maximum carbon (MtCO$_2$e)
Low-density grazing	2,180	3,043	7,863
Mixed agroforestry	3,332	7,777	16,627
Moderate to high-density crops and livestock	5,727	9,915	22,204
Unprotected virgin forest	5,368	14,535	30,730
Unmanaged	4,760	13,555	29,443
Other	2,348	6,497	13,811
Total	**23,715**	**55,322**	**120,678**

Source: Larson, Dinar, and Blankespoor 2012 based on data from the Food and Agriculture Organization (of the UN) and the United Nations Educational, Scientific, and Cultural Organization.
Note: km^2 = square kilometers. MtCO$_2$e = million tons of carbon dioxide equivalent.

MAP 10.2

Organic Carbon Content of Topsoils in Europe and Central Asia, 2010

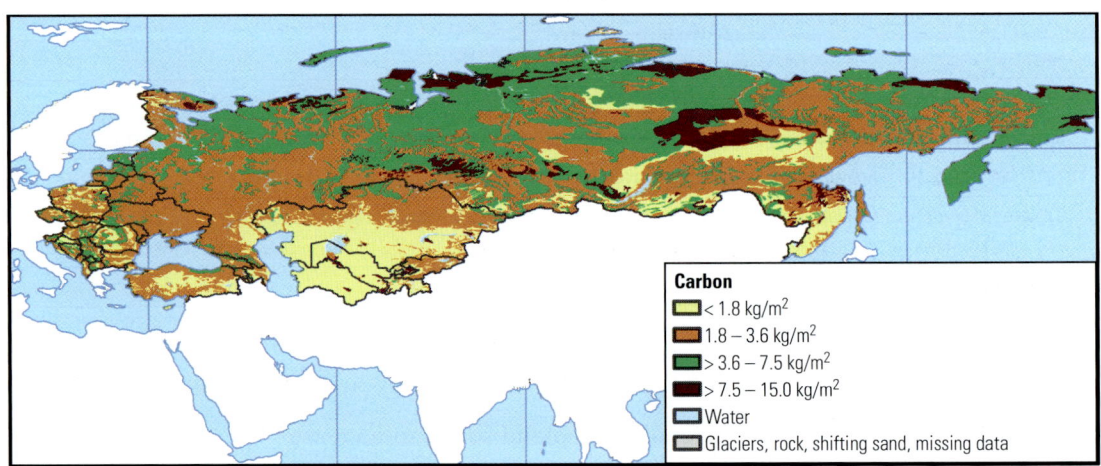

Source: Larson, Dinar, and Blankespoor 2012, based on Food and Agriculture Organization data.
Note: kg/m^2 = kilograms per square meter.

in soils beneath unprotected virgin forests. Policies aimed at protecting forests would therefore preserve not only aboveground carbon stores but also carbon contained in forest floors. However, such policies do not address the large reserves of soil carbon in the region (roughly 25 percent of the total) that are neither managed by farmers nor contained in forests.

Because restoring or enhancing soil carbon brings large productivity benefits, there is less of a need for extra funding that reflects the climate benefits. Local benefits to rural communities often far exceed global benefits. World Bank projects in Turkey (Anatolia watershed rehabilitation) and Uzbekistan (reduction of soil salinity) promoted new farming and land management practices that increased soil carbon (Sutton, Block, and Srivastava 2010). Even without accounting for carbon sequestration benefits, the projects generated rates of return of around 20 percent. Agricultural research and extension can yield similar rates of return by supplying improved and locally appropriate methods to conserve soil nutrients and increase soil moisture. Carbon sequestration and increased climate change resilience come as an additional bonus. Still, carbon finance can sometimes play a catalytic role in sequestration projects, as discussed in box 10.1.

BOX 10.1

Carbon Finance Can Assist in Financing Land Conservation and Restoration

Three land restoration and soil conservation projects in ECA illustrate how payments for environmental services (PES) systems[a] and carbon financing can be blended with standard development project financing. The projects support carbon sequestration while generating significant co-benefits, including improvements in agricultural productivity. Land degradation is a pervasive problem in the ECA region, and the sequestration of carbon through improved land management offers low-cost mitigation potential globally. Even so, land restoration and soil conservation projects are rare among the Clean Development Mechanism (CDM) and Joint Implementation projects that can offset emissions for signatories of global climate agreements.

The three projects are the Assisted Natural Regeneration of Degraded Lands Project in Albania; the Moldova Soil Conservation Project; and the Moldova Community Forestry Development Project. The first two projects are registered under the CDM, while the Moldova Community Forestry Development project awaits CDM registration.

Albania: Assisted Natural Regeneration of Degraded Lands

A core task of the project in Albania is to transform badly eroded lands into broadleaf forests.[b] The project is expected to benefit over 80,000 people through the sale of project carbon credits, employment, improved productivity, access to firewood and other forest products, cleaner water, reduced silt in reservoirs, and protection from flooding and erosion. Broader public goods include an improved habitat for native flora and fauna and reduced sediment runoff to the Adriatic Sea. Coordinating stakeholders is challenging, since 218 communes are involved in the project. The mitigation component of the Albanian project is relatively small. The project is expected to sequester 140,000 tCO_2e by 2012 and 250,000 tCO_2e by 2017 because the sequestration is

continued

BOX 10.1 *continued*

measured on only 6,000 ha managed by 30 communes, even though the project itself is expected
to result in better management of 660,000 ha of forest and pastureland. The expected project
costs totaled US$19.4 million, with contributions from the Government of Albania, the Swedish
Government, the Global Environment Facility (GEF), and the World Bank.

Moldova: Soil Conservation

The Soil Conservation project in Moldova is expected to establish forests on 15,000 ha of
degraded agricultural land on nearly 1,900 plots in 151 local communities. It aims to conserve
and improve the productivity of agricultural soils by planting shrubs and trees. Other benefits
include access to fuelwood and other forestry products to nearby communities, as well as global
biodiversity benefits. The project also was expected to sequester an estimated 1.22 $MtCO_2e$ by
2012 and 2.51 $MtCO_2e$ by 2017. The project includes financing from a World Bank loan, grants
from the Government of Japan, and funding from one of the World Bank's carbon funds.

Moldova: Community Forest Development

A follow-up project in Moldova provides protective forest belts to reduce erosion and landslides
and thereby protect agricultural soils, but co-benefits from access to forestry products and
restored habitats are expected as well. This project involves coordination among 264 local communities and was expected to sequester 229,000 tCO_2e by 2012 and 697,000 tCO_2e by 2017.

The projects have common features that the Kyoto Protocol project structure cannot easily
accommodate. To start, the projects deal with reforestation and soil sequestration for which climate benefits are hard to measure. The produced carbon credits are subject to special rules and
are not accepted in some carbon-trading schemes. Moreover, land ownership often remains
with public entities, some of which have little capacity to adequately manage and monitor project activities and the disbursement of credit funds. Both the large number of participants and
stakeholders and the weak capacity of the governing entities add to transaction costs. At the
same time, the projects generate significant direct and indirect benefits that are hard to value.

To make the projects work, the governments of Albania and Moldova, the World Bank, and
other donors funded the projects primarily from non-carbon-market sources, using the relatively small flow of revenue from carbon sales as a source of supplemental funding or as a
source of payments to communities as an incentive to pursue land management protocols.
Carbon finance remains the icing on the cake rather than the yeast that expands funding available for climate action.

Source: Larson, Dinar, and Blankespoor 2012.

a. Payments for environmental (or ecosystems) services provide funding to landowners or customary land users in return for
  preserving the ability of the land to offer benefits such as maintenance of water flows, soil productivity, or forest cover. In
  the context of climate action, payments could reduce overexploitation of land that would diminish its ability to store carbon.

b. See the World Bank's Carbon Finance Unit website: https://wbcarbonfinance.org, for project information.

The most effective practices that increase soil carbon sequestration depend on the ecozone and farming systems, as listed in table 10.3. These practices can reduce agriculture's contribution to global warming, but they should not be seen as a solution for removing excess atmospheric carbon emissions from fossil-fuel burning. Trying to sequester the geosphere (fossilized fuels) into the biosphere (soils and vegetation), as some scientists have put it, is no substitute for significant reductions in emissions from energy use. The contribution of sequestration is limited also because of characteristics that complicate its inclusion in carbon financing schemes: *saturation* (soils or trees can only capture a limited amount); *permanence* (human action can quickly reverse sequestration gains); *leakage* or *displacement* (land degradation or deforestation may simply shift elsewhere); and *verification* (problems in measurement and monitoring) (Smith, Nabuurs et al. 2008).

When successful, measures that restore and maintain soil carbon can provide vast areas for expanding agricultural production. Up-to-date estimates are scarce, but in 2000, as many as 685 million ha of land in ECA were degraded, of which 162 million ha were degraded by agricultural practices, as shown in table 10.4. With proper soil management and farming practices, 382 million ha could be suitable for production.

Reducing Direct GHG Emissions

Measures that enhance soil carbon sequestration, as discussed in the previous section, also prevent direct emissions from farming. Two additional direct sources of emissions are the energy used to operate farm machinery and emissions related to livestock production. The former is essentially an energy efficiency problem, and mitigation policies follow closely from the discussion in chapter 2 and the other sectoral chapters. Significant emission sources include oversized and outdated tractors, the energy consumed in farm buildings (such as

TABLE 10.3
Practices Promoting Soil Carbon Sequestration, by Ecosystem Type

Ecosystem	Practices benefiting soil sequestration
Cropland soils	Conservation tillage, cover crops, the use of manure, diverse cropping systems, mixed crop and livestock farming, agroforestry
Rangelands and grasslands	Grazing management, improved species, fire management, nutrient management
Restoration of degraded and desertified soils	Water and wind erosion control, creating new forests in poor farming land, water conservation and water harvesting
Irrigated soils	Using drip or sub irrigation, providing drainage, controlling salinity, enhancing water use efficiency or water conservation

Source: Lal 2004.

TABLE 10.4

Estimated Area of Degraded Lands in Europe and Central Asia as of 2000

hectares, thousands

Country	Degraded by agriculture	Degraded total area	Land suitable for agriculture
Albania	400	2,700	544
Armenia	300	300	241
Azerbaijan	4,700	4,900	2,434
Belarus	0	1,300	15,274
Bosnia and Herzegovina	0	5,100	1,922
Bulgaria	3,400	11,100	5,975
Croatia	300	5,600	2,934
Czech Republic	7,300	7,900	4,779
Estonia	0	200	1,997
Federal Republic of Yugoslavia[a]	2,000	12,000	4,605
Georgia	700	700	1,800
Hungary	2,700	6,000	6,929
Kazakhstan	3,500	47,300	3,107
Kyrgyz Republic	400	400	414
Latvia	0	4,400	5,258
Lithuania	0	1,200	5,287
Macedonia, FYR	0	2,200	634
Moldova	0	3,400	2,219
Poland	4,600	26,000	22,296
Romania	11,000	23,700	12,238
Russian Federation	100,800	380,800	219,696
Slovenia	100	1,600	665
Tajikistan	900	1,000	1,219
Turkey	3,000	77,000	14,577
Turkmenistan	6,400	6,600	312
Ukraine	3,100	45,900	42,886
Uzbekistan	6,000	6,000	2,027
Total	**161,600**	**685,300**	**382,269**

Source: FAO 2000.

a. At the time the data were gathered, the Federal Republic of Yugoslavia included the area now known as Kosovo, Serbia, and Montenegro.

heated buildings for livestock keeping), or inefficient irrigation equipment that wastes energy as well as water.

Livestock production is growing faster than any other agricultural commodity, which could significantly increase emissions. Meat and milk demand is expected to grow by 70 percent by 2050 as living standards in transition and developing countries increase. On the other hand, increased production efficiency is usually associated with lower emissions per unit of output as in the example of dairy production, further discussed in box 10.2. Factors improving efficiency are reduced losses by improving animal health services; improved pasture management; genetic improvement focused on disease resistance,

---

BOX 10.2

Mitigation Options in ECA's Livestock Production Systems

In ECA, all major types of livestock production systems exist, although their importance varies by country: for example, in Central Asia, traditional agropastoral systems in which livestock grazes across large areas predominate, while in Western Europe, more mixed crop-livestock and industrial systems are observed.

Agropastoral grazing system: The potential for mitigation is large, but difficult to attain, through carbon sequestration and rehabilitation of degraded pastures. The amount sequestered per unit area is relatively low (0.5–1.5 tCO_2e per ha under improved management) but is important given the large area of these rangelands. Better management includes plans for reducing the number of livestock in a given pasture; rotating the fields where they graze; and institutional measures such as reform of land tenure, community-based organizations, or pasture-user associations.

Ranching system: There is significant potential for mitigation through the use of improved management and, where appropriate, deeper-rooting grasses on pastures.[a] Increasing efficiency is the advocated mitigation strategy highlighted by the recent Low-Carbon Growth Study in Brazil.[b]

Silvopastoral system: Mitigation and adaptation effects are provided through the improvement of degraded pastures, with mixed vegetation from grasses, leguminous herbs and trees, and fodder shrubs. The approach has been piloted in a regional project in Central America through a program supported by the GEF, the World Bank, and the United Nations (UN) Food and Agriculture Organization.

Mixed crop-livestock production: This is the most widespread livestock system. Mitigation can be achieved through a focus on efficiency, as shown in figure B10.2.1. CO_2e per kilogram of milk output declines by a factor of 10 when production increases from a commonly found level of 500 liters per year to the Organisation for Economic Co-operation and Development (OECD) level of about 7,500 liters per year. This production increase requires improvements in veterinary services, livestock breeds, feeding and marketing, and crop residues management. Crop residues—the parts of the plant left behind after the harvest—are often in high

continued

BOX 10.2 *continued*

FIGURE B10.2.1

CO$_2$ Emissions per Dairy Unit of Output, Selected Countries, 2010

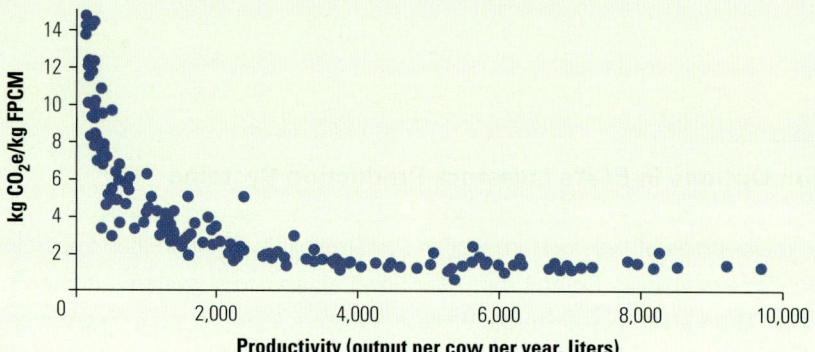

Source: Gerber et al. 2010.
Note: CO$_2$e = carbon dioxide equivalent; kg = kilogram; FPCM = fat and protein corrected milk (a standardized measure of milk production).

demand for animal feed, but this use can result in more CO$_2$ emissions from the bare soils and nitrous oxide emissions from the organic waste left behind. Partial grazing is therefore recommended.

Industrial system: This is the fastest growing subsector. Mitigation opportunities include dietary improvements that reduce the amount of land needed to grow feed. One-third of the global area cropped is used for feedgrain. A World Bank study looked at the feed conversion rate—the lower an animal's conversion rate, the less food it requires—in pigs and poultry in China. The study estimated that a decline of 1.5 percent in the feed conversion rate would save 25 million tons of feed in 2010, or about 5–7 million ha that would not need to be farmed. Another option is better manure management and biogas production from animal waste using methane digesters. This option will be economically attractive where energy costs are unsubsidized and surplus power can be sold into the grid.

Technology could soon enhance mitigation in livestock production. Examples of current applied research are (a) selecting for low-methane-emitting cattle genomes, as started in New Zealand; (b) identifying rumen microbes with enhanced capacity to digest fibrous feeds; (c) selecting grasses with low ammonia (NH$_3$) emission; and (d) enhancing knowledge on the use of biological nitrification inhibitors (BNI) in grasses.

Source: Contributed by Caroline Plante and Brian G. Bedard.

a. In the higher-rainfall areas, improved management on degraded pastures can sequester 1.5–4 tCO2e per year for up to eight years, at which time the organic soil equilibrium would be expected to be established.

b. The study projects a decline in demand for land and a reduction of the methane emissions from 23 kg to 18 kg of CO2e per kg of beef, through recovery of degraded areas and more intensive stocking and finishing systems (feedlots and crop-livestock systems) (World Bank 2010).

breeding livestock that can convert feed more efficiently, and developing biological resilience to climatic changes; and fine-tuning of feed rations to stock needs to avoid waste. Livestock production systems vary depending on ecological and socioeconomic conditions. Broadly there are grazing, mixed crop-livestock, and industrial systems, and the specific mitigation options vary by system and region, as box 10.2 also explains.

There are two primary links between livestock production and climate change. The first link has to do with how pasturelands are used. In cases where land property rights are weak and animal husbandry practices are poor, lands can be overgrazed, leading to degradation, erosion, and the depletion of the carbon stored in plants and the soil. Manure from grazing animals is also a source of methane; however, in well-managed pastures, livestock waste also contributes organic material to soils, facilitating sequestration and partly offsetting the methane-related effects on climate. With more commercialized livestock operation, feedlots and other production methods that concentrate animals on small areas of land become more prevalent. This process can lead to water contamination and overloaded soils, and manure is often stored in ways that release large amounts of methane. But concentrated livestock operations also make mitigation easier—for instance, by capturing methane to produce biogas or by using organic fertilization.

Bioenergy Production

Advanced bioenergy—as opposed to traditional, direct use of bioenergy such as fuelwood for cooking—includes electricity and heat produced from forest products, energy crops, agricultural residues, and manure or landfill gas as well as transportation fuels derived from organic matter such as bioethanol or biodiesel. There is a dazzling range of technologies to convert biological products into useful energy and heat, as figure 10.6 shows. The solid, gaseous, or liquid biofuels can generally be used in standard equipment or used in cofiring or fuel blending with fossil fuels. Although burning bioenergy releases carbon into the atmosphere, it is considered a renewable energy source because it emits only the carbon that the plants had previously captured. When it replaces fossil fuels, bioenergy therefore avoids net additions to GHG concentrations. In fact, life-cycle emissions from bioenergy are typically, but not always, much lower than those from fossil fuels (Chum et al. 2011).

Advanced bioenergy currently contributes a relatively small share to the global energy supply, perhaps around 1 percent. But in some countries its share is much bigger, such as in Sweden, where it provides almost a third or in Latvia with around 25 percent. In ECA,

FIGURE 10.6
The Many Ways of Producing Bioenergy

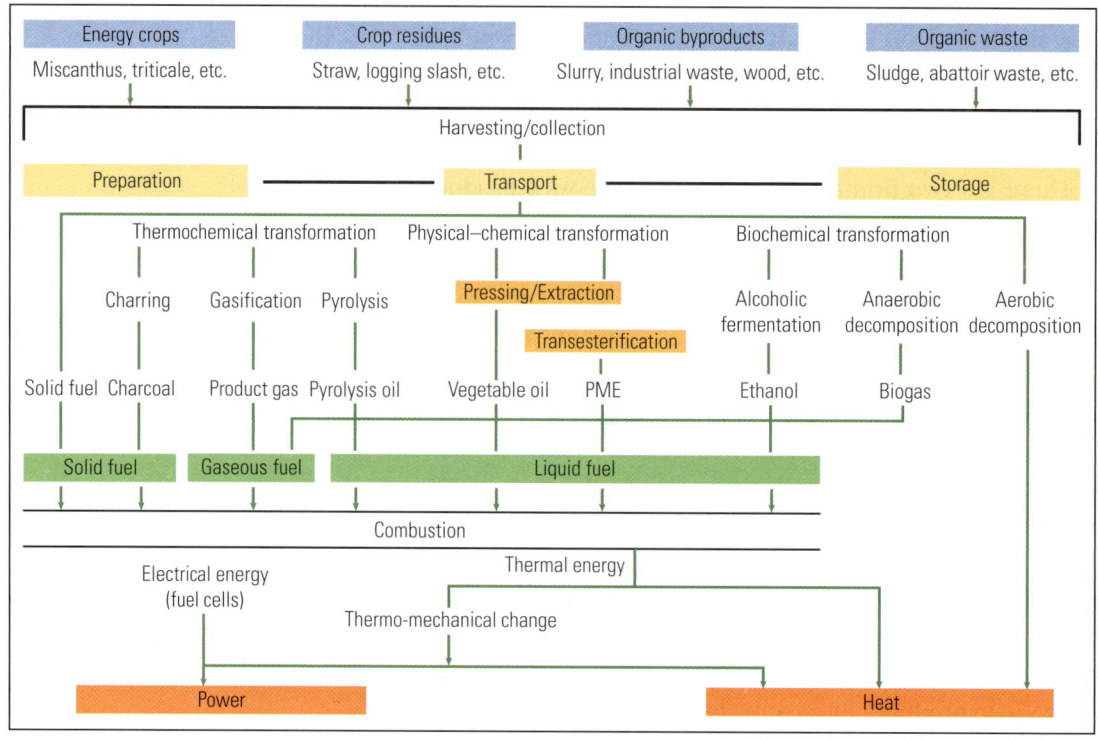

Source: Agency for Renewable Resources, German Federal Ministry of Food, Agriculture and Consumer Protection: http://www.nachwachsenderohstoffe.de.
Note: PME = palm methyl ester (biodiesel).

Poland consumes the largest amount of bioenergy, at more than 5 million tons of oil equivalent (mtoe), followed by Romania and Russia, as shown in table 10.5. Belarus, the Czech Republic, Hungary, and Latvia also all consume more than 1 mtoe. Ninety-three percent of bioenergy comes from solid biomass—mostly wood, charcoal, and agricultural residue. Biodiesel contributes almost 4 percent.

Analyses of bioenergy potential in selected countries show large expansion opportunities.[7] Croatia, for example, has significant technical potential for fast-growing broadleaved-species energy plantations on abandoned land or on land where agricultural production is not profitable. Croatia could technically produce about 0.2 mtoe of bioenergy annually from forests and 1.2 mtoe from agricultural lands. This is equivalent to about 16 percent of Croatia's total primary energy supply in 2009. Ukraine could produce as much as 12 percent of its 2009 energy supply, including 1.6 mtoe annually from forests (stem wood, primary forest residues, and secondary forest residues), 12 mtoe from agriculture (manure, primary agricultural residues, and secondary agricultural residues), and 0.4 mtoe from waste (landfill gas, sewage sludge, and sewage gas).

TABLE 10.5

Bioenergy Consumption in Europe and Central Asia, 2008

tons of oil equivalent

Country	Solid biomass	Municipal waste	Bio or landfill gas	Ethanol	Biodiesel	Total biomass
Albania	215.0	0.0	0.0	0.0	0.0	215.0
Armenia	1.0	0.0	0.0	0.0	0.0	1.0
Azerbaijan	4.4	0.0	0.0	0.0	0.0	4.4
Belarus	1,286.1	0.0	0.4	0.0	7.0	1,293.5
Bosnia and Herzegovina	183.4	0.0	0.0	0.0	0.0	183.4
Bulgaria	684.2	0.0	0.0	0.0	1.8	686.0
Croatia	319.4	0.0	5.0	0.0	1.8	326.2
Czech Republic	1,753.3	42.6	90.0	34.2	75.3	1,995.4
Estonia	629.3	0.0	2.8	0.0	0.0	632.2
Georgia	377.9	0.0	0.1	0.0	0.0	378.0
Hungary	1,219.5	46.1	21.8	46.4	121.8	1,455.6
Kazakhstan	164.2	0.0	0.0	0.0	0.0	164.2
Kyrgyz Republic	3.6	0.0	0.0	0.0	0.0	3.6
Latvia	1,099.2	0.0	8.8	0.0	1.8	1,109.8
Lithuania	737.6	0.0	3.0	15.4	46.6	802.5
Macedonia, FYR	171.8	0.0	0.0	0.0	0.9	172.7
Moldova	71.5	0.0	0.0	0.0	0.0	71.5
Poland	4,749.8	19.1	131.7	126.3	310.7	5,337.5
Romania	3,709.8	0.0	0.6	0.0	210.1	3,920.4
Russian Federation	3,158.4	0.0	0.0	0.0	0.0	3,158.4
Serbia	804.3	0.0	0.0	0.0	0.0	804.3
Slovak Republic	472.9	20.9	10.2	25.7	100.7	630.3
Slovenia	468.7	0.0	14.0	2.6	22.0	507.3
Tajikistan	0.0	0.0	0.0	0.0	0.0	0.0
Turkmenistan	0.0	0.0	0.0	0.0	0.0	0.0
Ukraine	900.0	0.0	0.0	0.0	0.0	900.0
Uzbekistan	0.2	0.0	0.0	0.0	0.0	0.2
Total	**23,185.5**	**128.7**	**288.4**	**250.4**	**900.4**	**24,753.3**

Source: IEA Bioenergy website, http://www.ieabioenergy.com.

Although the potential is large, the expansion of bioenergy production needs to proceed carefully for two reasons. The first is often called the "food versus fuel" trade-off. High prices for bioenergy can crowd out traditional agricultural production, thus raising food prices. This has happened globally in recent years when demand for bioenergy increased in response to renewable-energy support policies in industrialized countries. This trade-off is clearly an important concern, and the implications for the ECA region warrant further study. Reasonable increases in prices for agricultural products raise farmers' incomes, which encourages intensification and reintroduction of abandoned land into production. Some of the larger countries such as Russia and Ukraine also have vast land resources relative to their populations. Large abandoned or degraded areas could be restored to farming (as described above), and the efficiency of agricultural land use is far from the frontier. As outlined earlier, efficiency in livestock production and agricultural yields, while improving, are still well below Western European levels in parts of ECA. For instance, Belgium produced about 9.5 tons of wheat per ha compared with Georgia's 1.1 tons. Intensification could free up land both to expand food production and for bioenergy production on land already devoted to agriculture. So, in principle, land scarcity is a less significant barrier to the increased production of energy crops than in other regions, although the first priority for the region is to close the efficiency gap for food crops rather than expand agricultural land area for bioenergy production.

The second concern about expansion of bioenergy production is that the climate change mitigation benefits of bioenergy also depend on the sustainability of farming or forestry practices used to grow biomass for energy. Heavy use of fertilizers and other agrochemicals will negate some of the emission reductions besides having harmful impacts on biodiversity and land and water quality. Expansion of bioenergy production areas by clearing land, especially carbon-rich land such as forests or peat lands, imposes a "carbon debt" that can take many years to offset. Therefore, bioenergy strategies need to be considered within the context of sustainable agriculture more generally.

A useful way to think about the potential side effects of bioenergy is the "CLAW" approach proposed by the venture capital investor Vinod Khosla (Khosla 2011). It asks four questions when evaluating bioenergy projects:

- Are full life-cycle *carbon* emissions lower than for fossil-fuel alternatives like gasoline or diesel?

- Is the impact on *land* use beneficial or detrimental, especially in terms of competition with food crops or in affecting fragile or carbon-rich ecosystems?

- Does the bioenergy product cause *air* pollution during processing or consumption, and have external costs been incorporated in cost-benefit analysis?

- Finally, how much *water* is required relative to that required in equivalent fossil-fuel production?

Economics of Bioenergy

The cost of bioenergy depends on the feedstock, type of energy outputs (electricity versus heat), type of combustion, and location of end use. Recent estimates compiled for a Special Report by the IPCC (Chum et al. 2011) are based on a sample of country cases (see tables 10.6 and 10.7). For comparison, electricity generation costs about US$11–US$23 per gigajoule (GJ) for coal, US$15–US$35 per GJ for combined cycle gas turbine, US$7–US$26 per GJ for hydropower, and US$19–US$38 per GJ for nuclear power. Biofuels for transport tend to be somewhat more expensive than gasoline derived from oil (about US$16 per GJ) with the exception of Brazilian ethanol from sugar cane.

TABLE 10.6
Costs of Bioenergy for Heat and Power Production, 2005

Production process and energy output	Cost range (2005 US$ per GJ)	Remarks
Wood fuel and agricultural wastes for cofiring with coal (worldwide)	8–15	5–100 MW unit capacity with 30–40 percent efficiency
Wood fuel and agricultural wastes for direct combustion (worldwide)	20–25	10–100 MW unit capacity with 20–30 percent efficiency
Municipal solid waste (worldwide)	9–26	50–400 MW unit capacity with 22 percent efficiency
Wood fuel and agricultural wastes for small-scale gas engines (worldwide)	29–38	5–10 MW unit capacity with 15–30 percent efficiency
Wood pellets (EU countries) for cofiring with coal or cogasification	14–6	12.5–300 MW unit capacity
Biogas from manure (Finland, Sweden)	48–110	Small-scale biogas digesters and power plants

Source: Chum et al. 2011.
Note: GJ = gigajoule; MW = megawatts; EU = European Union.

TABLE 10.7
Costs of Biofuels for Transportation

Feedstock type	Cost range (2005 US$ per GJ)	Remarks
Ethanol		
Sugarcane	14.8 (Brazil); 31.8 (Australia)	Without bagasse revenues
Corn	20–31 (United States); 34.8 (France)	With revenues from using by-products for animal feed
Wheat	22.8 (United Kingdom); 40.7 (Australia)	With revenues from using by-products for animal feed
Sugar beets	24.4 (United Kingdom)	With coproduct revenues
Cassava	21–26 (Thailand, China, Australia)	No coproducts
Biodiesel		
Rapeseed	31–50 (Germany); 41.5 (France); 28.5 (United Kingdom)	Without coproducts in Germany, and without co-products in France and United Kingdom
Palm oil	26.1 (Indonesia and Malaysia)	With coproduct (residues for power generation) revenues

Source: Chum et al. 2011.
Note: GJ = gigajoule.

ECA's EU members are subject to EU directives regarding renewable energy in general and bioenergy in particular. Among non-EU ECA countries, several host CDM projects related to bioenergy, although these represent a small share of carbon finance activities.[8] At the end of 2011, ECA hosted 11 CDM projects related to bioenergy (in Armenia, Georgia, the former Yugoslav Republic of Macedonia, Moldova, Serbia, and Uzbekistan), although many of those were for landfill projects rather than agricultural bioenergy. Joint Implementation (JI) projects are larger in number, but as with CDM bioenergy projects, they also account for a relatively small amount of emission reductions.

Although there have been some assessments of bioenergy potential in parts of the ECA region, particularly for the EU-10 countries[9] (see, for example, Van Dam et al. 2007), there is a need for a more comprehensive study that explores what role bioenergy could play in replacing a share of fossil fuels used for electricity and heat production as well as for transport fuels. Sustainable energy production requires careful governance and monitoring of land use and productive but environmentally benign agricultural practices. The best options will be site specific, so there is a need for local research and experimentation, in particular to determine the best options for abandoned lands that are often of lower quality. If done well, bioenergy can not only contribute to climate change mitigation but also help raise agricultural incomes by increasing domestic and international demand—for instance, to supply countries that have bioenergy mandates. Given the growing awareness of bioenergy's risks, a credible certification system will be needed. In the short term, however, the focus should be on bioenergy options that pose fewer or no trade-offs with sustainability objectives, such as the use of agricultural waste products, residue, and manure for biogas production.

Promoting Mitigation in Agriculture

Productivity in ECA's agricultural sector has generally increased, although it still has some way to go to reach Western European levels. As agriculture both intensifies and expands production areas, GHG emissions could rise as well. But this does not have to be the case. Well-designed policies that encourage higher productivity can be carbon neutral or even increase carbon sequestration. The priority areas follow from the three main ways to reduce the sector's climate impacts:

- Increase incentives for sustainable land management, land conservation, and restoration

- Reduce direct emissions, most importantly from livestock production and the use of farm equipment

- Ensure that the emission-reducing benefits from bioenergy production are not outweighed by its negative effects.

EU policies have been successful in making agriculture more sustainable in member states. Much can be learned from that experience. However, it relies heavily on the incentives provided by the direct payment system of the Common Agricultural Policy to change farming practices (as discussed in box 10.3), which is not an option where rural subsidies are not as generous.

Some of the required actions, especially those that encourage greater farm productivity, require institutional reform and not

---

BOX 10.3

EU Policies for Promoting Environmental and Climate Benefits in the Agricultural Sector

Western Europe is the only major world region in which, according to IPCC estimates, emissions from agriculture are declining (Smith, Martino et al. 2008). This trend reflects strong national actions but also the harmonization of policies through the EU. These policies use a system of sticks and carrots to promote environmentally friendly practices among its farmers.

Sticks: EU agricultural policies condition direct income support to compliance with a minimum set of obligations (environmental, but also regarding food safety and animal welfare). "Cross-compliance" defines the baseline, or bare minimum, environmental obligations that a farmer must comply with. This cross-compliance involves two layers of obligations:

- First, ensuring "good agricultural and environmental conditions," which include mandatory practices such as preventing soil erosion, maintaining soil organic matter, ensuring a minimum level of maintenance, protecting and managing water, and so on. In addition, the ratio of permanent pastures must be maintained within certain limits on the national level.

- Second, meeting "statutory management requirements"—a set of directives and regulations including those that protect water against nitrate pollution from agricultural sources and those promoting nature and wildlife conservation. These requirements are mandatory for farmers whether they receive EU support or not.

Farmers who fail to comply with these obligations are penalized through reductions or total loss of EU direct income support. The EU is using a system of strict checks to ensure compliance but is also financially supporting farm advisory systems in each member state to help farmers meet their obligations.[a]

Carrots: The EU rewards farmers who voluntarily undertake environmental commitments that go beyond the mandatory requirements mentioned above or in terms of the minimum

continued

BOX 10.3 *continued*

requirements for fertilizer and plant protection product use. Rewards are administered through one of the measures of the EU's rural development menu, called agri-environment. It consists of area-based payments made to farmers who undertake voluntary environmental commitments over a period of at least five years. Payments are made every year, and there is a system for checking compliance (involving a certain percentage of on-the-spot controls). These payments are subject to a per-hectare ceiling regulated at the EU level, but their actual levels differ by country or region.

In addition, the EU aims to more explicitly integrate climate change into rural development measures. For instance, it is using existing rural development support schemes to support projects that have eligible mitigation or adaptation benefits. This means, for example, that farmers who make investments to introduce precision agriculture, improve manure storage, or introduce water-saving technologies can recover up to half of their costs from EU-funded rural development programs.

Source: Contributed by Irina Ramniceanu.

a. For more information about the EU cross-compliance mechanism, see http://ec.europa.eu/agriculture/direct-support/cross-compliance.

predominantly climate policies. Removal or reduction of price distortions for water and energy increases efficiency and lowers fiscal burdens—with emissions reductions as a side benefit. Where land ownership is uncertain, farmers have fewer incentives to invest in sustainable farming practices for the long term. Where ownership is scattered among many small and fragmented land holdings, efficiency will be limited; land consolidation on 7.5 million ha in Turkey reduced fuel consumption in agriculture by 25 percent (Republic of Turkey 2011). Restrictive land policies also discourage foreign direct investment (FDI) that could bring needed resources and expertise (Swinnen and van Herck 2011). Given the countries' size, inward FDI in agriculture could be much larger in Russia, Ukraine, and especially in Kazakhstan. Trade policy distortions—such as export bans during global food price spikes—further deter investors.

Complementary changes in access to credit, infrastructure (including grain handling, storage, and export facilities), and labor mobility are required for farmers to realize the benefits from these institutional reforms. Regulatory frameworks for sustainable bioenergy production are an example of climate-oriented institutional policies. These should be a prerequisite for introduction of bioenergy targets or inclusion of bioenergy production in feed-in tariff policies. They should favor low-risk options such as bioenergy from waste products but be open to opportunities for sustainable bioenergy crop production.

More direct policies target farming practices through information programs such as extension services or through regulations. For instance, using nitrogen more efficiently—through slow-release fertilizers or precision farming techniques—reduces emissions from fertilizer manufacturing as well as nitrous dioxide emissions in the fields. Encouraging more efficient animal feeding and regulations on manure management reduces emissions from livestock production. Agriculture is not the largest source of GHG emissions in the ECA region. But good policies could reduce the sector's impact on the global climate, while at the same time generating economic benefits in rural areas.

Forests[10]

The ECA region has the largest forest cover of all the World Bank's regions, accounting for more than 25 percent of the world's forests and storing more than 44 billion tons of carbon. The forestry sector also currently employs 2.2 million people and contributes 1.1–1.5 percent of GDP. If sustainably managed, the sector could potentially employ an additional 3.3 million workers and double its contribution to GDP. Unlike other regions, ECA has forests that have expanded steadily (1.6 percent over the past 20 years), and the potential for additional growth (both in terms of area and volume per hectare) is substantial. However, if ECA is to fully develop this economic and employment potential and further increase its carbon sequestration capacity, the region needs to

- Protect the forest from current and projected threats;

- Improve the enabling policy environment by deepening current reform processes; and

- Invest in forest infrastructure such as access roads, better harvesting techniques, and sustainable forest management practices.

Growing Forest Cover

Forests in ECA play an integral role in the global climate change mitigation agenda. By 2010, more than 1 billion ha were forested. However, these forests are not uniformly distributed. Most are in Russia, which is home to close to 20 percent of the world's forests and more than 60 percent of the world's standing softwood. Siberia is more than 50 percent forest, while forest cover in Europe ranges from 20 percent to 40 percent. Kazakhstan is the least forested country in the region with 1.2 percent. The average growing stock varies from

10 cubic meters per ha in Central Asia up to 202 cubic meters per ha in Central and Eastern Europe. In terms of forest area as a percentage of total land area, Slovenia, Latvia, and Estonia are leading in the ECA region, followed by Russia, Bosnia and Herzegovina, and Belarus, as shown in figure 10.7.

In the past 20 years, the planet lost more than 135 million ha of forests, whereas in ECA, the forest cover has increased by 15.7 million ha—an area the size of Croatia and Hungary combined. In the past 10 years, the forest area in ECA surpassed that of Latin America and the Caribbean for the first time, as table 10.8 shows. Overall, the forest area in ECA has increased at a steady rate of 1.6 percent over the past 20 years, and 39 percent of the region is now forest. This growth has frequently not been driven by a plan or a careful design process but rather is mainly the result of reductions in the rural population and the associated change in land use from agriculture and pasture to

FIGURE 10.7
Forest Area Trends in Europe and Central Asia, 1990–2010

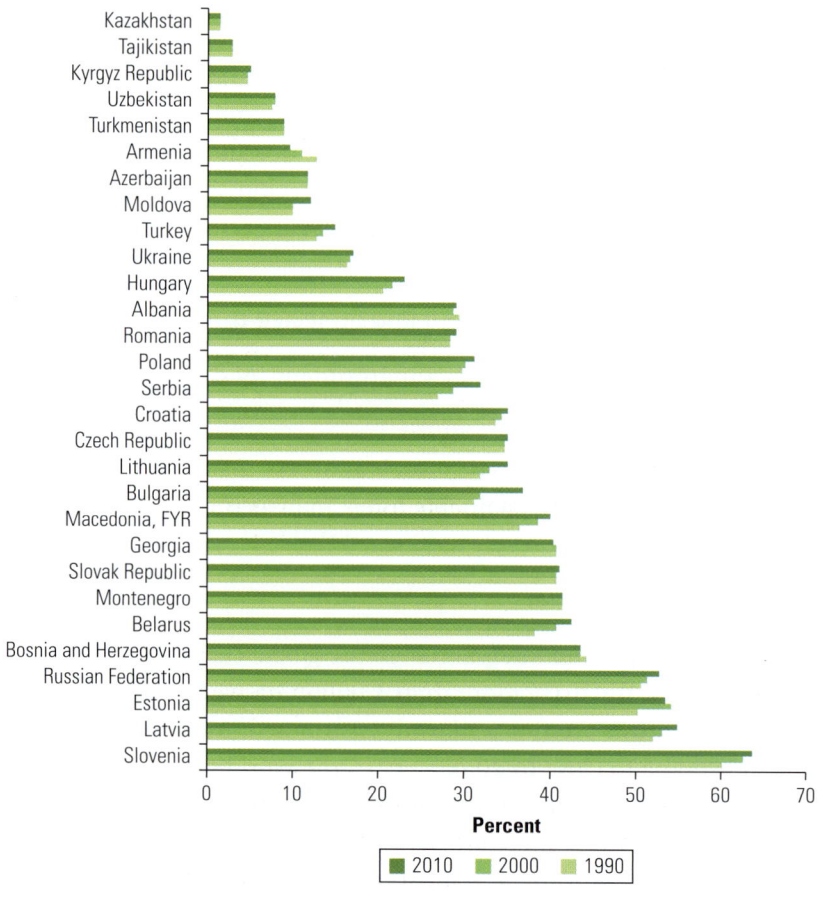

Source: Shvidenko et al. 2011.

TABLE 10.8
Change in Forest Cover in Selected Regions, 1990–2010

Region	Area (ha, thousands)		Annual change (ha, thousands)	Percentage change (%)
	1990	2010	1990–2010	1990–2010
Sub-Saharan Africa	664,112	595,605	(3,425.35)	−10.3
East Asia and Pacific	655,202	660,074	243.60	0.7
South Asia	79,513	81,659	107.30	2.7
Latin America and the Caribbean	1,048,363	955,584	(4,638.95)	−8.8
Europe and Central Asia	1,005,372	1,021,017	782.25	1.6
Middle East and North Africa	109,404	104,961	(222.15)	−4.1
North America	606,469	614,156	384.35	1.3
World Total	**4,168,435**	**4,033,056**	**(6,768.95)**	**−3.2**

Source: Shvidenko et al. 2011.
Note: ha = hectares.

land that has become forested through natural processes. This process is in direct contrast to the trend in tropical areas, where deforestation has been mainly caused by encroaching agriculture (commercial and subsistence).

The increase in forest cover is distributed across most ECA countries and has been larger than any other region. Unlike other regions—particularly Latin America and the Caribbean, Sub-Saharan Africa, and the Middle East and North Africa—most ECA countries have not had a deforestation issue. In the past 20 years, forest cover has grown in almost all ECA countries with the exception of Albania, Armenia, Bosnia and Herzegovina, Georgia, and Kazakhstan. Forests in East Asia and the Pacific and South Asia have also expanded; however, their growth has been concentrated in one major country in each region (that is, China in East Asia and the Pacific, and India in South Asia), with almost all the other countries experiencing varying degrees of deforestation. The forest area expansion in ECA is larger than that of East Asia and the Pacific, South Asia, and North America combined.

Currently, the carbon stock—or the amount of carbon stored—in ECA forests is second only to Latin America and the Caribbean. This growth in forest cover has also led to an increase in the amount of carbon stock in the trees. The overall carbon stock locked in ECA forest biomass is now estimated to exceed 44 billion tons of carbon. Even though ECA's forest cover has exceeded that of Latin America and the Caribbean, the carbon stock of the latter's forests is still the largest in the world, as shown in map 10.3. ECA's carbon stock has grown by 0.87 billion tons during the 1990s and by another 2.29 billion tons between 2000 and 2010, as table 10.9 shows. Annually, the

MAP 10.3

The World's Forest Carbon Stock, by Region, 2009

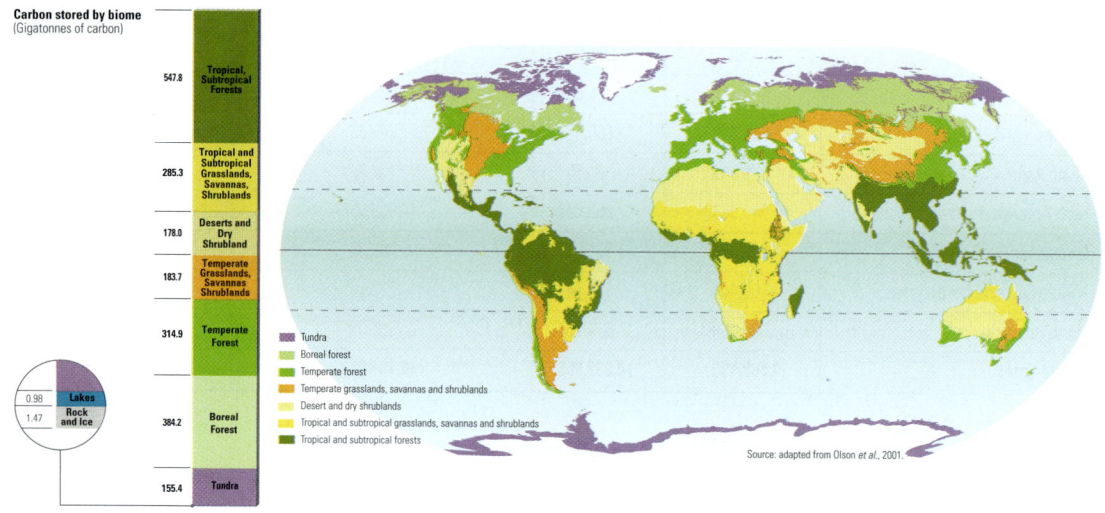

Source: UNEP/GRID-Arendal (http://www.grida.no/).

TABLE 10.9

Carbon Stock in Living Forest Biomass in Europe and Central Asia, by Subregion, 1990–2010

Tons of carbon, millions

Region	1990	2000	2005	2010
Southeastern Europe	1,479	1,666	1,766	1,955
Central and Eastern Europe	1,688	1,929	2,018	2,117
Baltics	1,572	1,837	1,989	2,169
South Caucasus	263	272	275	279
Central Asia	49	62	70	90
Kazakhstan	137	137	137	137
European–Ural Russian Federation North[a]	5,339	5,516	5,600	5,737
European–Ural Central Russian Federation [a]	4,112	4,248	4,312	4,418
European–Ural South Russian Federation [a]	357	369	374	383
Western Siberia[a]	4,073	4,208	4,272	4,377
Siberian and Far Eastern North[a]	5,566	5,751	5,838	5,981
Siberian and Far Eastern South[a]	15,453	15,964	16,207	16,604
Russian Federation total	34,900	36,055	36,602	37,500
Total ECA	**40,088**	**41,958**	**42,857**	**44,247**

Source: Shvidenko et al. 2011.
a. Included in Russian Federation total.

ECA region produces around 3.5 billion tons of carbon emissions from fossil-fuel combustion. However, 6.5 percent of these emissions are annually captured and stored by ECA forests.

Potential for Increased Employment and GDP Contribution

The ECA forest sector employs about 2.2 million people (FAO 2011). This represents a little over 1 percent of the region's total labor force, a share slightly higher than in the EU-15 countries.[11] In ECA countries, value added in forestry in 2010 was about US$26 billion versus US$113 billion in the EU-15. Although the contribution of the forestry sector to GDP was slightly higher—1.3 percent in ECA versus 0.9 percent in the EU-15—the percentage translates into very different labor productivities. The value added per forest sector employee ranged from less than US$5,000 in much of the eastern part of ECA to more than US$25,000 in Slovenia, as shown in figure 10.8. In the EU-15, the value added per employee was around US$70,000.

Globally, approximately one person is employed in forestry for every 1,000 ha of forest, and this employment is supplemented by another two jobs in forest processing (the wood and pulp and paper industries). In developing countries, there is one job in forest processing for every job in forestry, while in North America this ratio is 1 to 7. In Western Europe, one job in forestry leads to 4.4 jobs in forest processing, while in Eastern Europe this ratio is only 1 to 2, as figure 10.9 illustrates (FAO 2004).

More than 1 billion ha of forests in ECA should therefore be able to sustain around 1 million jobs in the forestry sector, and if wood and timber processing is further developed to reach Western European standards, 4.4 million people could be employed in wood processing. Hence, more than 3 million additional jobs could potentially be sustainably created in ECA's forestry sector. Even accounting for the inaccessibility of some of Russia's forest areas, the job creation potential will be great.

The forestry sector also contributes to GDP. This contribution varies greatly by country. In most ECA countries, the forestry sector contributes 1–1.5 percent of GDP. In Estonia and Latvia, where the forestry sector is more developed, the sector contributes around 4–5 percent of GDP. In comparison, in Canada and Sweden, this percentage is around 3 percent, and in Austria it is around 2.1 percent. Hence, many ECA countries could potentially double the forest sector's contribution—from 1–1.5 percent of national GDP to 2–3 percent of GDP.

FIGURE 10.8

Labor Productivity in the Forest Sector in Europe and Central Asia, 2010

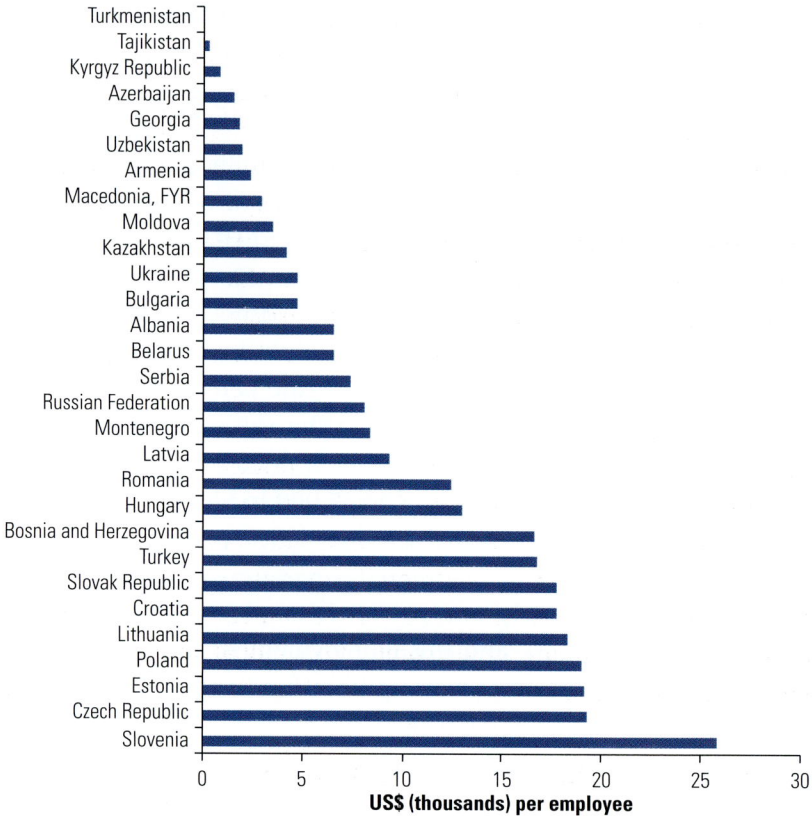

Source: FAO 2011.

FIGURE 10.9

Forest Industry Employment Ratios, Selected Regions

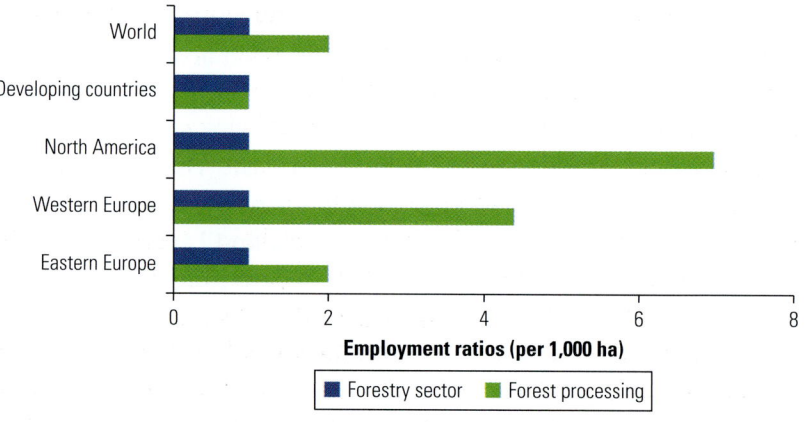

Source: FAO 2011.
Note: ha = hectare.

Additional Value of ECA's Forests

In addition to the economic contribution from sustainable wood harvesting and processing and the global climate benefits, forests provide an array of direct nontimber forest benefits such as mushrooms, berries, honey, hunting and wildlife management, tourism, and recreation. Indirect benefits include watershed protection, nutrient cycling, and ecological functions. Unfortunately, almost all of these global and local benefits are rarely captured as revenue to forest owners or managers. If these side effects were priced and captured, this would create additional incentives that would encourage forest area to increase faster, and the contributions from employment as well as timber, fuelwood, and other forest products would be even larger.

Even though EU accession has helped some countries in the region to benefit from payments for environmental services (for example, Natura 2000[12] and agri-environment payments), many ECA countries still need to develop domestic policies to pay for and encourage the delivery of additional environmental services. The GEF and other international donor environment programs should also be used to help internalize some of the global and local positive benefits from forests.[13]

Challenges and Opportunities for Forests in ECA

ECA's forestry sector is grappling with a diverse set of challenges, in particular, climate change and poor forest governance (described further below). However, the sector also offers several significant opportunities: there is further scope to increase the forested area as well as the yield, productivity, and in-country value added. Investing in the forestry sector would significantly increase both income and employment generation, environmental services, and global climate change mitigation benefits. The considerable direct economic benefits (income and employment) from sustainable, economic use of forest resources alone are sufficient to justify the protection of ECA's forests as a global carbon sink.

A priority for ECA countries is to create both the enabling environment and the incentives for investments in the forest sector. The former includes investments in market development and infrastructure (such as forest roads) that raise the profitability of the sector. The latter means improving the investment climate in the forestry sector through sensible regulation and support policies (such as energy policies that encourage sustainable forest bioenergy), governance of access, and improved investment finance.

Currently, the forests in ECA are consistently being harvested at less than the annual allowable cut. Better management could increase both the harvest and the growth and sink potential. The expanding forest area will allow for an even greater sustainable yield over time. The yield can be further enhanced by

- Preventing forest degradation (for example, from forest fires, poor maintenance, or lack of thinning);

- Increasing and improving the forests through afforestation and interplanting desirable tree species (enrichment planting); and

- Making better use of technology to improve yields in wood harvesting and processing.

The forest sector therefore has the potential to sustain a larger timber industry employing more people in rural areas and providing a larger contribution to GDP.

The forest cover in ECA is projected to continue to grow in the foreseeable future. Assuming that the growth experienced continues over the next 20 years, the region could annually gain 370,000 ha of forests, and the size of the carbon sink would increase by about 20 percent to around 48 billion tons of carbon. To put this in perspective: ECA's annual forest sink potential could be roughly equivalent to the annual GHG emissions of over 135 million passenger vehicles or the CO_2 emissions from 163 coal-fired power plants for one year.[14] Better management, improved local policies, and infrastructure investment are projected to further improve forest cover and increase the size of the carbon sink.

Forests in ECA can also contribute further toward climate change mitigation by fossil-fuel substitution with forest-derived biomass. Sweden now obtains almost 30 percent of its energy (120 terawatt-hours [TWh])[15] from biomass, much of it from by-products from logging in well-managed forests.[16] In 2009, Eastern Europe produced 81.3 million cubic meters of fuelwood while consuming only 78.6 million cubic meters. The rest was exported mainly to Western Europe, which produced 71.2 million cubic meters and consumed 73.1 million cubic meters. (Central Asia produced almost exactly what it consumed, 0.8 million cubic meters.)

A larger, sustainably managed ECA forest would allow for a larger maximum sustainable cut and increased sequestration. In 2008, northwestern Russia (including St. Petersburg and the surrounding area, which holds a population of around 13 million) consumed around 943 TWh of fossil fuels, out of which only 1 percent originated from biomass sources, as shown in figure 10.10. The same

FIGURE 10.10
Fuel Consumption Mix, Northwestern Russian Federation, 2008

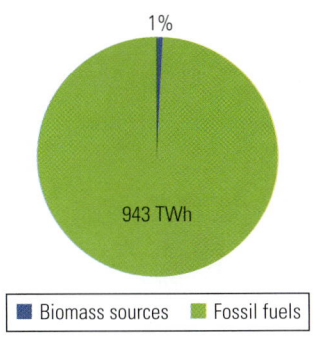

1%

943 TWh

■ Biomass sources   ■ Fossil fuels

Source: Gerasimov and Karjalainen 2009.
Note: TWh = terawatt hour.

FIGURE 10.11
Wood Energy Potential, Northwestern Russian Federation, 2008

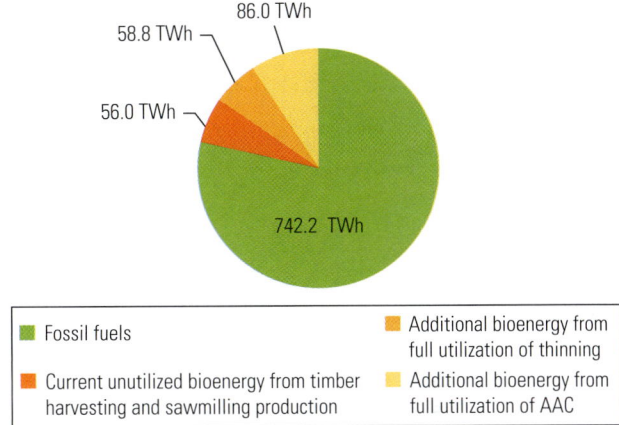

86.0 TWh

58.8 TWh

56.0 TWh

742.2 TWh

■ Fossil fuels

■ Current unutilized bioenergy from timber harvesting and sawmilling production

■ Additional bioenergy from full utilization of thinning

■ Additional bioenergy from full utilization of AAC

Source: Gerasimov and Karjalainen 2009.
Note: TWh = terawatt hour; AAC = annual allowable cut.

region has an estimated 56 TWh supply of energy wood that is currently unused.[17] Based on the full annual allowable cut in that area, an additional 86 TWh could be developed. If full use of thinning were also included, around 200.8 TWh could be produced from wood sources, as shown in figure 10.11 (Gerasimov and Karjalainen 2009). Hence, northwestern Russia has the potential to use wood energy to supply around 22 percent of its energy needs. This amount would be equivalent to reducing consumption of natural gas by around 20 billion cubic meters every year, which could be exported instead.

Hence there is scope for ECA forests to continue to grow, provide more harvested wood, sequester more carbon, increase employment and contribution to GDP, allow for more fuel switches, and provide more environmental services. For forests in ECA to achieve this potential, they have to be better managed and protected from existing and growing threats. In particular, ECA forests will require investments in sustainable forest management, including improving governance, strengthening institutions, and adapting to the changing climate.

Protecting ECA Forests

Climate change is projected to significantly affect the region and increase the risk of forest fires, pests, and diseases. Forest fires have become a growing problem in ECA (see map 10.4). In Bulgaria, for

MAP 10.4

Estimated Vegetation Area Destroyed Annually by Fire, by Region, 2006

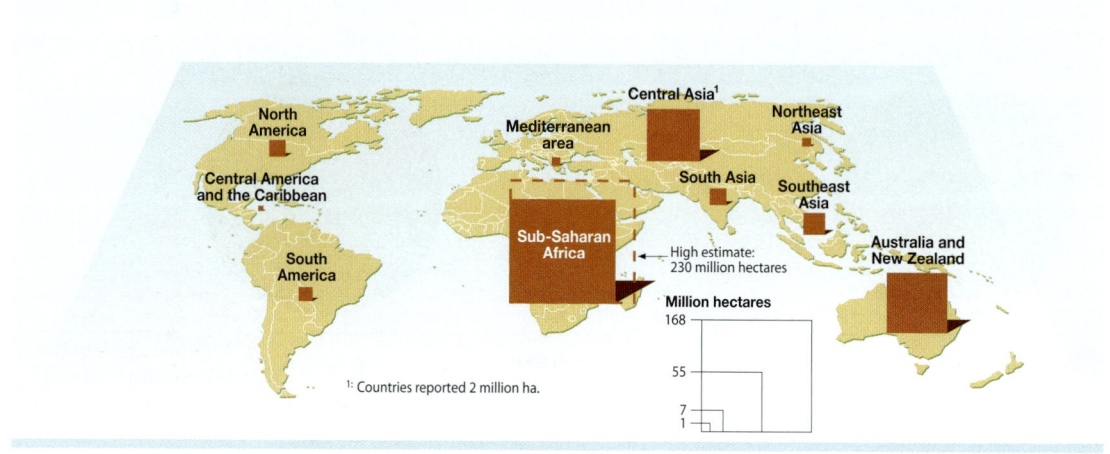

Source: UNEP/GRID-Arendal (http://www.grida.no/).

instance, the average forest area lost annually to fire has increased from 324 ha in the 1980s to 3,868 ha in the 1990s and further to 7,163 ha in the 2000s. Today, 40–50 percent of the forests in Bulgaria are classified as high risk in terms of fire. In Russia, a vast area of forest burns every year, ranging from 4 million to 17 million ha according to satellite imagery. Direct emissions from forest fires in the country are estimated at around 120 million tons of carbon annually.

These losses are of global concern in terms of carbon emissions into the atmosphere; for example, in the forest fires of 2010, plumes of smoke rose to the stratosphere over western Russia, and by August, coalesced into one massive cloud 3,000 kilometers (km) long. The economic costs in terms of lost timber, damage to property, carbon emissions, and people's health may have exceeded US$1 billion.[18] Almost all climate change models predict that the dry season in many ECA countries will become longer and dryer, creating the conditions for more fires. It is difficult to categorize concretely the direct causes of forest fires; however, it is generally accepted that 80–95 percent of forest fires are of human origin, caused through carelessness, escapes from burning agricultural residue, or arson. There is a clear link between forest governance and poor policy having unintended negative impacts (for example, banning of agricultural fires and lack of access to legitimate sources of timber).

Even though natural ecosystems, including forests, have coped with changing climates in the past, the current and projected climatic changes are estimated to be occurring faster than ecosystems' capacity to adapt. For example, the major tree species in the Northern Hemisphere cannot migrate by more than 0.3–0.5 km per year. It is projected that movement of the boundaries between ecological zones can occur at twice that order of magnitude (Gustafson et al. 2010). In other words, tree species will not migrate fast enough to more appropriate climates, and large forested areas would have tree species that are not completely compatible with new climatic conditions and therefore are more susceptible to disease, pests, and especially, forest fires.

Climate change is also predicted to have an impact on the productivity and carbon storage capacity of ECA's forests. Forests in the southern part of ECA will be under substantial risk of higher temperatures, increased droughts, and changes in precipitation. If these risks materialize, they could lead to a moderate decline in the productivity and health of forests in Southeastern Europe and European south Russia, and some scenarios predict a decrease in the current sink by approximately one-fourth. The carbon budget of Central Asia will likely not change significantly, although some decrease in productivity (and carbon sink) is expected in mountain forests. The Baltic States could see productivity gains of up to 30 percent with the sink function increasing by around a third. Although Russia's forests are expected to remain a net sink, there will be variations in the subregions. In Siberia and the far eastern north, for instance, the thawing of permafrost and the climate becoming drier could lead to a decrease in the expected net carbon budget of 30–50 percent. Forest fire prevention, anticipatory strategies to address climate change (particularly in taiga landscapes), and the introduction of adaptive forest management will be important to mitigate future impacts and to improve the resilience of Russia's forests.

Continuing the Policy Reform Process

ECA's forestry sector is frequently subject to poor governance characterized by a weak legislative and policy framework, policies with unintended negative consequences, and an uneven distribution of the benefits from forest resources, rather than the inadequate monitoring and enforcement of laws alone. Poor governance undermines sustainable economic growth, societal equity, and forest conservation and management. Illegal logging and trade, for instance, distorts timber markets and causes a loss of revenue for responsible forest enterprises that could be invested in sustainable forest management or economic development. Illicit activities also put at risk poor and forest-dependent populations who rely on timber and nontimber forest products for their livelihoods. Consequently, good forest governance is essential for achieving development outcomes.

Many ECA countries are struggling with how to transform their forestry institutions. A particular challenge in the ECA region has been how to transform formerly financially secure and centrally planned forestry institutions with strong regulatory functions to respond to a changing environment. This need for change has grown out of the political and economic reform process in Europe's transition countries, which created a significant incentive for forestry institutions to be more responsive to local demands. In the same way, forestry organizations are fundamentally service delivery–oriented and need to be supported through public expenditures in an environment where resources are increasingly constrained. This requires forestry organizations to ensure that forests are sustainably managed and to deliver services for forest industries (such as timber) and households (such as firewood), while continuing to perform their public-good functions (such as carbon sequestration, watershed protection, and biodiversity conservation).

For the ECA forestry sector to remain competitive and capture higher-value European and North American markets, timber harvesting has to comply with new international forestry standards. Many countries in the region have embraced international forest certification schemes to demonstrate compliance with environmental and forest conservation and management requirements, consistent with international and EU commitments. The new EU Timber Regulation (to become effective 2013) and the U.S. Lacey Act (recently amended to include forest products) are particularly relevant for the ECA region. Both address legality of timber and timber products by requiring the importer and timber trader to exercise "due diligence" or "due care" to ensure the legal sourcing of their wood. Enforcement will require greater transparency and a more sophisticated documentation of the legality of harvesting forest products along the entire supply chain of timber and timber products. Compliance with these new regulations will pose a challenge for ECA countries, especially the forest-rich but governance-poor wood-exporting countries (Belarus, Russia, Ukraine, and to a lesser extent Georgia). Russia, in particular, will be affected, considering that a large volume of Russian timber is processed and refined in China before being exported to the United States. The EU is helping Russia and the Eastern partnership countries[19] to address many of these policy reform issues, as box 10.4 discusses.

BOX 10.4

Improving Forest Governance in ECA

An EU– and Austrian Development Cooperation–funded program (the ENPI [European Neighborhood Policy East] FLEG [Forest Law Enforcement and Governance] Program: Improving Forest Law Enforcement and Governance in the European Neighborhood Policy East Countries and Russia) has developed a unique approach to contributing to legal and sustainable forest management. It has also improved local livelihoods in the participating countries (Armenia, Azerbaijan, Belarus, Georgia, Moldova, Russia, and Ukraine) by addressing the underlying drivers of poor governance.

Through legislative changes, stakeholder involvement, education and training, and other activities that strengthen governance and anticorruption measures, the ENPI FLEG Program has successfully created an environmentally, socially, and economically sound approach that engages and links governments with the business, academic, civil society, and rural communities. Progress has been made in improving forest governance, strengthening local capacity, creating transparency and greater understanding of specific forestry issues, and building inclusive relationships. The ENPI FLEG Program is working to institutionalize and transform these successes into changes in behavior that become engrained into everyday life.

In Russia, for instance, the Forest Code was recently amended, based in part on the ENPI FLEG Program's analysis, which identified measures to improve forest legislation and forest management and to combat illegal logging. In Ukraine, training manuals have been developed on forest crime processing and on the classification of wood species to educate forest guards. Other achievements include improved transparency through better state-level forestry websites (Belarus); increased public awareness of forestry issues, including training of journalists (Georgia, Moldova); assessment of the scale of firewood removal (Armenia); and the preparation of a long-term strategy on development of the forest sector (Azerbaijan).

The ENPI FLEG Program (http://www.enpi-fleg.org) arose out of the Europe and North Asia FLEG Ministerial conference held in St. Petersburg in 2005. The program has been implemented by the World Bank, in partnership with the International Union for Conservation of Nature (IUCN) and the World Wildlife Fund (WWF) from 2008–12 with support from the EU and the Austrian Development Cooperation.

Investing in Sustainable Forest Management

The forestry sector in the ECA region has almost universally suffered from underinvestment. Even some of the Eastern European countries that have since joined the European Union continue to have difficulties investing adequately in forest infrastructure. There is no specific EU Forest Directive, and there are few opportunities for externally supported investments in state forests. In addition, ECA countries have been unable to benefit from international efforts aimed at creating financial value for the carbon stored in forests, including mechanisms such as the UN's REDD program (Reducing Emissions from Deforestation and Forest Degradation) and REDD+, which includes the role of conservation, sustainable management of forests, and enhancement of forest carbon stocks.

Carbon prices frequently do not reflect the true cost of avoiding CO_2 emissions, and by early 2013, the price of carbon per ton fell below US$5 in EU emissions trading. The incentives for carbon financing were therefore limited. However, there is potential to reward countries whose carbon stock has increased by further developing price signals and market instruments that encourage investment in forest operations, not least through domestic policies that support payments for environmental services or by including forests in national carbon trading schemes that are being developed (for example, in Turkey and Ukraine). Initial modeling shows that at a price of US$10 per ton of CO_2, an additional 28 million tons of CO_2e could be sequestered annually. If this price increased to US$30 or US$50, an additional 68 and 110 million tons of CO_2e, respectively, could be stored annually (Shvidenko et al. 2011).

Overall, appropriate interventions in sustainable forest management could include support for

- *The forest policy, legislation, and institutional reform process*—creating laws that encourage effective and transparent forest management institutions;

- *Continued governance improvement*—increasing transparency through stakeholder participation, strengthening anticorruption measures and law enforcement, strengthening land ownership structures, expanding forest certification, and using modern information and communication technology;

- *Improving forest infrastructure*—rehabilitating forest road networks, improving design and maintenance standards to increase the efficiency of sustainable timber harvesting and transportation, improving selective seed collection and treatment, increasing nursery capacity and silvicultural systems, and better maintenance and tending of both plantations and natural forests;

- *Forest fire prevention and protection*—improving policies targeting the causes of forest fires as well as forest fire detection, monitoring, management, and control; and better forest pest detection and treatment systems;

- *Protecting global benefits*—mitigating climate change (for example, through fuel switching and enhancing carbon stocks), protecting biodiversity by developing payments for environmental services, adapting to the changing climate, and supporting domestic policies that include forests as part of carbon trading;

- *Rural development*—working with forest-dependent and rural communities to address their sustainable development needs in the broader rural landscape; and

- *New technologies*—improving forest harvesting and processing; creating higher-value wood products; developing forest information systems that support forest cadastre, inventory, growth, and yield monitoring; annual harvesting plans; carbon accounting; the tracking of timber origin, ownership, and legality; and forest certification.

The Way Forward

Although ECA's forests already play a crucial role in global GHG mitigation, they have the potential to make a significantly larger contribution. However, such a contribution will require targeted investments to address existing challenges and seize opportunities in the sector. Better management and governance of forests could create up to 3.3 million additional jobs and could double the sector's contribution to GDP. The full economic value of ECA's forests—including biodiversity, recreational value, and other direct and indirect benefits—should be at least accounted for if not completely captured. Doing so would allow ECA countries to benefit from payments for environmental services and take advantage of possible conservation and environmental finance opportunities. In the context of improving forest management, important opportunities exist for creating a more inclusive, transparent, and sustainable forest sector. Achieving this requires commitment to improving the regulatory environment, developing better price instruments, and investing in sustainable forest management.

On the regulatory side, land ownership structure, forest certification, and forest law enforcement and governance policies should support the development of wood industries and wood energy. The capacity of institutions should be developed to transparently manage and supervise the use of state and, where appropriate, privately owned forest resources, according to international best practices. There is also the need to address forest policy issues, which can unintentionally encourage poor governance and forest degradation. Even though demand for support in these areas is strong, these opportunities have been constrained by the recent global economic crises and countries' reluctance to invest in the environment.

Notes

1. ECA's soils contain an estimated 55 to 121 billion tons of carbon, forests 44 billion tons of carbon. One ton of carbon is equivalent to 3.67 tons of CO_2e (Larson, Dinar, and Blankspoor 2012; Shvidenko et al. 2011).

2. This section draws on background papers by Larson, Dinar, and Blankespoor (2012) and Timilsina (2012) and contributions from Benoit Blarel, Brian G. Bedard, Caroline Plante, and Irina Ramniceanu.

3. Although consistent estimates of emissions from fossil-fuel burning are available from the International Energy Agency (IEA), estimates on other emission sources are more dated and less clear. This section used data from the World Resources Institute's Climate Analysis Indicators Tool (WRI-CAIT) and the Joint Research Centre of the European Commission's Emission Database for Global Atmospheric Research (JRC-EDGAR).

4. The Balkan countries include Albania, Bosnia and Herzegovina, Croatia, Kosovo, FYR Macedonia, Montenegro, Serbia, and Slovenia. The CIS countries include Armenia, Azerbaijan, Belarus, Georgia, Kazakhstan, the Kyrgyz Republic, Moldova, the Russian Federation, Tajikistan, Turkmenistan, Ukraine, and Uzbekistan.

5. For more about the Clean Development Mechanism, see http://unfccc. int/kyoto_protocol/mechanisms/clean_development_mechanism/ items/2718.php.

6. The South Caucasus countries include Armenia, Azerbaijan, and Georgia. Central Asian countries include Kazakhstan, the Kyrgyz Republic, Tajikistan, Turkmenistan, and Uzbekistan

7. See the Biomass Energy Europe (BEE) project of the European Commission: http://www.eu-bee.info.

8. ECA countries that are not EU members have an opportunity to sell credits through Kyoto Protocol compliance vehicles, including the Clean Development Mechanism (CDM) and Joint Implementation (JI). These initiatives can help fund renewable energy projects in developing countries in exchange for credits that count toward meeting buying countries' emissions targets. JI is relevant for countries (like Belarus and Turkey) listed in Annex B of the Kyoto Protocol, which sets binding emissions targets. CDM is designed for developing countries without firm targets.

9. The EU-10 countries include Bulgaria, the Czech Republic, Estonia, Hungary, Latvia, Lithuania, Poland, Romania, the Slovak Republic, and Slovenia.

10. This section was prepared by Ahmad Slaibi, Nina Rinnerberger, and Andrew Mitchell, and builds on Shvidenko et al. (2011).

11. The EU-15 countries include Austria, Belgium, Denmark, Finland, France, Germany, Greece, Ireland, Italy, Luxembourg, the Netherlands, Portugal, Spain, Sweden, and the United Kingdom.

12. Natura 2000 is an EU-wide network of nature protection areas established under the European Commission's 1992 Habitats Directive (which, together with the Birds Directive, is the cornerstone of Europe's nature conservation policy). For more information about Natura 2000, see http://ec.europa.eu/environment/nature/natura2000/index _en.htm or http://www.natura.org/.

13. GEF 5, which started in July 2010 and ends in June 2014, includes US$317.5 million allocated to ECA countries. More than 90 percent of these funds have not been used.

14. U.S. Environmental Protection Agency, Greenhouse Gas Equivalencies Calculator: http://www.epa.gov/cleanenergy/energy-resources /calculator.html.

15. One terawatt-hour is equivalent to 114 megawatts.

16. The Swedish Bioenergy Association presents an overview of wood fuel use for energy generation: http://svebio.agriprim.com/attachments /33/118.pdf.

17. This supply includes the use of spruce stumps, nonindustrial wood, unused branches, and damaged wood during harvesting. There is a possibility that, in certain cases, full use of forest cut may lead to decreased forest soil fertility.

18. In Moscow on August 7, 2010, air samples showed 6.6 times the normal levels of carbon monoxide, and illness and pollution due to extreme air pollution reached high levels.

19. The Eastern partnership countries (also known as the European Neighborhood Policy East [ENPI] countries) include Armenia, Azerbaijan, Belarus, Georgia, Moldova, and Ukraine.

References

Burney, Jennifer A., Steven J. Davis, and David B. Lobell. 2010. "Greenhouse Gas Mitigation by Agricultural Intensification." *Proceedings of the National Academy of Sciences* 107 (26): 12052–57.

Chum, H., A. Faaij, J. Moreira, G. Berndes, P. Dhamija, H. Dong, B. Gabrielle, A. Goss Eng, W. Lucht, M. Mapako, O. Masera Cerutti, T. McIntyre, T. Minowa, and K. Pingoud. 2011. "Bioenergy." In *IPCC Special Report on Renewable Energy Sources and Climate Change Mitigation*, ed. by O. Edenhofer, R. Pichs-Madruga, Y. Sokona, K. Seyboth, P. Matschoss, S. Kadner, T. Zwickel, P. Eickemeier, G. Hansen, S. Schlömer, and C. von Stechow. Cambridge and New York: Cambridge University Press.

DB Climate Change Advisors. 2009. "Investing in Agriculture: Far-Reaching Challenge, Significant Opportunity. An Asset Management Perspective." Climate Change Investment Research paper, Deutsche Bank Group, New York.

FAO (United Nations Food and Agriculture Organization). 2000. *Land Resource Potential and Constraints at Regional and Country Levels*. World Soil Resources Reports. Rome: FAO.

———. 2004. "Trends and Current Status of Contribution of Forestry Sector to National Economies." Working paper FSFM/ACC/07, FAO, Rome.

———. 2011. *State of the World's Forests*. Rome: FAO.

Gerasimov, Yuri, and Timo Karjalainen. 2009. "Energy Wood Potential in Northwest Russia." Working Papers of the Finnish Forest Research Institute 108, Vantaa, Finland.

Gerber, Pierre, Theun Vellinga, Carolyn Opio, and Henning Steinfeld. 2010. "Productivity Gains and Greenhouse Gas Emissions Intensity in Dairy Systems." *Livestock Science* 139 (1–2): 100–108.

Gustafson, Eric J., Anatoly Z. Shvidenko, Brian R. Sturtevant, and Robert M. Scheller. 2010. "Predicting Global Change Effects on Forest Biomass and Composition in South-Central Siberia." *Ecological Applications* 20 (3): 700–15.

IPCC (Intergovernmental Panel on Climate Change). 2007. *Fourth Assessment Report: Climate Change 2007.* Geneva: IPCC of the United Nations Environment Programme and the World Meteorological Organization.

Khosla, Vinod. 2011. "What Matters in Biofuels and Where Are We?" Online article, greentechmedia.com, Greentech Media, San Francisco. http://www.greentechmedia.com/articles/read/guest-post-vinod-khosla-on-what-matters-in-biofuels.

Lal, R. 2004. "Soil Carbon Sequestration Impacts on Global Climate Change and Food Security." *Science* 304 (5677): 1623–27.

Larson, Donald F., Ariel Dinar, and Brian Blankespoor. 2012. "Aligning Climate Change Mitigation and Agricultural Policies in Eastern Europe and Central Asia." Background paper, Europe and Central Asia Region, World Bank, Washington, DC.

Pretty, Jules. 2008. "Agricultural Sustainability: Concepts, Principles and Evidence." *Philosophical Transactions of the Royal Society B* 363 (1491): 447–65.

Republic of Turkey. 2011. "National Climate Change Action Plan 2011–2023." Ministry of Environment and Urbanization, Republic of Turkey, Ankara.

Schierhorn, Florian, Daniel Mueller, Alexander V. Prishchepov, and Alfons Balmann. 2012. "Grain Potentials on Abandoned Cropland in European Russia." Paper presented at the "Annual World Bank Conference on Land and Poverty," World Bank, Washington, DC, April 23–26.

Shvidenko, Anatoly, Dmitry Schepaschenko, Hannes Böttcher, Mykola Gusti, Florian Kraxner, Michael Obersteiner, and Sylvain Leduc. 2011. "The Role of ECA's Forest Resources for Climate Change Mitigation." Background paper, International Institute for Applied Systems Analysis, Laxenburg, Austria.

Smith, Pete, Daniel Martino, Zucong Cai, Daniel Gwary, Henry Janzen, Pushpam Kumar, Bruce McCarl, Stephen Ogle, Frank O'Mara, Charles Rice, Bob Scholes, Oleg Sirotenko, Mark Howden, Tim McAllister, Genxing Pan, Vladimir Romanenkov, Uwe Schneider, Sirintornthep Towprayoon, Martin Wattenbach, and Jo Smith. 2008. "Greenhouse Gas Mitigation in Agriculture." *Philosophical Transactions of the Royal Society B* 363 (1492): 789–813.

Smith, P., G.-J. Nabuurs, I. Janssens, S. Reis, G. Marland, J.-F. Soussana, T. R. Christensen, L. Heath, M. Apps, V. Alexeyev, J. Y. Fang, J.-P. Gattuso, J. P. Guerschman, Y. Huang, E. Jobbagy, D. Murdiyarso, J. Ni, A. Nobre, C. H. Peng, A. Walcroft, S. Q. Wang, Y. Pan, and G. S. Zhou. 2008. "Sectoral Approaches to Improve Regional Carbon Budgets." *Climatic Change* 88 (3–4): 209–49.

Sutton, William R., Rachel I. Block, and Jitendra P. Srivastava. 2010. "The Unbuilt Environment: Agriculture and Forestry." In *Adapting to Climate Change in Eastern Europe and Central Asia*, ed. by M. Fay, R. I. Block, and J. O. Ebinger. Washington, DC: World Bank.

Swinnen, Johan, and Kristine van Herck. 2011. "Food Security and the Transition Region." Study report, United Nations Food and Agriculture Organization Investment Centre, Rome.

Swinnen, Johan F. M., and Liesbet Vranken. 2010. "Reforms and Agricultural Productivity in Central and Eastern Europe and the Former Soviet Republics: 1989–2005." *Journal of Productivity Analysis* 33 (3): 241–58.

Timilsina, Govinda R. 2012. "Bioenergy for Climate Change Mitigation in Eastern Europe and Central Asia." Background paper, Europe and Central Asia Region, World Bank, Washington, DC.

Van Dam, J., A. P. C. Faaij, I. Lewandowski, and G. Fischer. 2007. "Biomass Production Potentials in Central and Eastern Europe under Different Scenarios." *Biomass and Bioenergy* 31 (6): 345–66.

World Bank. 2010. "Brazil Low-Carbon Country Case Study." Low-Carbon Growth Country Studies Program, World Bank, Washington, DC.

———. 2012. "Looking Beyond the Horizon: Adapting Agriculture to Climate Change in Four Europe and Central Asia Countries." Report AAA71-7E, Europe and Central Asia Region, World Bank, Washington, DC.